IFCoLog Journal of Logics and their Applications

Volume 3, Number 4

October 2016

Disclaimer

Statements of fact and opinion in the articles in IfCoLog Journal of Logics and their Applications are those of the respective authors and contributors and not of the IfCoLog Journal of Logics and their Applications or of College Publications. Neither College Publications nor the IfCoLog Journal of Logics and their Applications make any representation, express or implied, in respect of the accuracy of the material in this journal and cannot accept any legal responsibility or liability for any errors or omissions that may be made. The reader should make his/her own evaluation as to the appropriateness or otherwise of any experimental technique described.

ISBN 978-1-84890-219-0
ISSN (E) 2055-3714
ISSN (P) 2055-3706

College Publications
Scientific Director: Dov Gabbay
Managing Director: Jane Spurr

http://www.collegepublications.co.uk

Printed by Lightning Source, Milton Keynes, UK

Editorial Board

Scope and Submissions

This journal considers submission in all areas of pure and applied logic, including:

- pure logical systems
- proof theory
- constructive logic
- categorical logic
- modal and temporal logic
- model theory
- recursion theory
- type theory
- nominal theory
- nonclassical logics
- nonmonotonic logic
- numerical and uncertainty reasoning
- logic and AI
- foundations of logic programming
- belief revision
- systems of knowledge and belief
- logics and semantics of programming
- specification and verification
- agent theory
- databases
- dynamic logic
- quantum logic
- algebraic logic
- logic and cognition
- probabilistic logic
- logic and networks
- neuro-logical systems
- complexity
- argumentation theory
- logic and computation
- logic and language
- logic engineering
- knowledge-based systems
- automated reasoning
- knowledge representation
- logic in hardware and VLSI
- natural language
- concurrent computation
- planning

This journal will also consider papers on the application of logic in other subject areas: philosophy, cognitive science, physics etc. provided they have some formal content.

Submissions should be sent to Jane Spurr (jane.spurr@kcl.ac.uk) as a pdf file, preferably compiled in LaTeX using the IFCoLog class file.

Contents

EDITORIAL

HANNES LEITGEB, IOSIF PETRAKIS, PETER SCHUSTER, AND
HELMUT SCHWICHTENBERG
Munich and Verona

This special issue contains selected papers from the summer school "Proof, Truth, Computation (PTC)" which was held from 21st to 25th July 2014 in Frauenwörth, Chiemsee, Germany, within the Volkswagen Foundation's programme Symposia and Summer Schools. With its focus on the interactions between modern foundations of mathematics and contemporary philosophy, the topics of the summer school included truth theories, predicativity, constructivity, proof theory, formal epistemology and set-theoretic truth.

Mathematical methods are about to shape some branches of contemporary philosophy just as they have formed most of the natural and many of the social sciences. The thread of the school was to mirror this development, known as mathematical philosophy or formal epistemology; to highlight the challenges that arise from it; and in particular to display its repercussions in mathematics. As for theoretical computer science, a quite comparable spin-off of mathematics, the principal counterpart within mathematics is mathematical logic.

Since many of the objects of study lie beyond the typical commitment of contemporary mathematics, it is decisive to include non-classical issues such as predicativity and constructivity. Proof theory does indeed play a pivotal role: as the area of mathematical logic that is closest to the understanding of logic as the science of formal languages and reasoning, it is predestined for interaction both with philosophical and computer science logic.

A hot topic that crossed over wide ranges of the school is whether axiomatic theories of truth and of related notions, such as provability and knowledge, are possible at all in the stress field between syntax and semantics. Rational belief and rational choice, epistemic issues of principal philosophical relevance, are currently put under mathematical scrutiny by applying probabilism: that is, the thesis that a rational agent's degrees of belief should conform to the axioms of probability theory.

The summer school was thought to help to bridge the gap between two of the most fundamental faculties of human intellect, mathematics and philosophy, right at their natural point of contact: that is, logic. The gap, which would have been inconceivable for Leibniz, say, has opened when mathematics went abstract in the

19th century; and has widened considerably with the resulting crisis in the foundations of mathematics. Allowing this divide to remain would prevent the wealth of knowledge obtained since from being transferred across.

Times have changed, however: new generations have arrived with fresh mindsets; and challenges emerging from scientific and technological progress can now only be met by joining forces over the borders of the disciplines. Closing the divide between mathematics and philosophy has thus become both possible and necessary. Moreover, it will contribute to bridging the general split between sciences and humanities, which has been a drawback for too long.

Among other things, the advent of the computer, and the predominant role it plays in the modern world, have made it indispensable to take much more seriously than ever before the foundations of mathematics at large, and logic in particular. In computer science, for example, the need for secure computer software prompted the use of formal methods for program verification, or even for extracting programs directly from mathematical proofs; in any case, logic is heavily employed, and the need for appropriate foundations of mathematics is manifest.

The increasing complexity of proofs almost beyond the reach of a human mathematician (e.g. Wiles's proof of Fermat's Last Theorem) has further suggested to use computers also for proving theorems in mathematical practice. In fact a computer-assisted formalisation of mathematics has been started fairly successfully, and more and more man-made proofs have already been checked by machine; examples include the Four-Colour Theorem and the Feit-Thompson Theorem. A widely used proof assistant is Coq, named after Coquand.

For formalising mathematics proof theory is indispensable. For example, the formalisation in Coq of ever more abstract mathematics has turned out to require a deep revision of the concept of identity in Martin-Löf type theory. As a way out Voevodsky has put forward the Univalence Axiom. It is not by coincidence that identity has ever constituted one of the most puzzling issues in the foundations of mathematics and the exact sciences.

This special issue features the following five articles:
Jacob Cook and Michael Rathjen, "A classification of the provably total set functions of KP".
Sy Friedman, "Evidence for set-theoretic truth and the hyperuniverse programme".
Maria Emilia Maietti and Samuele Maschio, "A predicative realizability tripos for the minimalist foundation".
Graham E. Leigh, "Reflecting on truth".
Joan Rand Moschovakis, "A translation theorem for restricted R-formulas".

Acknowledgements

First and foremost the organisers of the summer school and guest editors of this special issue would like to express their gratitude to the Volkswagen Foundation for offering the patronage of the summer school and for generously giving financial support. Dr. Christoph Kolodziejski provided considerate guidance throughout the whole process, from application to reporting.

Next we thank all speakers, disputants and participants of the summer school. It is only them who made up the scholarly value of the school, and thus layed the cornerstone of this special issue.

Hosting the almost 100 participants would have been impossible without the organisational skills of Sister Scholastica McQueen OSB. A helping hand often came from Basil Karádais and Kenji Miyamoto.

When preparing this special issue Schuster received funding by the Alexander-von-Humboldt Foundation, in the form of a Research Fellowship at the Munich Center for Mathematical Philosophy (MCMP) hosted by Leitgeb; and by the John Templeton Foundation within the project "Abstract Mathematics for Actual Computation: Hilbert's Program in the 21st Century". The opinions expressed in this publication are those of the authors and do not necessarily reflect the views of the John Templeton Foundation.

Dov Gabbay has kindly established our contact with the journal editors. Special thanks are owed to Jane Spurr, executive editor of this journal, for diligently handling the refereeing process and for her advice and patience. Last but not least, the guest editors are indebted to the authors and to the anonymous peer reviewers, without whose contributions this special issue would hardly have emerged.

Munich and Verona, September 2016

Hannes Leitgeb
E-mail: hannes.leitgeb@lmu.de
Iosif Petrakis
E-mail: petrakis@math.lmu.de
Peter Schuster
E-mail: peter.schuster@univr.it
Helmut Schwichtenberg
E-mail: schwicht@math.lmu.de

Received 23 September 2016

Evidence for Set-Theoretic Truth and the Hyperuniverse Programme

Sy-David Friedman
Kurt Gödel Research Center, University of Vienna
sdf@logic.univie.ac.at

I discuss three potential sources of evidence for truth in set theory, coming from set theory's roles as a branch of mathematics and as a foundation for mathematics as well as from the intrinsic maximality feature of the set concept. I predict that new *non first-order* axioms will be discovered for which there is evidence of all three types, and that these axioms will have significant first-order consequences which will be regarded as true statements of set theory. The bulk of the paper is concerned with the *Hyperuniverse Programme*, whose aim is to discover an optimal mathematical principle for expressing the maximality of the set-theoretic universe in height and width.

1 Introduction

The truth of the axioms of ZFC is commonly accepted for at least two reasons. One reason is *foundational*, as they endow set theory with the ability to serve as a remarkably good foundation for mathematics as a whole, and another is *intrinsic*, as (with the possible exception of AC, the axiom of choice) they can be seen to be derivable from the concept of set as embodied by the maximal iterative conception.

In fact a little bit more than ZFC is justifiable on intrinsic and perhaps also foundational grounds. I refer here to *reflection principles* and their related *small large cardinals*, which are also derivable from the maximal iterative conception through *height (ordinal) maximality* and, at least in the case of inaccessible cardinals, are occasionally useful for the development of certain kinds of highly abstract mathematics (such as *Grothendieck universes*). These extensions of ZFC are *mild* in the sense that they are compatible with the *powerset-minimality* principle $V = L$.

But finding strong evidence for the truth of axioms that contradict $V = L$ has been exceedingly difficult. There are a number of reasons for this. One is the fact that mild extensions of ZFC have been in a sense *too good*, in that they alone have until recently been sufficient to serve the needs of set theory as a foundation for mathematics. Another is the difficulty of squeezing more out of the maximal iterative conception through a *width (powerset) maximality* analogue of the height maximality principles that give rise to reflection. And the development of set theory as a branch of mathematics has been so dramatic, diverse and ever-changing that it has been impossible to select those perspectives on the subject whose choices of new axioms can be regarded as "the most true".

My aim in this article is to provide evidence for the following three predictions.

The Richness of Set-Theoretic Practice. The development of set theory as a branch of mathematics is so rich that there will never be a consensus about which first-order axioms (beyond ZFC plus small large cardinals) best serve this development.

A Foundational Need. Just as AC is now accepted due to its essential role for mathematical practice, a systematic study of independence results across mathematics will uncover first-order statements contradicting CH (and hence also $V = L$) which are best for resolving such independence.

An Optimal Maximality Criterion. Through the *Hyperuniverse Programme* it will be possible to arrive at an optimal *non first-order* axiom expressing the maximality of the set-theoretic universe in height and width; this axiom will have first-order consequences contradicting CH (and hence also $V = L$).

And as a synthesis of these three predictions I propose the following optimistic scenario for making progress in the study of set-theoretic truth.

Thesis of Set-Theoretic Truth. There will be first-order statements of set theory that well serve the needs of set-theoretic practice and of resolving independence across mathematics, and which are derivable[1] from the maximality of the set-theoretic universe in height and width. Such statements will come to be regarded as *true statements of set theory.*

This *Thesis* has a converse: In order for a first-order statement contradicting $V = L$ to be regarded as *true*, in my view it must well serve the needs of set-theoretic practice and of resolving independence in mathematics, and it must at least be compatible with the maximality of the set-theoretic universe as expressed

[1]For a discussion of this notion of *derivability* see the final Subsection 4.12.

by the optimal maximality criterion. Indeed the strength of the evidence for such a statement's truth is in my view measured by the extent to which it fulfills these three requirements.

An important consequence of the *Thesis* is the failure of CH. Thus part of my prediction is that CH will be regarded as false.

Note that in the *Thesis* I do not refer to *true first-order axioms* but only to *true first-order statements*. The reason is the following additional claim.

Beyond First-Order. There will never be a consensus about the truth of proposed first-order axioms that contradict $V = L$; instead true first-order statements will arise solely as consequences of true *non first-order* axioms.

One reason for this claim is the inadequancy of first-order statements to capture the maximality of the set-theoretic universe.

The plan of this paper is as follows. First I'll review some of the popular first-order axioms that well serve the needs of set-theoretic practice and argue for the *Richness* prediction above. Second I'll discuss what little is known about independence across mathematics, discussing the role of forcing axioms as evidence for the *Foundational* prediction above. And by far the bulk and central aim of the paper is the third part, in which I present the *Hyperuniverse Programme*, including its philosophical foundation and most recent mathematical developments.

2 Set-Theoretic Practice

Set theory is a burgeoning subject, rife with new ideas and new developments, constantly leading to new perspectives. Naturally certain of these perspectives stand out among the chaotic mass of new results being proved, and it is worth focusing on a few of these to expose the difficulty of settling on particular new axioms as being "the true ones".

I have emphasized the need to find evidence for the truth of axioms that contradict $V = L$, but purely in terms of the value of an axiom for the development of good set theory, what I will refer to as *Type 1 evidence*, this is not possible. Jensen's deep work unlocking the power of this axiom reveals the power of $V = L$, indeed it appears to give us, when combined with small large cardinals, a theory that is complete for all natural set-theoretic statements! That is a remarkable achievement and speaks volumes in favour of declaring $V = L$ to be *true* based on Type 1 evidence.

A natural Type 1 objection to $V = L$ is that it doesn't take forcing into account, a fundamental method for building new models of set theory. Admittedly, even in L one has forcing extensions of countable models, but it is more natural to force over the full L and not just over some small piece of it. So now we contradict $V = L$ in favour of "V contains many generic extensions of L" or something similar.

Having lots of forcing extensions of L sounds good, but then what is our canonical universe now? Shouldn't we also have a sentence that is true only in V, and not in any of its proper inner models, while at the same time having many generic extensions of L? Indeed this is possible with class forcing (see [11]). So now we have a nice Type 1 axiom: V is a canonical universe which is class-generic over L, containing many set-generic extensions of L. This is an excellent context for doing set theory, as the forcing method is now available.

In fact we can do even better and take V to be $L[0^\#]$. Not only does this model contain many generic extensions of L, it is also a canonical universe and we recover all of the powerful methods that Jensen developed under $V = L$, relativised now to the real $0^\#$. So our Type 1 evidence leads us to the superb axiom $V = L[0^\#]$.

Objection! What about measurable cardinals? Recall the important hierarchy of consistency strengths: Natural theories are wellordered (up to bi-interpretability) by their consistency strengths and the consistency strengths of large cardinal axioms provide a nice collection of consistency strengths which is cofinal in a large initial segment of (if not all of) this hierarchy. This does not mean that large cardinals must exist but at the very least there should be inner models having them. So now based on Type 1 evidence we get some version of "There are inner models with large cardinals", an attractive environment in which to do good set theory.

Moreover, notice that if we have inner models for large cardinals we haven't lost the option of looking at L or its generic extensions, they are still available as inner models. So we seem to have reached the best Type 1 axiom yet.

But we could ask for even more. Recall that L has a nice internal structure, very powerful for deriving consequences of $V = L$. Can V not only have inner models for large cardinals but also an L-like internal structure? Of course the answer is positive, as we can adopt the axiom "There are inner models with large cardinals and $V = L[x]$ for some real x". A better answer is provided in [14], where it is shown that V can be L-like together with arbitrary large cardinals, not only in inner models but in V itself. However, as attractive as this may sound, it fails to address a key problem, and this is where we see the multiple perspectives of set theory, with no single perspective having a claim to being "the best".

Even if we produce a nice axiom[2] of the form "There are large cardinals and V is a canonical generalisation of L", doing so commits us to an L-like environment in which to do set theory. Indeed there are other compelling perspectives on set theory which lead us to non L-like environments and correspondingly to entirely different Type 1 axioms. I will mention two of them. (Further information about the notions mentioned below is available in [22]).

Forcing axioms have a long history, dating back to Martin's axiom (MA), a special case of which asserts the existence of generics for ccc partial orders (i.e. partial orders with only countable antichains) over models of size $\aleph_1$. This simple axiom can be used to establish in one blow the relative consistency of a huge range of set-theoretic statements. Naturally there has been interest in strengthenings of MA, and a popular one is the Proper Forcing Axiom (PFA), which strengthens this[3] to the wider class of *proper* partial orders.

Now with regard to Type 1 evidence the point is that PFA has even more striking consequences than MA, qualifying it as a central and important tool for solving combinatorial problems in set theory. A powerful case can be made for its truth based on Type 1 evidence. But of course PFA conflicts with any axiom which asserts that V is L-like, as it implies the negation of CH. In fact PFA implies that the size of the continuum is $\aleph_2$.

The diversity of Type 1 evidence goes beyond just L-likeness and forcing axioms; there are also cardinal characteristics. These are natural and heavily-investigated cardinal numbers that arise when studying definability-theoretic and combinatorial properties of sets of real numbers. Each of these cardinal characteristics is an uncountable cardinal number of size at most the continuum. Now given the variety of such characteristics together with the fact that they can consistently differ from each other, isn't it compelling to adopt the axiom that cardinal characteristics provide a large spectrum of distinct uncountable cardinals below the size of the continuum and therefore the continuum is indeed quite large, in contradiction to both L-likeness and forcing axioms?[4]

[2]Woodin has in fact proposed such an axiom which he calls *Ultimate L*.

[3]For the experts, to get PFA one must allow non-transitive models of size $\aleph_1$.

[4]As a specific example, let $\mathfrak{a}$ denote the least size of an infinite almost disjoint family of subsets of ω, and $\mathfrak{b}$ ($\mathfrak{d}$) the least size of an unbounded (dominating) family of functions from ω to ω ordered by eventual domination. Then $\mathfrak{b} < \mathfrak{a} < \mathfrak{d}$ is consistent; shouldn't it in fact be true?

Thus we have three distinct types of axioms with excellent Type 1 evidence: L-likeness with large cardinals, forcing axioms and cardinal characteristic axioms. They contradict each other yet each is consistent with the existence of inner models for the others. In my view, this makes a clear case that Type 1 evidence is insufficient to establish the truth of axioms of set theory; it is also insufficient to decide whether or not CH is true.

3 Set Theory as a Foundation for Mathematics

Of course axiomatic set theory can be heartily congratulated for its success in providing a foundation for mathematics. An overwhelming case can be made that when theorems are proved in mathematics they can be regarded as theorems of a mild extension of ZFC (compatible with $V = L$). In particular, we routinely expect questions in mathematics to be answerable (perhaps with great difficulty!) in a mild extension of ZFC.

A consequence is that an independence result for such mild extensions is indeed an independence result for mathematics as a whole. This is of course of minor importance if the independence result in question is a statement of set theory, as set theory is just a small part of mathematics. But this is of considerable importance when independence arises with questions of mathematics outside of set theory, as is the case with the the Borel, Kaplansky and Whitehead Conjectures of measure theory, functional analysis and group theory, respectively. Let us not forget the great mathematician David Hilbert's thesis that the questions of mathematics can be resolved using the powerful tools of the subject. An understanding of how to deal with independence is needed to restore the status of mathematics as the complete and definitive field of study that Hilbert envisaged.

The time is ripe for set-theorists to focus on this problem. The central question is:

Foundational or Type 2 Evidence: Are there particular axioms of set theory which best serve the needs of resolving independence in other areas of mathematics?

Recently there are signs that a positive answer to this question is emerging, as new applications of set theory to functional analysis, topology, abstract algebra and model theory (a field of logic, but still outside of set theory) are being found. The *Foundational Need* that I expressed earlier is precisely the prediction that a pattern will emerge from these applications to reveal that particular axioms of set theory are best for bringing set theory closer to the complete foundation that Hilbert was hoping for.

Now where are these *foundationally advantageous* axioms of set theory to be found? Consider the following list of candidates with good Type 1 evidence:

$V = L$
V is a canonical and rich class-generic extension of L
Large Cardinal Axioms (like supercompacts)
Forcing Axioms like MA, PFA
Determinacy Axioms like AD in $L(R)$
Cardinal Characteristic Axioms like $\mathfrak{b} < \mathfrak{a} < \mathfrak{d}$

As already said, each of these axioms is important for the development of set theory, providing a unique perspective on the subject. But perhaps it is surprising to discover that only two of them, $V = L$ and Forcing Axioms, have had any significant impact on mathematics outside of set theory! The impact of Large Cardinal Axioms (like supercompacts) and Cardinal Characteristic Axioms has been minimal and that of Determinacy Axioms non-existent so far.

To give a bit more detail, both $V = L$ and Forcing Axioms can be used to answer the following questions (in different ways):

Functional Analysis: Must every homomorphism from $C(X)$, X compact Hausdorff, into another Banach algebra be continuous (the Kaplansky Problem)? Is the ideal of compact operators on a separable Hilbert space in the ring of all bounded operators the sum of two smaller ideals?; Are all automorphisms of the Calkin Algebra inner?

Topology and Measure Theory: Is every normal Moore Space metrizable? Are there S-spaces (regular, hereditarily separable spaces where some open cover has no countable subcover)? Is every strong measure 0 set of reals countable (the Borel conjecture)?

Abstract Algebra: Is every Whitehead group free (the Whitehead Problem)? What is the homological dimension of $R(x, y, z)$ as an $R[x, y, z]$-module where R is the field of real numbers? Does the direct product of countably many fields have global dimension 2?

One could also mention the field of Model Theory (part of Logic, but not part of Set Theory), where new axioms of Set Theory may play an important role in the study of Morley's theorem for Abstract Elementary Classes or perhaps even in the resolution of Vaught's Conjecture.

My prediction is that $V = L$ and Forcing Axioms will be the definite winners among choices of axioms of Set Theory that resolve independence across mathematics as a whole. But as $V = L$ is in conflict with the maximality of the set-theoretic universe in width, it is not suitable as a realization of the *Thesis of Set-Theoretic Truth*, leaving Forcing Axioms as the current leading candidate for that.

4 The Maximality of the Set-Theoretic Universe and the HP

The letters HP stand for the *Hyperuniverse Programme*, which I now discuss in detail.

4.1 The Iterative Conception of Set

As Gödel put it, the *iterative conception* of set expresses the idea that a set is something obtainable from well-defined objects by iterated application of the powerset operation. In more detail (following Boolos [7]; also see [27]): Sets are formed in *stages*, where only the empty set is formed at stage 0 and at any stage greater than 0, one forms collections of sets formed at earlier stages. (Said this way, a set is re-formed at every stage past where it is first formed, but that is OK.) Any set is formed at some least stage, after its elements have been formed. This conception excludes anomalies: We can't have $x \in x$, there is no set of all sets, there are no cycles $x_0 \in x_1 \in \cdots \in x_n \in x_0$ and there are no infinite sequences $\cdots \in x_n \in x_{n-1} \in x_{n-2} \in \cdots \in x_1 \in x_0$, as there must be a least stage at which one of the x_n's is formed. We'll assume that there are infinite sets[5], so the iteration process leads to a limit stage ω, which is not 0 and is not a successor stage.

The iterative conception yields that the universe of sets is a model of the axioms of Zermelo Set Theory, i.e. ZFC without Replacement and without the Axiom of Choice. The standard model for this theory is $V_{\omega+\omega}$.

Nevertheless, Replacement and AC (the Axiom of Choice) are included as part of the standard axioms of Set Theory, for very different reasons. The case for AC is typically made on *extrinsic grounds*, citing its *fruitfulness* for the development

[5]This is derivable once we add *maximality* to the iterative conception, but is convenient to assume already as part of the iterative conception.

of mathematics and its corresponding necessity for Set Theory as a foundation for mathematics (a case of what I have called *Type 2 evidence*). It is not clear to me that Choice is derivable from the iterative conception, nor from its necessity for doing good Set Theory (*Type 1 evidence*).

Replacement, on the other hand, is derivable from the concept of set. To see this, we need to extend the iterative conception to the stronger *maximal iterative conception*, also implicit in the set-concept.

4.2 Maximality and the Iterative Conception

The term *maximal* is used in many different senses in Set Theory, what I have in mind here is a very specific use associated to the iterative conception (IC). Recall that according to the IC, sets appear inside levels indexed by the ordinal numbers, where each successor level $V_{\alpha+1}$ is the powerset of the previous. As Boolos explained, the IC alone takes no stand on how many levels there are (the *height* of the universe V) or on how fat the individual levels are (the *width* of V). However it is generally regarded as implicit in the set-concept that both of these should be *maximal*:

Height (or Ordinal) maximality: The universe V is *as tall as possible*, i.e., the sequence of ordinals is *as long as possible.*

Width (or Powerset) maximality: The universe V is *as wide (or thick) as possible*, i.e., the powerset of each set is *as large as possible.*

If we conjunct the IC with maximality we arrive at the MIC, the *maximal iterative conception*, also part of the set-concept but more of a challenge to explain than the simple IC.

It is natural to see a *comparative* aspect to maximality, as to be *as large as possible* suggests *as large as possible within the realm of possibilities.* Thus a natural way to explain height and width maximality would be to compare V to other possible universes.

But now we face a serious problem. If V is the fixed universe of *all* sets, then there are no universes other than those already included in V. In other words V is maximal by default, as no other universe can threaten its maximality, and therefore we are limited in what we can say about this concept.

I will postpone this problem for now, and instead discuss an easier one: Let M denote a countable transitive model of ZFC (ctm). What could it mean to say that M is maximal?

Now we have a different problem. The natural way to express the maximality of M is to say that M cannot be expanded to a larger universe. Let us call this *structural maximality*. But under a very mild assumption (there is a set-model of ZFC containing all of the reals) this is impossible: Any ctm M is an element (and therefore proper subset) of a larger ctm.

So instead we move to a milder form of maximality, called *syntactic maximality*, expressed as follows.

In the case of (syntactic-) height maximality, we consider *lengthenings* of M, i.e. ctm's M^* of which M is a rank-initial segment (the ordinals of M form an initial segment of the ordinals of M^* and the powerset operations of these two universes agree on the sets in M).

In the case of width maximality, we consider *thickenings* of M, i.e. ctm's M^* of which M is an inner model (M and M^* have the same ordinals and M is included in M^*).

In this way we can produce forms of *height maximality* and *width maximality* for ctm's as follows.

If M is *height maximal* then a property of M also holds of some rank-initial segment of M. This is the typical formulation of *reflection*. (However we will see that *height maximality* is stronger than *reflection*.) Of course specific realizations of height maximality must specify which properties are to be taken into account.

If M is *width maximal* then a property of a thickening of M also holds of some inner model of M. In the case of first-order properties this is called the *Inner Model Hypothesis*, or *IMH* (introduced in [12]).

The above discussion of maximality for ctm's, although brief, will suffice for establishing the strategy of the HP.

We return now to the problem of maximality for V. Can the above discussion for ctm's also be applied to V? Does it make sense to talk about *lengthenings* and *thickenings* of V in the way we talk about them for ctm's? There are differences of opinion about this, which I'll take up next.

4.3 Actualism and Potentialism

Recall that in the *IC* we describe V, the universe of sets, via a process of iteration of the powerset operation. Does this process come to an end, or is it indefinite, always extendible further to a longer iteration? The former possibility, that there is a "limit" to the iteration process is referred to as *height actualism* and the latter view is called *height potentialism.* Analogously there is a question of the definiteness of the powerset operation: For a given set, is its powerset determined or is it always possible to extend it further by adding more subsets? The former is called *width actualism* and latter *width potentialism.*

There is a vast literature on this topic ([4, 19, 20, 21, 23, 24, 25, 26, 29, 31, 32]). However as the Hyperuniverse Programme is very flexible on the choice of ontology, we will not engage here in a lengthy discussion of the actualism/potentialism debate, but only mention some points in favour of a *Zermelian view*, combining height potentialism with width actualism, the view which we choose to adopt for our analysis of maximality via the HP.

We can summarize the situation as follows. Without difficulty, height potentialism facilitates an analysis of height maximality. Surprisingly, we will show that even with width actualism, it also facilitates an analysis of width maximality, using the method of *V-logic*. A further benefit of height potentialism is that we can reduce the study of maximality for V to the study of maximality for ctm's[6]. Our arguments also show that height actualism is viable for our analysis of width maximality, provided it is enhanced with a strong enough fragment of MK (Morse-Kelley class theory; one only needs Σ^1_1 comprehension). Thus the only problematic ontology for the HP is height actualism supported by only a weak class theory; otherwise the choice of ontology is not critical for the HP (although the programme develops slightly differently with width potentialism than it does with width actualism) [7].

I will now present some arguments due to Geoffrey Hellman ([33][8]) in favour of

[6]The set of ctm's is called the *Hyperuniverse*; hence we arrive at the *Hyperuniverse Programme.*

[7]Height actualism with just GB (Gödel-Bernays) appears inadequate for a fruitful analysis of maximality. A referee has informed us about *agnostic Platonism*, the view that there is a well-determined universe V of all sets but without taking a position on whether ZFC holds in it. But as this perspective allows for the possibility of height actualism with just GB, it is problematic for the HP.

[8]These comments were made during a lively e-mail exchange among numerous set-theorists and philosophers of set theory from August until November 2014, triggered by my response to Sol Feferman's preprint *The Continuum Hypothesis is neither a definite mathematical problem nor a definite logical problem.* Some of this discussion is documented at <http://logic.harvard.edu/blog/?cat=2>, but regrettably Hellman's comments do not appear there.

height potentialism and width actualism, *the Zermelian view*. Hellman says:

"The idea that any universe of sets can be properly extended (in height, not width) is extremely natural, endorsed by many mathematicians (e.g. MacLane, seemingly by Gödel, et. al.) ... As Maddy and others say, if it's possible that sets beyond some (putatively maximal) level exist, then they do exist ... Thus, if 'imaginable' (end) extensions of V are not incoherent, then they are possible, and then, on an actualist, platonist reading, they are actual, and V wasn't really maximal after all. ... such extensions are always possible, so that the notion of a single fixed, absolutely maximal universe V of sets is really an incoherent notion."

And again:

"I have no earthly or heavenly idea what 'as high as possible' could mean, since the notion of a set domain that absolutely could not in logic be extended seems to me incoherent (or at any rate empty). As Putnam put it in his controversial paper, 'Mathematics without Foundations' (1967), 'Even God couldn't make a universe for Zermelo set theory that it would be impossible to extend.' And I agree, theology aside."

Regarding width potentialism, Hellman says ([33]):

"I have a good idea, I think, about 'as thick as possible', since the notion of full power set of a given set makes perfect sense to me ... Granted that forcing extensions can be viewed as 'thickenings' of the cumulative hierarchy, as usually described, when we assert the standard Power Sets axiom, we implicitly build in bivalence, i.e. that either x belongs to y or it doesn't, i.e. we are in effect ruling forcing extensions or Boolean-valued generalizations as *non-standard* [my italics], i.e. 'full power set' is to be understood only in the standard way."

And further:

"Thus, to my way of thinking, there is an important disanalogy between 'all ordinals' ... and 'all subsets of a given set'. The latter is 'already relativized'; there is nothing implicit in the notion of 'subset' that allows for indefinite extensions, so long as we are speaking of 'subsets of a fixed, given set' ... In contrast, 'all ordinals' cries out for relativization (a point I find in Zermelo's [1930]); without it, it does allow for indefinite extensibility, by the very operations that we use to describe ordinals"

I do appreciate Hellman's point here, and indeed will (for the most part) adopt the *Zermelian perspective*, height potentialism with width actualism, in this paper. Another strong point in favour of this view is that although we have a clear and coherent way of generating the ordinals through a process of iteration, there is currently no analogous iteration process for generating increasingly rich power sets[9].

In light of this adoption of potentialism in height, I will now use the symbol V ambiguously, not to denote the fixed universe of all sets (which does not exist) but as a variable to range over universes within the *Zermelian multiverse* in which each universe is a rank initial segment of the next.

Despite my adoption of the Zermelian view, I will for expository purposes also consider a form of potentialism in both height and width which I will call *radical potentialism.* The HP can be run with either point of view. Although it is simpler with radical potentialism, there are interesting issues (both mathematical and philosophical) which arise when employing the Zermelian view which are worth exploring.

To describe radical potentialism, let me begin with something less radical, *width potentialism.* First as motivation, consider a Platonist view, so that V is the fixed universe of all sets, and consider the method of forcing for producing generic sets. If M is a ctm we can easily build a generic extension $M[G]$ of M using the countability of M. But of course generic extensions $V[G]$ of V do not exist, as our "real V" has all the sets. Despite this we can talk definably in V about what can be true in such a *generic extension* without actually having such extensions in V, by constructing the Boolean universe V^B within V and taking *true in a generic extension of* V to just mean *of nonzero Boolean truth value in* V^B. Thus the Platonist view is in fact dualistic: It allows for the possibility of making sense of truth in universes (generic extensions) without allowing these universes to actually exist.

Width potentialism is a view in which any universe can be thickened, keeping the same ordinals, even to the extent of making ordinals countable. Thus for example it allows for the existence of the generic extensions of V (now a variable ranging over the multiverse of all possible universes) that are prohibited by the Platonist. So for any ordinal α of V we can thicken V to a universe where α is countable; i.e., any ordinal is *potentially countable.* But that does not mean that every ordinal of V *is* countable in V, it is only countable in a larger universe. So this *potential countability* does not threaten the truth of the powerset axiom in V.

[9]But I am not 100% sure that there could not be such an analogous iteration process, perhaps provided by a wildly successful theory of inner models for large cardinals.

Now *radical potentialism* is in effect a unification of width and height potentialism. It entails that any V (in the multiverse of possible universes) looks countable inside a larger universe: We allow V to be lengthened and thickened simultaneously. Note that even just width potentialism (allowing universes to be thickened) forces us also into height potentialism: If we were to keep thickening to make every ordinal of V countable then after $\text{Ord}(V)$ steps we are forced to also lengthen to reach a universe that satisfies the powerset axiom. In that universe, the original V looks countable. But then we could repeat the process with this new universe until it too is seen to be countable. The height potentialist aspect is that we cannot end this process by taking the union of all of our universes, as this would not be a model of ZFC (the powerset axiom will fail) and therefore would have to be lengthened. Note that once again, the *potential countability of* V does not threaten the truth of the axioms of ZFC in V.

4.4 Maximality in Height and #-Generation

The analysis of height maximality is the first major success of the HP. The programme has produced a robust principle expressing the maximality of V in height which appears to encompass all prior height maximality principles, including reflection, and to constitute the definitive expression of the height maximality of V in mathematical terms.

For our discussion of height maximality, height potentialism will suffice (radical potentialism is not needed). Thus we allow ourselves the option of lengthening V to universes V^* which have V as a rank-initial segment. Of course we can also consider shortenings of V, replacing V by one of its own rank-initial segments. Let us now make use of lengthenings and shortenings to formulate a height maximality principle for V, expressing the idea that the sequence of ordinals is *as long as possible.*

But before embarking on our analysis of height maximality we should take note of the following: No *first-order* statement φ can be adequate to fully capture height maximality. This is simply because a first-order statement true in V will reflect to one of its rank initial segments and we are then naturally led from φ to the stronger first-order statement "φ holds both in V and in some transitive set model of ZFC". We will also see that no first-order statement is adequate to capture width maximality. This is an instance of the *Beyond First-Order* claim of the introduction: True first-order statements contradicting $V = L$ only arise as consequences of true *non first-order* axioms.

But how do we capture height maximality with a non first-order axiom? We do this via a detailed analysis of the relationship between V and its lengthenings and shortenings.

Standard Lévy reflection tells us that a single first-order property of V with parameters will hold in some V_κ which contains those parameters. It is natural to strengthen this to the simultaneous reflection of all first-order properties of V to some V_κ, allowing arbitrary parameters from V_κ. Thus we have reflected V to a V_κ which is an elementary submodel of V.

Repeating this process leads us to an increasing, continuous sequence of ordinals $(\kappa_i \mid i < \infty)$, whee ∞ denotes the ordinal height of V, such that the models $(V_{\kappa_i} \mid i < \infty)$ form a continuous chain $V_{\kappa_0} \prec V_{\kappa_1} \prec \cdots$ of elementary submodels of V whose union is all of V.

Let C be the proper class consisting of the κ_i's. We can apply reflection to V with C as an additional predicate to infer that properties of (V, C) also hold of some $(V_\kappa, C \cap \kappa)$. But the unboundedness of C is a property of (V, C) so we get some $(V_\kappa, C \cap \kappa)$ where $C \cap \kappa$ is unbounded in κ and therefore κ belongs to C. As a corollary, properties of V in fact hold in some V_κ where κ belongs to C. It is convenient to formulate this in its contrapositive form: If a property holds of V_κ for all κ in C then it also holds of V.

Now note that for all κ in C, V_κ can be *lengthened* to an elementary extension (namely V) of which it is a rank-initial segment. By the contrapositive form of reflection of the previous paragraph, V itself also has such a lengthening V^*.

But this is clearly not the end of the story. For the same reason we can also infer that there is a continuous increasing sequence of such lengthenings $V = V_{\kappa_\infty} \prec V^*_{\kappa_{\infty+1}} \prec V^*_{\kappa_{\infty+2}} \prec \cdots$ of length the ordinals. For ease of notation, let us drop the *'s and write W_{κ_i} instead of $V^*_{\kappa_i}$ for $\infty < i$ and instead of V_{κ_i} for $i \leq \infty$. Thus V equals W_∞.

But which tower $V = W_{\kappa_\infty} \prec W_{\kappa_{\infty+1}} \prec W_{\kappa_{\infty+2}} \prec \cdots$ of lengthenings of V should we consider? Can we make the choice of this tower *canonical*?

Consider the entire sequence $W_{\kappa_0} \prec W_{\kappa_1} \prec \cdots \prec V = W_{\kappa_\infty} \prec W_{\kappa_{\infty+1}} \prec W_{\kappa_{\infty+2}} \prec \cdots$. The intuition is that all of these models resemble each other in the sense that they share the same first-order properties. Indeed by virtue of the fact that they form an elementary chain, these models all satisfy the same first-order sentences. But again in the spirit of "resemblance", the following should hold:

For $i_0 < i_1$ regard $(W_{\kappa_{i_1}}, W_{\kappa_{i_0}})$ as the structure $(W_{\kappa_{i_1}}, \in)$ together with $W_{\kappa_{i_0}}$ as a unary predicate. Then it should be the case that any two such pairs $(W_{\kappa_{i_1}}, W_{\kappa_{i_0}})$, $(W_{\kappa_{j_1}}, W_{\kappa_{j_0}})$ (with $i_0 < i_1$ and $j_0 < j_1$) satisfy the same first-order sentences, even allowing parameters which belong to both $W_{\kappa_{i_0}}$ and $W_{\kappa_{j_0}}$. Generalising this to triples, quadruples and n-tuples in general we arrive at the following situation:

$(*)$ V occurs in a continuous elementary chain $W_{\kappa_0} \prec W_{\kappa_1} \prec \cdots \prec V = W_{\kappa_\infty} \prec W_{\kappa_{\infty+1}} \prec W_{\kappa_{\infty+2}} \prec \cdots$ of length $\infty + \infty$, where the models W_{κ_i} form a *strongly-indiscernible chain* in the sense that for any n and any two increasing n-tuples $\vec{i} = i_0 < i_1 < \cdots < i_{n-1}$, $\vec{j} = j_0 < j_1 < \cdots < j_{n-1}$, the structures $W_{\vec{i}} = (W_{\kappa_{i_{n-1}}}, W_{\kappa_{i_{n-2}}}, \cdots, W_{\kappa_{i_0}})$ and $W_{\vec{j}}$ (defined analagously) satisfy the same first-order sentences, allowing parameters from $W_{\kappa_{i_0}} \cap W_{\kappa_{j_0}}$.

We are getting closer to the desired axiom of #-generation. Surely we can impose higher-order indiscernibility on our chain of models. For example, consider the pair of models $W_{\kappa_0} = V_{\kappa_0}$, $W_{\kappa_1} = V_{\kappa_1}$. We can require that these models satisfy the same second-order sentences; equivalently, we require that $H(\kappa_0^+)^V$ and $H(\kappa_1^+)^V$ satisfy the same first-order sentences. But as with the pair $H(\kappa_0)^V$, $H(\kappa_1)^V$ we would want $H(\kappa_0^+)^V$, $H(\kappa_1^+)^V$ to satisfy the same first-order sentences *with parameters.* How can we formulate this? For example, consider κ_0, a parameter in $H(\kappa_0^+)^V$ that is second-order with respect to $H(\kappa_0)^V$; we cannot simply require $H(\kappa_0^+)^V \models \varphi(\kappa_0)$ iff $H(\kappa_1^+)^V \models \varphi(\kappa_0)$, as κ_0 is the largest cardinal in $H(\kappa_0^+)^V$ but not in $H(\kappa_1^+)^V$. Instead we need to replace the occurence of κ_0 on the left side with a "corresponding" parameter on the right side, namely κ_1, resulting in the natural requirement $H(\kappa_0^+)^V \models \varphi(\kappa_0)$ iff $H(\kappa_1^+)^V \models \varphi(\kappa_1)$. More generally, we should be able to replace each parameter in $H(\kappa_0^+)^V$ by a "corresponding" element of $H(\kappa_1^+)^V$. It is natural to solve this parameter problem using embeddings.

Definition 1. *(See [10])*

A structure $N = (N, U)$ is called a # with critical point κ, *or just a #, if the following hold:*

(a) N is a model of ZFC$^-$ (ZFC minus powerset) in which κ is both the largest cardinal and strongly inaccessible.

(b) (N, U) is amenable (i.e. $x \cap U \in N$ for any $x \in N$).

(c) U is a normal measure on κ in (N, U).

(d) N is iterable, i.e., all of the successive iterated ultrapowers starting with (N, U) are well-founded, yielding iterates (N_i, U_i) and Σ_1 elementary iteration maps $\pi_{ij} : N_i \to N_j$ where $(N, U) = (N_0, U_0)$.

We let κ_i denote the largest cardinal of the i-th iterate N_i.

If N is a # and λ is a limit ordinal then $\mathrm{LP}(N_\lambda)$ denotes the union of the $(V_{\kappa_i})^{N_i}$'s for $i < \lambda$. (LP stands for *lower part.*) $\mathrm{LP}(N_\infty)$ is a model of ZFC.

Definition 2. *We say that a transitive model V of ZFC is #-*generated *iff there is $N = (N, U)$, a # with iteration $N = N_0 \to N_1 \to \cdots$, such that V equals $LP(N_\infty)$ where ∞ denotes the ordinal height of V.*

#-generation fulfills our requirements for vertical maximality, with powerful consequences for reflection. L is #-generated iff $0^\#$ exists, so this principle is compatible with $V = L$. If V is #-generated via (N, U) then there are elementary embeddings from V to V which are canonically-definable through iteration of (N, U): In the above notation, any order-preserving map from the κ_i's to the κ_i's extends to such an elementary embedding. If $\pi : V \to V$ is any such embedding then we obtain not only the indiscernibility of the structures $H(\kappa_i^+)$, for all i but also of the structures $H(\kappa_i^{+\alpha})$ for any $\alpha < \kappa_0$ and more. Moreover, #-generation evidently provides the maximum amount of vertical reflection: If V is generated by (N, U) as $\mathrm{LP}(N_\infty)$ where ∞ is the ordinal height of V, and x is any parameter in a further iterate $V^* = N_{\infty^*}$ of (N, U), then any first-order property $\varphi(V, x)$ that holds in V^* reflects to $\varphi(V_{\kappa_i}, \bar{x})$ in N_j for all sufficiently large $i < j < \infty$, where $\pi_{j,\infty^*}(\bar{x}) = x$. This implies any known form of vertical reflection and summarizes the amount of reflection one has in L under the assumption that $0^\#$ exists, the maximum amount of reflection in L. This is reinforced by a Jensen's *#-generated coding theorem* (Theorem 9.1. of [6]) which states that if V is #-generated then V can be coded into a #-generated model $L[x]$ for a real x where the given # which generates V extends to the natural generator $x^\#$ for the model $L[x]$.

From this we can conclude that #-generated models have the same large cardinal and reflection properties as does L when $0^\#$ exists.

#-generation also answers our question about which *canonical* tower of lengthenings of V to look at in reflection, namely the further lower parts of iterates of any # that generates V. This tower of lengthenings is independent of the choice of generating # for V and is therefore entirely *canonical.* And #-generation fully realizes the idea that V should look exactly like closed unboundedly many of its rank initial segments as well as its *canonical* lengthenings of arbitrary ordinal height.

In summary, #-generation stands out as the correct formalization of the principle of *height maximality*, and we shall refer to #-generated models as being *maximal in height.* It is not first-order (we have argued that no optimal height maximality principle can be), however it is second-order in a very restricted way: For a countable V, the property of being a $\#$ that generates V is expressible by quantifying universally over the models $L_\alpha(V)$ as α ranges over the countable ordinals.

4.5 Maximality in Width and the IMH

Whereas in the case of *maximality in height* we can use height potentialism (i.e., the option of lengthening V to taller universes) to arrive at an optimal principle, the case of *maximality in width* is of a very different nature. Unlike in the case of height maximality, we will see that there are many distinct criteria for width maximality and will not easily arrive at an optimal criterion. Moreover, to get a fair picture of maximality in both height and width, it is necessary to *synthesise* or *unify* width maximality criteria with $\#$-generation, the optimal height maximality criterion.

A thorough analysis of the different possible width maximality criteria and their synthesis with $\#$-generation, with an aim towards arriving at an optimal criterion, is the principal aim of the *Hyperuniverse Programme.*

I'll begin with a discussion of width maximality in the context of radical potentialism, as this offers a simpler theory than that provided by the Zermelian view. Thus we use the symbol V to be a variable ranging not over the Zermelian multiverse (in which universes are ordered by the relation of rank-initial segment) but over elements of the rich multiverse provided by radical potentialism, in which each universe is potentially countable. We begin with the fundamental:

Inner Model Hypothesis (IMH, [12]) If a first-order sentence holds in some outer model of V then it holds in some inner model of V.

For the current presentation, we may take *outer model* to mean a transitive set V^* containing V, with the same ordinals as V, which satisfies ZFC. An *inner model* in this presentation is a V-definable subclass of V with the same ordinals as V which satisfies ZFC. By radical potentialism, any transitive model of ZFC is countable in a larger such model and from this we can infer the existence of a rich collection of outer models of V.

The consistency of $\#$-generation follows from the existence of $0^\#$. But the consistency of the IMH, i.e. the assertion that there are universes V satisfying the IMH, requires more.

Consistency of the IMH

Theorem 3. *([18]) Assuming large cardinals there exists a countable transitive model M of ZFC such that if a first-order sentence φ holds in an outer model N of M then it also holds in an inner model of M.*

Proof. For any real R let $M(R)$ denote the least transitive model of ZFC containing R. We are assuming large cardinals so indeed such an $M(R)$ exists (the existence of just an inaccessible is sufficient for this). We will need the following consequence of large cardinals:

$(*)$ There is a real R such that for any real S in which R is recursive, the (first-order) theory of $M(R)$ is the same as the theory of $M(S)$.

One can derive $(*)$ from large cardinals as follows. Large cardinals yield Projective Determinacy (PD). A theorem of Martin is that PD implies the following *Cone Theorem*: If X is a projective set of reals closed under Turing-equivalence then for some real R, either S belongs to X for all reals S in which R is recursive or S belongs to the complement of X for all reals S in which R is recursive.

Now for each sentence φ consider the set $X(\varphi)$ consisting of those reals R such that $M(R)$ satisfies φ. This set is projective and closed under Turing-equivalence. By the cone theorem we can choose a real $R(\varphi)$ so that either φ is true in $M(S)$ for all reals S in which $R(\varphi)$ is recursive or this holds for $\sim \varphi$. Now let R be any real in which every $R(\varphi)$ is recursive; as there are only countably-many φ's this is possible. Then R witnesses the property $(*)$.

We claim that if N is an outer model of $M(R)$ satisfying ZFC and φ is a sentence true in N then φ is true in an inner model of $M(R)$. For this we need the following deep theorem of Jensen.

Coding Theorem (see [6]) Let α be the ordinal height of N. Then N has an outer model of the form $L_\alpha[S]$ for some real S which satisfies ZFC and in which N is Δ_2-definable with parameters.

As R belongs to $M(R)$ it also belongs to N and hence to $L_\alpha[S]$ where S codes N as above. Also note that since α is least so that $M(R) = L_\alpha[R]$ models ZFC, it is also least so that $L_\alpha[S]$ satisfies ZFC and therefore $L_\alpha[S]$ equals $M(S)$.

Clearly we can choose S to be Turing above R (simply replace S by its join with R). But now by the special property of R, the theories of $M(R)$ and $M(S)$ are the same. As N is a definable inner model of $M(S)$, part of the theory of $M(S)$ is the statement "There is an inner model of φ which is Δ_2-definable with parameters" and therefore there is an inner model of $M(R)$ satisfying φ, as desired. □

Note that the model that we produce above for the IMH, $M(R)$ for some real R, is the minimal model containing the real R and therefore satisfies "there are no inaccessible cardinals". This is no accident:

Theorem 4. *[12] Suppose that M satisfies the IMH. Then in M: There are no inaccessible cardinals and in fact there is a real R such that there is no transitive model of ZFC containing R.*

Proof. A theorem of Beller and David (also in [6]) extends Jensen's Coding Theorem to say that any model M has an outer model of the form $M(R)$ for some real R, where as above $M(R)$ is the minimal transitive model of ZFC containing R. Now suppose that M satisfies the IMH and consider the sentence "There is no inaccessible cardinal". This is true in an outer model $M(R)$ of M and therefore in an inner model of M. It follows that there are no inaccessibles in M. The same argument with the sentence "There is a real R such that there is no transitive model of ZFC containing R" gives an inner model M_0 of M with this property for some real R; but then also M has this property as any transitive model of ZFC containing R in M would also give such a model in the $L[R]$ of M and therefore in M_0, as M_0 contains the $L[R]$ of M. □

It follows that if M satisfies the IMH then some real in M has no # and therefore boldface Π^1_1 determinacy fails in M (although $0^\#$ does exist and lightface Π^1_1 determinacy does hold).

Width actualism

So far I have presented the IMH in the context of radical potentialism, which allows us to talk freely about *outer models (thickenings)* of the universe V. This is of course unacceptable to the width actualist, who sees a fixed meaning to V_α for each ordinal α (although possibly an unfixed, potentialist view of what the ordinals are). Is it possible to nevertheless talk about the *maximality of V in width* from a width actualist perspective (where V is now a variable ranging over the Zermelian multiverse)? Can we express the idea that V is *as thick as possible* without actually comparing V to thicker universes (which do not exist)?

A positive answer to the latter question emerges through a study of V-logic, to which I turn next. A useful reference for this material is Barwise's book [5].

V-Logic

Let's start with something simpler, V_ω-logic. In V_ω-logic we have constant symbols $\bar{a}$ for $a \in V_\omega$ as well as a constant symbol $\bar{V}_\omega$ for V_ω itself (in addition to $\in$ and the other symbols of first-order logic). Then to the usual logical axioms and the rule of *Modus Ponens* we add the rules:

For $a \in V_\omega$: From $\varphi(\bar{b})$ for each $b \in a$ infer $\forall x \in \bar{a}\varphi(x)$.

From $\varphi(\bar{a})$ for each $a \in V_\omega$ infer $\forall x \in \bar{V}_\omega\ \varphi(x)$.

Introducing the second of these rules generates new provable statements via proofs which are now infinite. The idea of V_ω-logic is to capture the idea of a model *in which V_ω is standard.* By the *ω-completeness theorem*, the logically provable sentences of V_ω-logic are exactly those which hold in every model in which $\bar{a}$ is interpreted as a for $a \in V_\omega$ and $\bar{V}_\omega$ is interpreted as the (real, standard) V_ω. Thus a theory T in V_ω-logic is consistent in V_ω-logic iff it has a model in which V_ω is the real, standard V_ω.

Now the set of logically-provable formulas (i.e. validities) in V_ω-logic, unlike in first-order logic, is not arithmetical, i.e. it is not definable over the model V_ω. Instead it is definable over a larger structure, a *lengthening of* V_ω. Let me explain.

As proofs in V_ω-logic are no longer finite, they do not naturally belong to V_ω. Instead they belong to *the least admissible set* $(V_\omega)^+$ containing V_ω as an element, this is known to higher recursion-theorists as $L_{\omega_1^{ck}}$, where ω_1^{ck} is the least non-recursive ordinal. Something very nice happens: Whereas proofs in first-order logic belong to V_ω and therefore provability is Σ_1 definable over V_ω (*there exists a proof* is Σ_1), proofs in V_ω-logic belong to $(V_\omega)^+$ and provability is Σ_1 definable over $(V_\omega)^+$.

For our present purposes the point is that $(V_\omega)^+$ is a lengthening, not a thickening of V_ω and in this lengthening we can formulate theories which describle arbitrary models in which V_ω is standard. For example the existence of a real R such that (V_ω, R) satisfies a first-order property can be formulated as the consistency of a theory in V_ω-logic. As the structure (V_ω, R) can be regarded as a "thickening" of V_ω, we have described what can happen in "thickenings" of V_ω by a theory in $(V_\omega)^+$, a lengthening of V_ω. This is even more dramatic if we start not with V_ω but with $(V_\omega)^+ = L_{\omega_1^{ck}}$ and introduce $L_{\omega_1^{ck}}$*-logic*, a logic for ensuring that the recursive ordinals are standard. Then in the lengthening $(L_{\omega_1^{ck}})^+$ of $L_{\omega_1^{ck}}$, the least admissible set containing $L_{\omega_1^{ck}}$ as an element, we can express the existence of a thickening of

$L_{\omega_1^{ck}}$ in which a first-order statement holds, and such thickenings can contain new reals and more as elements.

V-logic is analogous to the above. It has the following constant symbols:

1. A constant symbol $\bar{a}$ for each set a in V.
2. A constant symbol $\bar{V}$ to denote the universe V.

Formulas are formed in the usual way, as in any first-order logic. To the usual axioms and rules of first-order logic we add the new rules:

($*$) From $\varphi(\bar{b})$ for all $b \in a$ infer $\forall x \in \bar{a}\varphi(x)$.

($**$) From $\varphi(\bar{a})$ for all $a \in V$ infer $\forall x \in \bar{V}\varphi(x)$.

This is the logic to describe models in which V is standard. The proofs of this logic appear in V^+, the least admissible set containing V as an element; this structure V^+ is a special lengthening of V of the form $L_\alpha(V)$, the α-th level of Gödel's L-hierarchy built over V. We refer to such lengthenings as *Gödel lengthenings*. Recall that with our height potentialist perspective, we can lengthen V to models V^* with V as a rank-initial segment, and therefore surely lengthen V to the Gödel lengthening V^+. (This is also the case with a height actualist perspective, provided we allow our classes to satisfy MK (Morse-Kelley), as in MK we can construct a class coding V^+.)

The Inner Model Hypothesis for a Width Actualist

As width actualists we cannot talk directly about outer models or even about sets that do not belong to V. However using V-logic we can talk about them indirectly, as I'll now illustrate. Consider the theory in V-logic where we not only have constant symbols $\bar{a}$ for the elements of V and a constant symbol $\bar{V}$ for V itself, but also a constant symbol $\bar{W}$ to denote an "outer model" of V. We add the new axioms:

1. The universe is a model of ZFC (or at least the weaker KP, admissibility theory).
2. $\bar{W}$ is a transitive model of ZFC containing $\bar{V}$ as a subset and with the same ordinals as V.

So now when we take a model of our axioms which obeys the rules of V-logic, we get a universe modelling ZFC (or at least KP) in which $\bar{V}$ is interpreted correctly as V and $\bar{W}$ is interpreted as an outer model of V. Note that this theory in V-logic has been formulated without "thickening" V, indeed it is defined inside V^+, the least admissible set containing V, a Gödel lengthening of V. Again the latter makes sense thanks to our adoption of height (not width) potentialism.

So what does the *IMH* really say for a width actualist? It says the following:

IMH: Suppose that φ is a first-order sentence and the above theory, together the axiom "$\bar{W}$ satisfies φ" is consistent in V-logic. Then φ holds in an inner model of V.

In other words, instead of talking directly about "thickenings" of V (i.e. "outer models") we instead talk about the consistency of a theory formulated in V-logic and defined in V^+, a (mild) Gödel lengthening of V.

Note that this also provides a powerful extension of the Definability Lemma for set-forcing. The latter says that definably in V we can express the fact that a sentence with parameters holds in a "set-generic extension" (for sentences of bounded complexity, such as Σ_n sentences for a fixed n). The above shows that we can do the same for arbitrary "thickenings" of V, but where the definability takes place not in V but in V^+. (In the case of *omniscient* universes V, we can in fact obtain definability in V, and under mild large cardinal assumptions, V will be omniscient. See Subsection 4.11 for a discussion of this.)

So far we have worked with V, its lengthenings and its "thickenings" (via theories expressed in its lengthenings). We next come to an important step, which is to reduce this discussion to the study of certain properties of countable transitive models of ZFC, i.e., to the *Hyperuniverse* (the set of countable transitive models of ZFC). The net effect of this reduction is to show that our width actualist discussion of maximality is in fact equivalent to a radical potentialist discussion in which all models under consideration belong to the Hyperuniverse.

4.6 The Reduction to the Hyperuniverse

Of course it would be much more comfortable to remove the quotes in "thickenings" of V, as we could then dispense with the need to reformulate our intuitions about outer models via theories in V-logic. Indeed, if we were to have this discussion not about V but about a countable transitive ZFC model *little-V*, then our worries evaporate, as genuine thickenings become available. For example, if P is a forcing notion in little-V then we can surely build a P-generic extension to get a little-$V[G]$.

Of course we can't do this for V itself as in general we cannot construct generic sets for partial orders with uncountably many maximal antichains.

But the way we have analysed things with V-logic allows us to *reduce* our study of maximality criteria for V to a study of countable transitive models. As the collection of countable transitive models carries the name *Hyperuniverse*, we are then led to what is known as the *Hyperuniverse Programme.*

I'll illustrate the reduction to the Hyperuniverse with the specific example of the IMH. Suppose that we formulate the IMH as above, using V-logic, and want to know what first-order consequences it has.

Lemma 5. *Suppose that a first-order sentence φ holds in all countable models of the IMH. Then it holds in all models of the IMH.*

Proof. Suppose that φ fails in some model V of the IMH, where V may be uncountable. Now notice that the IMH is first-order expressible in V^+, a lengthening of V. But then apply the downward Löwenheim-Skolem theorem to obtain a countable little-V which satisfies the IMH, as verified in its associated little-V^+, yet fails to satisfy φ. But this is a contradiction, as by hypothesis φ must hold in all *countable* models of the IMH. □

So *without loss of generality*, when looking at first-order consequences of maximality criteria as formulated in V-logic, we can restrict ourselves to countable little-V's. The advantage of this is then we can dispense with the little-V-logic and the quotes in "thickenings" altogether, as by the Completeness Theorem for little-V-logic, consistent theories in little-V-logic do have models, thanks to the countability of little-V. Thus for a countable little-V, we can simply say:

IMH for little-V's: Suppose that a first-order sentence holds in an outer model of little-V. Then it holds in an inner model of little-V.

This is exactly the radical potentialist version of the IMH with which we began. Thus the width actualist and radical potentialist versions of the IMH coincide on countable models.

#-Generation Revisited

The reduction of maximality principles to the Hyperuniverse is however not always so obvious, as we will now see in the case of #-generation. This reveals a difference in the development of the HP form a Zermelian perspective versus a radical potentialist perspective.

First consider the following encouraging analogue for #-generation of our earlier reduction claim for the IMH.

Lemma 6. *Suppose that a first-order sentence φ holds in all countable models which are #-generated. Then it holds in all models which are #-generated.*

Proof. Suppose that φ fails in some #-generated model V, where V may be uncountable. Let (N, U) be a generating # for V and place both V and (N, U) inside some transitive model of ZFC minus powerset T. Now apply Löwenhiem-Skolem to T to produce a countable transitive $\bar{T}$ in which there is a $\bar{V}$ which $\bar{T}$ believes to be generated by $(\bar{N}, \bar{U})$ with an elementary embdding of $\bar{T}$ into T, sending $\bar{V}$ to V and $(\bar{N}, \bar{U})$ to (N, U). But the fact that (N, U) is iterable and $(\bar{N}, \bar{U})$ is embedded into (N, U) is enough to conclude that also $(\bar{N}, \bar{U})$ is iterable. So we now have a countable $\bar{V}$ which is #-generated (via $(\bar{N}, \bar{U})$) in which φ fails, contrary to hypothesis. □

However the difficulty is this: How do we express #-generation from a width actualist perspective? Recall that to produce a generating # for V we have to produce a set of rank less than $\mathrm{Ord}(V)$ which does not belong to V, in violation of width actualism.

And recall that a # is a structure (N, U) meeting certain first-order conditions which is in addition *iterable*: For any ordinal α if we iterate (N, U) for α steps then it remains wellfounded. V is #-generated if there is a # which generates it. But notice that to express the iterability of a generating # for V we are forced to consider theories T_α formulated in $L_\alpha(V)$-logic for *arbitrary* Gödel lengthenings $L_\alpha(V)$ of V: T_α asserts that V is generated by a *pre-#* (i.e. by a structure that looks like a # but may not be fully iterable) which is *α-iterable*, i.e. iterable for α-steps. Thus we have no fixed theory that captures #-generation but only a tower of theories T_α (as α ranges over ordinals past the height of V) which capture closer and closer approximations to it.

Definition 7. *V is* weakly #-generated *if for each ordinal α past the height of V, the theory T_α which expresses the existence of an α-iterable pre-# which generates V is consistent.*

Weak #-generation is meaningful for a width actualist (who accepts enough height potentialism to obtain Gödel lengthenings) as it is expressed entirely in terms of theories internal to Gödel lengthenings of V.

For a countable little-V, weak $\#$-generation can be expressed semantically. First a useful definition:

Definition 8. *Let little-V be a countable transitive model of ZFC and α an ordinal. Then little-V is* α-generated *if there is an α-iterable pre-$\#$ which generates little-V (as the union of the lower parts of its first γ iterates, where γ is the ordinal height of little-V).*

Then a countable little-V is weakly $\#$-generated if it is α-generated for each countable ordinal α (where the witness to this may depend on α). Little-V is $\#$-generated iff it is α-generated when $\alpha = \omega_1$ iff it is α-generated for all ordinals α.

Just as a syntactic approach is needed for a width actualist fomulation of $\#$-generation, the reduction of this weakened form of $\#$-generation to the Hyperuniverse takes a syntactic form:

Lemma 9. *Suppose that a first-order sentence φ holds in all countable little-V which are weakly $\#$-generated, and this is provable in ZFC. Then φ holds in all models which are weakly $\#$-generated.*

Proof. Let W be a weakly $\#$-generated model (which may be uncountable). Thus for each ordinal α above the height of W, the theory $T_\alpha + \sim \varphi$ expressing that φ fails in W and W is generated by an α-iterable pre-$\#$ is consistent. If we choose α so that $L_\alpha(W)$ is a model of ZFC (or enough of ZFC where the truth of φ in countable $\#$-generated models provable) then $L_\alpha(W)$ is a model of (enough of) ZFC in which W is weakly $\#$-generated. Apply Löwenheim-Skolem to obtain a countable $\bar{W}$ and $\bar{\alpha}$ such that $L_{\bar{\alpha}}(\bar{W})$ embeds elementarily into $L_\alpha(W)$ and therefore satisfies (enough of) ZFC plus "$\bar{W}$ is weakly $\#$-generated". Now let g be generic over $L_{\bar{\alpha}}(\bar{W})$ for the Lévy collapse of (the height of) $\bar{W}$ to ω; then $L_{\bar{\alpha}}(\bar{W})[g]$ is a model of (enough of) ZFC in which $\bar{W}$ is both countable and weakly $\#$-generated. By hypothesis $L_{\bar{\alpha}}(W)[g]$ satisfies "$\bar{W}$ satisfies φ" and therefore $\bar{W}$ really does satisfy φ. Finally, by elementarity W satisfies φ as well, as desired. □

To summarise: As radical potentialists we can comfortably work with full $\#$-generation as our principle of height maximality. But as width actualists we instead work with weak $\#$-generation, expressed in terms of theories inside Gödel lengthenings $L_\alpha(V)$ of V. Weak $\#$-generation is sufficient to maximise the height of the universe. And properly formulated, the reduction to the Hyperuniverse applies to weak $\#$-generation: To infer that a first-order statement follows from weak $\#$-generation it suffices to show that in ZFC one can prove that it holds in all weakly #-generated countable models.

Weak #-generation is indeed strictly weaker than #-generation for countable models: Suppose that $0^{\#}$ exists and choose α to be least so that α is the α-th Silver indiscernible (α is countable). Now let g be generic over L for Lévy collapsing α to ω. Then by Lévy absoluteness, L_α is weakly #-generated in $L[g]$, but it cannot be #-generated in $L[g]$ as $0^{\#}$ does not belong to a generic extension of L.

In what follows I will primarily work with #-generation, as at present the mathematics of weak #-generation is poorly understood. Indeed, as we'll see in the next section, a synthesis of #-generation with the IMH is consistent, but this remains an open problem for weak #-generation.

4.7 Synthesis

We introduced the IMH as a criterion for width maximality and #-generation as a criterion for height maximality. It is natural to see how these can be combined into a single criterion which recognises both forms of maximality. We achieve this in this section through *synthesis*. Note that the IMH implies that there are no inaccessibles yet #-generation implies that there are. So we cannot simply take the conjunction of these two criteria.

A #-generated model M satisfies the IMH# iff whenever a sentence holds in a #-generated outer model of M it also holds in an inner model of M.

Note that IMH# differs from the IMH by demanding that both M and M^*, the outer model, are #-generated (while the outer models considered in IMH are arbitrary). The motivation behind this requirement is to impose width maximality only with respect to those models which are height maximal.

Theorem 10. *[15] Assuming that every real has a # there is a real R such that any #-generated model containing R satisfies the IMH#.*

Proof. (Woodin) Let R be a real with the following property: Whenever X is a lightface and nonempty Π^1_2 set of reals, then X has an element recursive in R. We claim that any #-generated model M containing R as an element satisfies the IMH#.

Suppose that φ holds in M^*, a #-generated outer model of M. Let (m^*, U^*) be a generating # for M^*. Then the set X of reals S such that S codes such an (m^*, U^*) (generating a model of φ) is a lightface Π^1_2 set. So there is such a real recursive in

R and therefore in M. But then M has an inner model satisfying φ, namely any model generated by a # coded by an element of X in M. □

The argument of the previous theorem is special to the weakest form of IMH#. The original argument from [15], used *#-generated Jensen coding* to prove the consistency of a stronger principle, SIMH#(ω_1); see Theorem 15.

Corollary 11. *Suppose that φ is a sentence that holds in some V_κ with κ measurable. Then there is a transitive model which satisfies both the IMH# and the sentence φ.*

Proof. Let R be as in the proof of Theorem 10 and let U be a normal measure on κ. The structure $N = (H(\kappa^+), U)$ is a #; iterate N through a large enough ordinal ∞ so that $M = LP(N_\infty)$, the lower part model generated by N, has ordinal height ∞. Then M is #-generated and contains the real R. It follows that M is a model of the IMH#. Moreover, as M is the union of an elementary chain $V_\kappa = V_\kappa^N \prec V_{\kappa_1}^{N_1} \prec \cdots$ where φ is true in V_κ, it follows that φ is also true in M. □

Note that in Corollary 11, if we take φ to be any large cardinal property which holds in some V_κ with κ measurable, then we obtain models of the IMH# which also satisfy this large cardinal property. This implies the compatibility of the IMH# with arbitrarily strong large cardinal properties.

Question 12. *Reformulate IMH# using weak #-generation, as follows: V is weakly #-generated and for each sentence φ, if the theories expressing that V has an outer model satisfying φ with an α-iterable generating pre-# are consistent for each α, then φ holds in an inner model of V. Is this consistent?*

The above formulation of IMH# for weak #-generation takes the following form for a countable V: V is α-generated for each countable α and for all φ, if φ holds in an α-generated outer model of V for each countable α then φ holds in an inner model of V. It is not known if this is consistent.

Remark. An even weaker form of #-generation asserts that V is just Ord(V) + Ord(V)-generated, a sufficient amount of iterability to obtain ordinal maximality. However a synthesis of the IMH with this very weak #-generation yields a consistent principle that contradicts large cardinals (indeed the existence of #'s for arbitrary reals). These different forms of #-generation, and of their synthesis with the IMH, are in need of further philosophical discussion.

We have now laid the foundations for the HP and discussed the two most basic maximality principles, #-generation and the IMH. Most of the mathematical work in the HP remains to be done. Therefore what I will do in the remainder of this article is simply present a range of maximality criteria which are yet to be fully analysed and which give the flavour of how the HP is intended to proceed. These criteria are also referred to as *H-axioms*, formulated as properties of elements of the Hyperuniverse H, expressible as maximality properties within H.

4.8 The Strong IMH

Our discussion of the IMH has been always with regard to sentences, without parameters. Stronger forms result if we introduce parameters.

First note the difficulties with introducing parameters into the IMH. For example the statement

"If a sentence with parameter ω_1^V holds in an outer model of V then it holds in an inner model"

is inconsistent, as the parameter ω_1^V could become countable in an outer model and therefore the above cannot hold for the sentence "ω_1^V is countable". If we however require that ω_1 is preserved then we get a consistent principle.

Theorem 13. *Let SIMH(ω_1) be the following principle: If a sentence with parameter ω_1 holds in an ω_1-preserving outer model then it holds in an inner model. Then the SIMH(ω_1) is consistent (assuming large cardinals).*

Proof. Again use PD to get a real R such that the theory of $M(S)$, the least transitive ZFC model containing S, is fixed for all S Turing above R. Now suppose that $\varphi(\omega_1)$ is a sentence true in an ω_1-preserving outer model N of $M(R)$, where ω_1 denotes the ω_1 of $M(R)$. Then as in the proof of consistency of the IMH, we can code N into $M(S)$ for some real S Turing above R, and moreover this coding is ω_1-preserving. As $\varphi(\omega_1)$ holds in a definable inner model of $M(S)$ and ω_1 is the same in $M(R)$ and $M(S)$, it follows that $M(R)$ also has an inner model satisfying $\varphi(\omega_1)$. □

The above argument uses the fact that Jensen-coding is ω_1-preserving. It is however not ω_2-preserving unless CH holds, and therefore we have the following open question:

Question 14. *Let SIMH(ω_1, ω_2) be the following principle: If a sentence with parameters ω_1, ω_2 holds in an ω_1-preserving and ω_2-preserving outer model then it holds in an inner model. Then is the SIMH(ω_1, ω_2) consistent (assuming large cardinals)?*

The SIMH(ω_1, ω_2) implies that CH fails, as any model has a cardinal-preserving outer model in which there is an injection from ω_2 into the reals. Is there an analogue $M^*(R)$ of the minimal model $M(R)$ which does not satisfy CH? Is there a coding theorem which says that any outer model of $M^*(R)$ which preserves ω_1 and ω_2 has a further outer model of the form $M^*(S)$, also with the same ω_1 and ω_2? If so, then one could establish the consistency of the SIMH(ω_1, ω_2).

The most general from of the SIMH makes use of *absolute parameters.* A parameter p is *absolute* if some formula defines it in all outer models which preserve cardinals up to and including the hereditary cardinality of p, i.e. the cardinality of the transitive closure of p. Then SIMH(p) for an absolute parameter p states that if a sentence with parameter p holds in an outer model which preserves cardinals up to the hereditary cardinality of p then it holds in an inner model. The full SIMH (Strong Inner Model Hypothesis) states that this holds for every absolute parameter p.

The SIMH is closely related to strengthenings of Lévy absoluteness. For example, define Lévy(ω_1) to be the statement that Σ_1 formulas with parameter ω_1 are absolute for ω_1-preserving outer models; this follows from the $SIMH(\omega_1)$ and is therefore consistent. But the consistency of Lévy(ω_1, ω_2), i.e. Σ_1 absoluteness with parameters ω_1, ω_2 for outer models which preserve these cardinals, is open.

The SIMH#

A synthesis of the SIMH with #-generation can be formulated as follows: V satisfies the SIMH# if V is #-generated and whenever a sentence φ with absolute parameters holds in a #-generated outer model having the same cardinals as V up to the hereditary cardinality of those parameters, φ also holds in an inner model of V. A special case is SIMH#(ω_1), where the only parameter involved is ω_1 and we are concerned only with ω_1-preserving outer models.

Theorem 15. *[15] Assuming large cardinals, the SIMH#(ω_1) is consistent.*

Proof. Assume there is a Woodin cardinal with an inaccessible above. For each real R let $M^{\#}(R)$ be $L_\alpha[R]$ where α is least so that $L_\alpha[R]$ is #-generated. The Woodin cardinal with an inaccessible above implies enough projective determinacy to enable us to use Martin's Lemma to find a real R such that the theory of $M^{\#}(S)$ is constant for S Turing-above R. We claim that $M^{\#}(R)$ satisfies SIMH#(ω_1): Indeed, let M be a #-generated ω_1-preserving outer model of $M^{\#}(R)$ satisfying some sentence $\varphi(\omega_1)$. Let α be the ordinal height of $M^{\#}(R)$ (= the ordinal height of M). By the result of Jensen quoted before (Theorem 9.1 of [6]), M has a *#-generated* ω_1-preserving outer model W of the form $L_\alpha[S]$ for some real S with $R \leq_T S$. Of course α is least so that $L_\alpha[S]$ is #-generated. So W equals $M^\#(S)$ and the ω_1 of W equals the ω_1 of $M^\#(R)$. By the choice of R, $M^\#(R)$ also has a definable inner model satisfying $\varphi(\omega_1)$. □

However as with the SIMH(ω_1, ω_2), the consistency of SIMH#(ω_1, ω_2) is open.

4.9 A Maximality Protocol

This protocol aims to organise the study of height and width maximality into three stages.

Stage 1. Maximise the ordinals (height maximality).

Stage 2. Having maximised the ordinals, maximise the cardinals.

Stage 3. Having maximised the ordinals and cardinals, maximise powerset (width maximality).

Stage 1 is taken care of by #-generation. So we focus now on Stage 2, cardinal-maximisation.

In light of Stage 1, we assume now that V is #-generated and when discussing outer models of V we only consider those which are also #-generated.

We would like a criterion which says that for each cardinal κ, κ^+ is *as large as possible.* To get started let's consider the case $\kappa = \omega$, so we want to maximise ω_1. The basic problem of course is the following. As set-generic extensions of #-generated models are also #-generated:

Fact. V has a #-generated outer model in which ω_1^V is countable.

But surely we would want something like: $\omega_1^{L[x]}$ is countable for each real x. The reason for this is that $\omega_1^{L[x]}$, unlike ω_1^V in general, is *absolute* between V and all of its outer models.

Definition 16. *Let p be a parameter in V and P a set of parameters in V. Then p is* strongly absolute relative to P *if there is a formula φ with parameters from P*

that defines p in V and all #-generated outer models of V which preserve cardinals up to and including the hereditary cardinality of the parameters mentioned in φ[10].

Typically we will take P to consist of all subsets of some infinite cardinal κ, in which case the cardinal-preservation in the above definition refers to cardinals up to and including κ.

CardMax(κ^+) (for κ an infinite cardinal). Suppose that the ordinal α is strongly absolute relative to subsets of κ. Then α has cardinality at most κ.

It is possible to show that if κ is regular then there is a set-forcing extension in which CardMax(κ^+) holds.

Question 17. *Is CardMax consistent, where CardMax denotes CardMax(κ^+) for all infinite cardinals κ, both regular and singular?*

Internal Cardinal Maximality

Another approach to cardinal maximality is to relate the cardinals of V to those of its inner models. Two large inner models are HOD, the class of hereditarily ordinal-definable sets, and the smaller inner model $\mathbb{S}$, the *Stable Core* of [13]. V is class-generic over each of these models.

Let M denote an inner model.

M-cardinal Violation. For each infinite cardinal κ, κ^+ is greater than the κ^+ of M.

In [9] it is shown that HOD-cardinal violation is consistent. Can we strengthen this?

Question 18. *Is it consistent that for each infinite cardinal κ, κ^+ is inaccessible, measurable or even supercompact in HOD? Is this consistent with HOD replaced by the Stable Core $\mathbb{S}$?*

A result of Shelah states that all subsets of κ belong to HOD_x for some fixed subset x of κ when κ is a singular strong limit cardinal of uncountable cofinality. By [8] this need not be true at countable cofinalities.

Question 19. *Is it consistent that for each infinite cardinal κ, κ^+ is greater than κ^+ of $\mathbb{S}_x$ (the Stable Core relativised to x) for each subset x of κ?*

[10] We thank one of the referees for pointing out that an earlier version of cardinal-maximality with a weaker parameter-absoluteness assumption is inconsistent. A similar phenomenon with weakly absolute parameters occurs in Theorem 10 of [18].

A major difference between HOD and $\mathbb{S}$ is that while any set is set-generic over HOD, this is not the case for $\mathbb{S}$.

Question 20. *Is it consistent that for each infinite cardinal κ, some subset of κ^+ is not set-generic over $\mathbb{S}_x$ for any subset x of κ?*

A positive answer to any of these three questions would yield a strong internal cardinal-maximality principle for V.

Stage 3: Having maximised the ordinals and cardinals, maximise powerset.

This is where we revisit the SIMH, but only in the context of #-generation and cardinal-preservation. Again assume that V is #-generated.

A parameter p in V is *cardinal-absolute* if there is a parameter-free formula which defines p in all #-generated outer models of V which have the same cardinals as V.

SIMH#(CP) (Cardinal-preserving SIMH#). Suppose that p is a cardinal-absolute parameter, V^* is a #-generated outer model of V with the same cardinals as V and φ is a sentence with parameter p which holds in V^*. Then φ holds in an inner model of V.

Question 21. *Is the SIMH#(CP) consistent?*

Note that SIMH#(CP) implies a strong failure of CH.

4.10 Width Indiscernibility

An alternative to the Maximality Protocol (which ideally should be synthesised with it) is *Width Indiscernibility.* The motivation is to provide a description of V in width analogous to its description in height provided by #-generation.

Recall that with #-generation we arrive at the following:

$$V_0 \prec V_1 \prec \cdots \prec V = V_\infty \prec V_{\infty+1} \prec \cdots$$

where for $i < j$, V_i is a *rank-initial segment* of V_j. Moreover the models V_i form a collection of *indiscernible models* in a strong sense. This picture was the result of an analysis which began with *height reflection*, starting with the idea that V must have unboundedly many rank-initial segments V_i which are elementary in V.

Analogously, we introduce *width reflection.* We would like to say that V has proper inner models which are "elementary in V". Of course this cannot literally be true, as if V_0 is an elementary submodel of V with the same ordinals as V then it is easy to see that V_0 equals V. Instead, we use elementary embeddings.

Width Reflection. For each ordinal α, there is a proper elementary submodel H of V such that $V_\alpha \subseteq H$ and H is *amenable*, i.e. $H \cap V_\beta$ belongs to V for each ordinal β.

Equivalently:

Width Reflection. For each ordinal α, there is a nontrivial elementary embedding $j : V_0 \to V$ with critical point at least α such that j is *amenable*, i.e. $j \restriction (V_\beta)^{V_0}$ belongs to V for each ordinal β.

Let's write $V_0 < V$ if there is a nontrivial amenable $j : V_0 \to V$, as in the second formulation of width reflection. This relation is transitive.

Proposition 22. *(a) If $V_0 < V$ then V_0 is a proper inner model of V.*
(b) Width Reflection is consistent relative to the existence of a Ramsey cardinal.

Proof. (a) This follows from Kunen's Theorem that there can be no nontrivial elementary embedding from V to V.
(b) Suppose that κ is Ramsey. Then it follows that any structure of the form $\mathcal{M} = (V_\kappa, \in, \ldots)$ has an unbounded set of indiscernibles, i.e. an unbounded subset I of κ such that for each n, any two increasing n-tuples from I satisfy the same formulas in $\mathcal{M}$. Now apply this to $\mathcal{M} = (V_\kappa, \in, <)$ where $<$ is a wellorder of V_κ of length κ. Let J be any unbounded subset of I such that $I \setminus J$ is unbounded and for any $\alpha < \kappa$, let $H(J \cup \alpha)$ denote the Skolem hull of $J \cup \alpha$ in $\mathcal{M}$. Then $H(J \cup \alpha)$ is an elementary submodel of V_κ and is not equal to V_κ because no element of $I \setminus J$ greater than α belongs to it. As V_κ contains all bounded subsets of κ it follows that $H(J \cup \alpha)$ is amenable. □

A variant of the argument in (b) above yields the consistency of arbitrarily long finite chains $V_0 < V_1 < \cdots < V_n$. But obtaining infinite such chains seems more difficult, and even more ambitiously we can ask:

Question 23. *Is it consistent to have $V_0 < V_1 < \cdots < V$ of length Ord $+ 1$ such that the union of the V_i's equals V?*

The latter would be a good start on the formulation of a consistent criterion of *Width Indiscernibility*, as an analogue for maximality in width to the criterion of maximality in height provided by #-generation.

4.11 Omniscience

By OMT(V), the *outer model theory of* V, we mean the class of sentences with arbitrary parameters from V which hold in all outer models of V. We have seen using V-logic that OMT(V) is definable over V^+. However for many universes V, OMT(V) is in fact first-order definable over V. These universes are said to be *omniscient.*

Recall the following version of Tarski's result on the undefinability of truth:

Proposition 24. *The set of sentences with parameters from V which hold in V is not (first-order) definable in V with parameters.*

Surprisingly, Mack Stanley showed however that OMT(V) can indeed be V-definable.

Theorem 25. *(M.Stanley [30]) Suppose that in V there is a proper class of measurable cardinals, and indeed this class is V^+-stationary, i.e. Ord(V) is regular with respect to V^+-definable functions and this class intersects every club in Ord(V) which is V^+-definable. Then OMT(V) is V-definable.*

Proof. Using V-logic we can translate the statement that a first-order sentence φ (with parameters from V) holds in all outer models of V to the validity of a sentence φ^* in V-logic, a fact expressible over V^+ by a Σ_1 sentence. Using this we show that the set of φ which hold in all outer models of V is V-definable.

As Ord(V) is regular with respect to V^+-definable functions we can form a club C in Ord(V) such that for κ in C there is a Σ_1-elementary embedding from Hyp(V_κ) into V^+ (with critical point κ, sending κ to Ord(V)). Indeed C can be chosen to be V^+-definable.

For any κ in C let φ^*_κ be the sentence of V_κ-logic such that φ holds in all outer models of V_κ iff φ^*_κ is valid (a Σ_1 property of Hyp(V_κ)). By elementarity, φ^*_κ is valid iff φ^* is valid.

Now suppose that φ holds in all outer models of V, i.e. φ^* is valid. Then φ^*_κ is valid for all κ in C and since the measurables form a V^+-stationary class, there is a measurable κ such that φ^*_κ is valid.

Conversely, suppose that φ^*_κ is valid for some measurable κ. Now choose a normal measure U on κ and iterate $(H(\kappa^+), U)$ for Ord(V) steps to obtain a wellfounded structure $(H^*, U*)$. (This structure is wellfounded, as for any admissible set A, any measure in A can be iterated without losing wellfoundedness for α steps, for any ordinal α in A.) Then H^* equals $\mathrm{Hyp}(V^*)$ for some $V^* \subseteq V$. By elementarity, the sentence $\varphi^*_{V^*}$ which asserts that φ holds in all outer models of V^* is valid. But as V^* is an inner model of V, φ also holds in all outer models of V.

Thus φ belongs to $\mathrm{OMT}(V)$ exactly if it belongs to $\mathrm{OMT}(V_\kappa)$ for some measurable κ, and this is first-order expressible. □

Are measurable cardinals needed for omniscience? Actually, Stanley was able to use just Ramsey cardinals, but as far as the consistency of omniscience we have the following:

Theorem 26. *([16]) Suppose that κ is inaccessible and GCH holds. Then there is an omniscient model of the form $V_\kappa[G]$ where G is generic over V. Moreover, $V_\kappa[G]$ carries a definable wellorder.*

Omniscience demonstrates that it is possible to treat truth in arbitrary outer models internally in a way similar to how truth in set-generic extensions can be handled using the standard definability and truth lemmas of set-forcing. In fact, the situation is even better in that the entire outer model theory is first-order definable, not just the restriction of this theory to sentences of bounded complexity, as is the case for set-forcing. (The key difference is that in the case of set-forcing, the ground model V is uniformly definable in its set-generic extensions and therefore the full $\mathrm{OMT}(V)$ cannot be first-order definable in V by Proposition 24. An omniscient V cannot be uniformaly definable in its arbitrary outer models for the same reason.)

Note also that by Theorem 25, omniscience synthesises well with #-generation: We need only work with models that have sufficiently many measurable cardinals.

4.12 The Future of the HP

We have discussed evidence of Type 1, coming from set theory's role as a branch of mathematics, and evidence of Type 2, coming from set theory's role as a foundation for mathematics. In the first case, evidence is judged by its value for the mathematical development of set theory and in the second case it is judged by its value for resolving independence in (and providing tools for) other areas of mathematics. In both cases the weight of the evidence is measured by a consensus of researchers working in the field.

Type 3 evidence is also measured by a consensus of researchers working in set theory (and its philosophy) but emanates instead from an analysis of the intrinsic maximality feature of the set concept as expressed by the maximal iterative conception. The Hyperuniverse Programme provides a strategy for *deriving* mathematical consequences from this conception.

To illustrate more clearly how the HP derives consequences of the maximality of V I'll discuss the case of #-generation and the search for an *optimal maximality criterion.*

#-generation is a major success of the HP. It provides a powerful mathematical criterion for height maximality which implies all prior known height maximality principles and provides an elegant description of how the height of V is maximised in a way analogous to the way L is maximised in height by the existence of large cardinals (or equivalently, by the existence of $0^{\#}$). There are good reasons to beleive that #-generation will be accepted by the community of set-theorists and philosophers of set theory as the definitive expression of height maximality.

Width maximality is of course much more difficult than height maximality and the formulation, analysis and synthesis of the various possible width maximality criteria is at its early stages. The basic IMH is a good start, but must be synthesised with #-generation. The biggest challenge at the moment is dealing with formulations of width maximality which make use of parameters. The *maximality protocol* is a promising approach. But it is important to emphasize that the mathematical analysis of width maximality principles is challenging and there are sure to be some false turns in the development of the programme, leading to inconsistnet principles (this has already happened several times). Such false turns are not damaging to the programme, but rather provide valuable further understanding of the nature of maximality.

The aim of the HP is to arrive after extensive mathematical work at an *optimal criterion* of maximality for the height and width of the universe of sets, providing a full mathematical analysis of the maximal iterative conception. As already said, the validation of such a criterion as optimal depends on a consensus of researchers working in set theory and its philosophy. *Derivability* from the maximal iterative conceptions refers to formal derivability form this sought-after optimal criterion. Of greatest interest are the first-order statements derivable from maximality, but it is already clear that the criteria being developed in the programme, such as the ones mentioned in this paper, are almost exclusively *non first-order.* My prediction is that the optimal criterion will include some form of the SIMH and therefore imply the (first-order) failure of CH.

I remain optimistic that when the discoveries of this programme are combined with further work in set theory and its application to resolving problems of independence in other areas of mathematics, the prediction expressed by the *Thesis of Set-Theoretic Truth* will be satisfyingly realized. But there is first a lot of work to be done.

References

[1] C.Antos, S.Friedman, R.Honzik and C.Ternullo, Multiverse conceptions in set theory, Synthèse 192, no.8, pp. 2463–2488, 2015.

[2] T.Arrigoni and S.Friedman, Foundational implications of the Inner Model Hypothesis, Annals of Pure and Applied Logic, vol.163, pp.1360–66, 2012.

[3] T.Arrigoni and S.Friedman, The Hyperuniverse Program, Bulletin of Symbolic Logic, Volume 19, Number 1, March, 2013, pp.77–96.

[4] Barton, Neil, Multiversism and Concepts of Set: How much Relativism is acceptable? Objectivity, Realism, and Proof. FilMat Studies in the Philosophy of Mathematics, F. Boccuni, A. Sereni (eds.), Boston Studies in the Philosophy and History of Science, Springer, 2016.

[5] J.Barwise, *Admissible sets and structures*, Springer 1975.

[6] A.Beller, R.Jensen and P.Welch, *Coding the universe*, Cambridge University Press 1982.

[7] G.Boolos, The iterative conception of set, Journal of Philosophy 68 (8):215-231 (1971).

[8] J.Cummings, S.Friedman, M.Magidor, A.Rinot and D.Sinapova, Definable subsets of singular cardinals, submitted.

[9] J.Cummings, S.Friedman and M.Golshani, Collapsing the cardinals of HOD, Journal of Mathematical Logic, Vol. 15, No. 02, 2015.

[10] A.Dodd, *The Core Model*, Cambridge University Press 1982.

[11] S.Friedman, *Fine structure and class forcing*, de Gruyter 2000.

[12] S.Friedman, Internal consistency and the inner model hypothesis, Bulletin of Symbolic Logic, Vol.12, No.4 (2006), pp. 591–600.

[13] S.Friedman, The Stable Core, Bulletin of Symbolic Logic, vol.18, no.2, June 2012, pp. 261–267.

[14] S.Friedman and P.Holy, A quasi-lower bound on the consistency strength of PFA, Transactions AMS 366 (2014), 4021-4065.

[15] S.Friedman and R.Honzik, On strong forms of reflection in set theory, Mathematical Logic Quarterly, Volume 62, Issue 1-2, pages 52–58, February 2016.

[16] S.Friedman and R.Honzik, Definability of satisfaction in outer models, to appear, Journal of Symbolic Logic.

[17] S.Friedman and C.Ternullo, The search for new axioms in the hyperuniverse programme, Objectivity, Realism, and Proof. FilMat Studies in the Philosophy of Mathe-

matics, F. Boccuni, A. Sereni (eds.), Boston Studies in the Philosophy and History of Science, Springer, 2016.

[18] S.Friedman, P.Welch and H.Woodin, On the consistency strength of the Inner Model Hypothesis, Journal of Symbolic Logic, vol.73, no.2, pp. 391–400, 2008.

[19] Hamkins, J.D., A multiverse perspective on the axiom of constructibility, in *Infinity and Truth*, World Sci. Publ., Hackensack, NJ, 2014, vol. 25, pp. 25-45.

[20] Hellman, Geoffrey (1989). Mathematics Without Numbers: Towards a Modal-Structural Interpretation. Oxford University Press.

[21] Isaacson, Daniel (2011) The reality of mathematics and the case of set theory, Zsolt Novak and Andras Simonyi (eds), Truth, Reference and Realism, Central European University Press, Budapest, 2011, pp 1-76.

[22] T.Jech, *Set Theory*, Springer 2003.

[23] Koellner, Peter (2009). On reflection principles. Annals of Pure and Applied Logic 157 (2):206-219.

[24] Linnebo, Øystein (2013). The potential hierarchy of sets. Review of Symbolic Logic 6(2):205-228.

[25] Maddy, Penelope (2011). Defending the Axioms: On the Philosophical Foundations of Set Theory. Oxford University Press.

[26] Meadows, Toby (2015). Naive Infinitism: The Case for an Inconsistency Approach to Infinite Collections. Notre Dame Journal of Formal Logic 56 (1):191-212.

[27] C.Parsons, What is the iterative conception of set?. Logic, Foundations of Mathematics, and Computability Theory, Volume 9 of the University of Western Ontario Series in Philosophy of Science, pp. 335-367, 1977.

[28] Reinhardt, W (1974) Remarks on Reflection Principles, Large Cardinals, and Elementary Embeddings, Proceedings of Symposia in Pure Mathematics, vol 13, 189-205.

[29] Rumfitt, Ian (2015) Determinacy and bivalence. Forthcoming in Michael Glanzberg, ed., The Oxford Handbook of Truth (Oxford: Clarendon Press, 2015).

[30] M.Stanley, Outer model satisfiability, preprint.

[31] J.Steel, Gödel's Program, in *Interpreting Gödel*, Juliette Kennedy (ed.) Cambridge University Press, 2014.

[32] E.Zermelo, (1930) On Boundary Numbers and Domains of Sets, in William Bragg Ewald (ed.) (1996). From Kant to Hilbert: A Source Book in the Foundations of Mathematics. pp. 1208–1233.

[33] *The Thread*, an e-mail discussion during June-November 2014 (with extensive contributions by S.Feferman, H.Friedman, S.Friedman, G.Hellman, P.Koellner, P.Maddy, R.Solovay and H.Woodin).

Received 25 June 2015

Reflecting on Truth

Graham E. Leigh*
Department of Philosophy, Linguistics, and Theory of Science, University of Gothenburg, Box 200, 405 30, Gothenburg, Sweden
graham.leigh@gu.se

Abstract

What is implicit in the acceptance of the Tarskian truth biconditionals? In this article we expand and generalise results by Horsten and Leigh [14] to characterise the proof- and truth-theoretic content of iterated reflection over disquotational theories of truth. In particular, we confirm the conjecture that, modulo reflection, all there is to typed and Kripke–Feferman truth is captured by simple and natural collections of truth (and in the latter case falsity) biconditionals.

1 Introduction

Consider the theory of typed truth over arithmetic, that is the first-order theory of arithmetic expanded to the language $\mathscr{L}_T = \mathscr{L}_0 \cup \{T\}$ featuring a fresh unary predicate symbol T whose intended interpretation is the Gödel numbers of true sentences in the base language $\mathscr{L}_0$. The theory has an obvious standard model, given by expanding the standard model of arithmetic $\mathbb{N}$ to an $\mathscr{L}_T$ structure by interpreting the new predicate T as

$$\mathrm{Th}_{\mathbb{N}} := \{\#A \mid A \text{ is an } \mathscr{L}_0\text{-sentence and } \mathbb{N} \models A\},$$

where $\#\colon \mathscr{L}_T \to \omega$ is some fixed injective *Gödel numbering* for $\mathscr{L}_T$ formulæ.

Axiomatically, this expansion of the standard model can be characterised by a collection of *truth biconditionals.* As is usual, it is necessary to assume the Gödel numbering fulfils certain 'definability' assumptions (cf. section 2 below). Let

The author would like to thank Leon Horsten, Volker Halbach and the anonymous referees for their insightful comments and suggestions.

*Research supported by the Wiener Wissenschafts-, Forschungs- und Technologiefonds (WWTF), grant no. VRG12-04 and a Wallenberg Academy Fellowship from the Knut and Alice Wallenberg Foundation.

$[\cdot]\colon \mathcal{L}_T \to$ Terms be a Gödel coding that maps each $\mathcal{L}_T$ formula A to a closed $\mathcal{L}_0$-term $[A]$ whose value (in $\mathbb{N}$) is $\#A$; henceforth we identify $[A]$ and $\#A$, writing only the former. The *typed Tarskian truth biconditionals*, *TB*, is the collection of sentences

$$A \leftrightarrow T[A]$$

for A an $\mathcal{L}_0$-sentence.

Proposition 1.1. *Let $X \subseteq \operatorname{dom}\mathbb{N}$ be a set of natural numbers and let $\langle \mathbb{N}, X\rangle$ denote the expansion of the standard model of arithmetic to an $\mathcal{L}_T$-structure in which the predicate T is interpreted as X. The following are equivalent.*

1. $X \cap \mathrm{Sent}_{\mathcal{L}_0} = \mathrm{Th}_{\mathbb{N}}$,
2. $\langle \mathbb{N}, X\rangle \models TB$,
3. $\langle \mathbb{N}, X\rangle \models \mathsf{CT}$, *the theory extending* PA *by the* compositional axioms for typed truth *(see Definition 3.1).*

Proof. $2 \Rightarrow 1 \Rightarrow 3$ is obvious. Since the equivalences $A \leftrightarrow T[A]$ for A in $\mathcal{L}_0$ are derivable from the CT-axioms, also $3 \Rightarrow 2$. □

Unlike in the case of typed truth there is no obvious *standard* model for typed-free truth. On the one hand, the collection of all truth biconditionals $\{A \leftrightarrow T[A] \mid A \in \mathrm{Sent}_{\mathcal{L}_T}\}$ has no model. On the other, McGee's trick [22, Theorem 1] shows every consistent $\mathcal{L}_T$-theory extending a weak theory of arithmetic can be characterised by some consistent collection of biconditionals.

There are, however, some 'natural' theories of self-applicable truth that stand out. One such is *Kripkean truth*, introduced by Kripke in [18].[1] The theory is formulated in a language $\mathcal{L}_{T,F}$ extending $\mathcal{L}_0$ by two predicates, a truth predicate T and a falsity predicate F. T (respectively F) is interpreted as the set of $\mathcal{L}_{T,F}$ sentences whose truth (resp. falsehood) is *grounded* in $\mathrm{Th}_{\mathbb{N}}$. Formally this is given by an inductive definition. Let Γ be the function on $\mathrm{Pow}(\omega) \times \mathrm{Pow}(\omega)$ defined by $\Gamma(X_0, X_1) = (\Gamma_0(X_0, X_1), \Gamma_1(X_0, X_1))$ where

for each $i < 2$, $[A] \in \Gamma_i(X_0, X_1)$ iff one of the following conditions hold.

- A is an atomic $\mathcal{L}_0$-formula and $i = 0$ iff $\mathbb{N} \models A$,

[1] Other popular theories of type-free truth include the *revision theory*, or *stable truth*, and *supervaluationist truth*, though Kripkean truth is by far the most commonly considered theory (besides *naïve truth*).

- $A = (B \vee C)$ and $|\{[B],[C]\} \cap X_i| > i$,
- $A = (B \wedge C)$ and $|\{[B],[C]\} \cap X_i| > (1-i)$,
- $A = \forall x B$ and $i = 0$, or $A = \exists x B$ and $i = 1$, and for every term s, $[B(s)] \in X_i$,
- $A = \forall x B$ and $i = 1$, or $A = \exists x B$ and $i = 0$, and for some term s, $[B(s)] \in X_i$,
- $A = \neg B$ and $[B] \in X_{1-i}$,
- $A = T(s)$ and $s^{\mathbb{N}} \in X_i$,
- $A = F(s)$ and $s^{\mathbb{N}} \in X_{1-i}$.

If $X \subseteq X'$ and $Y \subseteq Y'$ then $\Gamma_i(X,Y) \subseteq \Gamma_i(X',Y')$, so Γ is a monotone function and has ($2^{\aleph_0}$) fixed points and, in particular, a *least* fixed point, which can also be defined by iterating Γ through the transfinite on $(\emptyset, \emptyset)$.

In the same vein as typed truth, Γ-fixed points over the standard model are characterised by collections of truth (and falsity) biconditionals and compositional axioms. Let $\mathscr{L}^+_{T,F}$ be the set of $\mathscr{L}_{T,F}$-formulæ in which predicates T and F appear strictly positively, that is not within the scope of negation symbols. For a formula A, define $\overline{A}$ to be the *dual* of A, given by exchanging each logical connective for its logical dual (with $\overline{\neg A} = A$), T for F and vice versa, and negating $\mathscr{L}_0$ atoms (see Section 5). Let *TFB* comprise the axioms

$$A \leftrightarrow T[A] \qquad\qquad \overline{A} \leftrightarrow F[A]$$

for A ranging over $\mathscr{L}^+_{T,F}$-sentences.

Proposition 1.2 (Kripke [18] and Feferman [9]). *Let $X, Y \subseteq \{[A] \mid A \in \mathrm{Sent}_{\mathscr{L}^+_{T,F}}\}$ be sets of natural numbers, and let $\langle \mathbb{N}, X, Y\rangle$ denote the expansion of $\mathbb{N}$ to an $\mathscr{L}_{T,F}$ structure with X interpreting the predicate T and Y the predicate F. The following statements are equivalent.*

1. $X = \Gamma_0(X,Y)$ *and* $Y = \Gamma_1(X,Y)$,
2. $\langle \mathbb{N}, X, Y\rangle \models$ *TFB*,
3. $\langle \mathbb{N}, X, Y\rangle \models \mathsf{KF}$, *the Kripke–Feferman theory of* positive, *or* strongly compositional, truth *(see Definition 5.1).*

Proof. 3 ⇒ 2 is on account of the truth and falsity biconditionals for $\mathscr{L}^+_{T,F}$-formulæ being derivable from the KF-axioms. 1 ⇒ 3 is easily verified and 2 ⇒ 1 results from combining the inductive definition of satisfaction with the truth and falsity biconditionals of *TFB*. □

1.1 Disquotational vs. compositional

The basic disquotational theories considered above are given by *local* truth (and falsity) biconditionals, collections $A \leftrightarrow T[A]$ for appropriate A. A natural generalisation of the local biconditionals is the inclusion of parameters. If A is a formula with free variable x the *uniform truth biconditional* for A is the axiom

$$\forall x(A \leftrightarrow T[A\dot{x}]) \tag{1}$$

where $[A\dot{x}]$ (unlike $[A]$) is an open term with exactly x free (with the dot above x marking it as a variable of the term). Whereas the local biconditionals axiomatise expressions *'A' is true iff A*, the uniform version formalises *for every y, 'λxA' is true of y iff A(y)*. Notice that in the presence of a background theory of sequences (such as in Elementary Arithmetic) the uniform biconditionals naturally expand to formulæ with more free variables and the single variable form presented above is equivalent to the multi-variable version:

$$\forall x_1 \cdots \forall x_k(Bx_1 \cdots x_k \leftrightarrow T[B\dot{x}_1 \cdots \dot{x}_k])$$

for formulæ B with at most $x_1, \ldots, x_k$ free. Let TB_0 and UTB_0 be the theories extending Elementary Arithmetic, EA (see Definition 2.1 below), by respectively the local and uniform truth biconditionals for $\mathscr{L}_0$ formulæ, and TFB_0 and UTFB_0 the extension of EA by the local and uniform truth and falsity biconditionals for the language $\mathscr{L}^+_{T,F}$.

Propositions 1.1 and 1.2 suggest an implicit link exists between collections of truth biconditionals and compositional theories, at least over the standard model. This link has not escaped attention and is clearly voiced (for example by Horwich [15] and Quine [23]) in the defense of *disquotationalism*, the general theory of truth that holds that the only functions of the concept of truth are *semantic ascent* ($A \to T[A]$) and *disquotation* ($T[A] \to A$).

When it comes to provability, however, disquotational theories are left wanting: they typically cannot derive even the most basic compositional axioms such as $\forall x \forall y(\mathrm{Sent}_{\mathscr{L}_0}(x \dot{\wedge} y) \to (T(x \dot{\wedge} y) \leftrightarrow Tx \wedge Ty))$ (*the conjunction of two $\mathscr{L}_0$ sentences is true iff each conjunct is true*), let alone reflection principles like *all theorems of Peano Arithmetic are true*, $\forall x(\mathrm{Bew}_{\mathsf{PA}}(x) \wedge \mathrm{Sent}_{\mathscr{L}_0}(x) \to Tx)$ (see, e.g. [12, Theorem 7.6]). Indeed, the standard disquotational theories are proof-theoretically trivial: UTFB_0 and UTB_0 conservatively extends Elementary Arithmetic (so if $\mathsf{UTFB}_0 \vdash A$ and A is an $\mathscr{L}_0$ formula, already $\mathsf{EA} \vdash A$) and the extension of TFB_0 by induction for *all* $\mathscr{L}_*$-formulæ conservatively extends Peano arithmetic (this result is a corollary of [4, Theorem 2.2] and a direct proof is presented in [14, Theorem 13]). On the contrary, theories such as CT and KF are proof-theoretically strong, equivalent to the extension of PA by transfinite induction (up to the ordinals $\varepsilon_{\varepsilon_0}$ and $\varphi\varepsilon_0 0$ respectively).

The *truth*-theoretic connection between disquotational and compositional theories, however, is a different matter and was examined in [14] building on earlier results of Halbach. In particular, the following was shown.

Theorem 1.3 (Halbach [11]; Horsten and Leigh [14]). *For a theory* S, *let* $\mathsf{R}(\mathsf{S})$ *denote the extension of* EA *by the uniform reflection principle for* S *(see Definition 2.11 below). Then*

1. CT *is identical to the theory* $\mathsf{R}(\mathsf{UTB}_0)$ *and a sub-theory of* $\mathsf{R}(\mathsf{R}(\mathsf{TB}_0))$*;*
2. KF *is identical to* $\mathsf{R}(\mathsf{UTFB}_0)$ *and a sub-theory of* $\mathsf{R}(\mathsf{R}(\mathsf{TFB}_0))$.

According to Feferman [9], acceptance of a theory implicitly commits one to accept a reflection principle of that theory. Modulo implicit commitment, therefore, Theorem 1.3 implies compositional truth à la CT and KF is captured by simple collections of truth (and falsity) biconditionals (see [14] for more on the discussion of implicit commitment in this context).

In the present paper we generalise Theorem 1.3 to iterations of the uniform reflection principle and thus characterise the principles implicit in our acceptance of disquotational theories (at least with respect to reflection hierarchies). Thus we are concerned not as much with the effect of reflection on the $\mathscr{L}_0$ consequences (i.e. its proof-theoretic strength) but rather with what new principles of truth are made available through reflection (its truth-theoretic content).

It might be expected that further acts of reflection permit the derivation of new truth-theoretic principles independent of CT or KF. This turns out not to be the case and iterations of the reflection principle (even into the transfinite) add no more theorems beyond (the unavoidable) transfinite induction. If transfinite induction is regarded as non-truth-theoretic then CT and KF can be said to be truth-theoretically complete. That is we obtain the following theorem.

Theorem 1.4. *For a theory* S *and (elementary) ordinal* κ, *let* S_κ *denote the extension of* S *by the schema of transfinite induction for ordinals below* κ *(for all formulæ), and* $\mathsf{R}_\kappa(\mathsf{S})$ *the theory* EA + *'κ-times iterated uniform reflection over* S*'. For all* κ *we have*

1. $\mathsf{CT}_{\varepsilon_\kappa} = \mathsf{R}_{1+\kappa}(\mathsf{UTB}_0)$ *and for* $\kappa > 0$ *also* $\mathsf{CT}_{\varepsilon_\kappa} = \mathsf{R}_{1+\kappa}(\mathsf{TB}_0)$*;*
2. $\mathsf{KF}_{\varepsilon_\kappa} = \mathsf{R}_{1+\kappa}(\mathsf{UTFB}_0)$ *and for* $\kappa > 0$ *also* $\mathsf{KF}_{\varepsilon_\kappa} = \mathsf{R}_{1+\kappa}(\mathsf{TFB}_0)$.

Theorem 1.4 is quite striking. Focusing on the theories of typed truth, observe that acceptance of CT boils down to accepting one act of (uniform) reflection over UTB_0 or two acts of reflection over the weaker theory TB_0. The compositional axioms of typed truth are not derivable in UTB_0 (nor, indeed, in UTB_α for any α in a *natural* ordinal notation system[2]), but are equivalent to the reflection principle for the theory. Furthermore, the case $\kappa > 1$ demonstrates that the entire reflection hierarchy above UTB_0 is captured by the compositional axioms and sufficient transfinite induction. In other words, modulo induction, the compositional axioms are the only truth principles revealed through reflection.

From Theorem 1.4 it becomes simple to also realise the proof-theoretic strength of the systems.

Theorem 1.5. *For each* $\kappa > 0$*, the theories* $\mathsf{R}_{1+\kappa}(\mathsf{TB}_0)$*,* $\mathsf{R}_{1+\kappa}(\mathsf{UTB}_0)$ *and* $\mathsf{CT}_{\varepsilon_\kappa}$ *derive the same* $\mathscr{L}_0$ *consequences as* $\mathsf{PA} + \mathrm{TI}(<\varepsilon_{\varepsilon_\kappa})$*, and the theories* $\mathsf{R}_{1+\kappa}(\mathsf{TFB}_0)$*,* $\mathsf{R}_{1+\kappa}(\mathsf{UTFB}_0)$ *and* $\mathsf{KF}_{\varepsilon_\kappa}$ *derive the same* $\mathscr{L}_0$ *consequences as* $\mathsf{PA} + \mathrm{TI}(<\varphi\varepsilon_\kappa 0)$*.*

1.2 Outline of paper

The paper is organised as follows. In Section 2 we fix the technical details required in later sections, including our assumptions on coding, background theory and the representation of transfinite induction and reflection hierarchies. Sections 3 to 5 explore the relation between compositional and disquotational truth theories case-by-case, considering *typed truth* (CT), *ramified truth* (RT), and *Kripke–Feferman truth* (KF). Section 6 concerns the technically challenging part of the main theorems, namely establishing that iterated reflection over disquotational theories is derivable in compositional theories extended by sufficient transfinite induction. The section introduces two infinitary calculi for theories of truth, establishes (partial) cut elimination for them and derives the desired result from an ordinal analysis of the calculi.

The proof of Theorem 1.4 is thus split into four parts. Theorem 3.10 establishes the inclusions $\mathsf{CT}_{\varepsilon_\kappa} \subseteq \mathsf{R}_{1+\kappa}(\mathsf{UTB}_0) \subseteq \mathsf{R}_{1+\kappa}(\mathsf{TB}_0)$ (the latter holding if $\kappa > 0$), and Theorem 5.6 proves the KF version. The converse inclusions, namely $\mathsf{R}_{1+\kappa}(\mathsf{UTB}_0) \subseteq \mathsf{CT}_{\varepsilon_\kappa}$ and $\mathsf{R}_{1+\kappa}(\mathsf{UTFB}_0) \subseteq \mathsf{KF}_{\varepsilon_\kappa}$ rely on an ordinal analysis of reflection hierarchies and are consequences of Theorems 6.22 (for CT) and 6.23 (for KF). Regarding Theorem 1.5, the first part is established by Lemma 3.11 and Theorem 6.24, and the second part by Theorems 5.7 and 6.24.

[2] Needless to say, naturality of notation systems for ordinals is a problematic concept; cf. Rathjen [24] and Feferman [6].

2 Syntax, ordinals and reflection

It is technically convenient to assume all theories (including the truth-free ones) are formulated in a single 'all encompassing' language $\mathscr{L}_*$ which includes, in addition to the standard logical symbols, the language $\mathscr{L}_0$ of *elementary arithmetic* (defined below), unary predicates F, T and also T_i for each $i < \omega$. If $X \subseteq \{T, F\} \cup \{T_i \mid i < \omega\}$, we write $\mathscr{L}_X$ for the sub-language of $\mathscr{L}_*$ extending $\mathscr{L}_0$ by only the predicate(s) in X. $\mathscr{L}_T$ is shorthand for $\mathscr{L}_{\{T\}}$.

Definition 2.1 (Elementary arithmetic). The language of elementary arithmetic, $\mathscr{L}_0$, comprises a constant symbol 0, successor function s, binary relations $=$, $<$ and $exp(\cdot,\cdot,)$, and ternary relations $add(\cdot,\cdot,\cdot)$, $mult(\cdot,\cdot,\cdot)$. The three additional predicates represent, respectively, binary exponentiation, addition and multiplication as relations. The class of Δ^0_0, or *elementary*, formulæ contains the $\mathscr{L}_0$ formulæ for which every quantified sub-formulæ takes one of the forms: $\forall z(z < t \to A')$, $\exists z(z < t \wedge A')$, $\forall z(exp(s,z) \to A')$, $\exists z(exp(s,z) \wedge A')$, $\forall z(P(s,t,z) \to A')$ or $\exists z(P(s,t,z) \wedge A')$, where $P \in \{add, mult\}$ and s, t are terms not containing z.

The theory *Elementary Arithmetic*, denoted EA, is the $\mathscr{L}_*$ theory with the following axioms.

EA1 Successor: $\neg(sx = 0) \wedge (sx = sy \to x = y)$;

EA2 Addition: $(add(x,0,z) \leftrightarrow z = x) \wedge (add(x,sy,z) \leftrightarrow \exists z'(z = sz' \wedge add(x,y,z')))$;

EA3 Multiplication: $(mult(x,0,z) \leftrightarrow z = 0) \wedge (mult(x,sy,z) \leftrightarrow \exists z'(mult(x,y,z') \wedge add(z',x,z)))$;

EA4 Binary exponentiation: $(exp(0,z) \leftrightarrow z = s0) \wedge (exp(sx,z) \leftrightarrow \exists z'(exp(x,z') \wedge add(z',z',z)))$;

EA5 Ordering: $\neg(x < 0) \wedge (x < sy \leftrightarrow x = y \vee x < y)$;

EA6 Δ^0_0-induction: $A(0) \wedge \forall x(A(x) \to A(sx)) \to A(x)$ for each Δ^0_0 formula A.

Lemma 2.2. EA *proves the that relations add, mult and exp define (total) functions.*

Proof. Consider, for example, $add(x,y,z)$, and let $A(x,y)$ be the elementary formula

$$\exists z\, add(x,y,z) \wedge \forall w(add(x,y,w) \wedge \forall z(add(x,y,z) \to w = z)).$$

We have $\mathsf{EA} \vdash A(x,0)$ and $\mathsf{EA} \vdash A(x,y) \to A(x,sy)$ so by Δ^0_0-induction we deduce $\mathsf{EA} \vdash \forall x \forall y A(x,y)$. Similarly for *mult* and *exp*. □

By definition the only $\mathscr{L}_0$-terms are numerals. However, as the previous lemma shows, functions defined from successor, addition, multiplication and binary exponentiation (the so-called *elementary functions*) can be represented and proved total in EA. For this reason we allow ourselves informal use of elementary terms built from these operations, such as the syntactic functions given below.

Peano Arithmetic, PA, is taken to be the extension of EA by induction for all $\mathscr{L}_0$ formulæ, and PA* is the extension by induction for *all* $\mathscr{L}_*$ formulæ. Let $\mathrm{Ind}(\mathscr{L})$ denote the schema of induction for all $\mathscr{L}$-formulæ. Thus $\mathsf{PA} = \mathsf{EA} + \mathrm{Ind}(\mathscr{L}_0)$ and $\mathsf{PA}^* = \mathsf{EA} + \mathrm{Ind}(\mathscr{L}_*)$.

We fix a Gödel coding $e \mapsto [e]$ of $\mathscr{L}_*$-expressions as closed $\mathscr{L}_0$-terms such that if A is a subformula of B then $\mathsf{EA} \vdash [A] < [B]$ and, moreover, the usual syntactic manipulations on formulæ, given by the functions

$$\dot\wedge\colon [A],[B] \mapsto [A \wedge B] \qquad \dot\forall\colon [x],[A] \mapsto [\forall x A] \qquad \dot=\colon m,n \mapsto [\bar m = \bar n]$$
$$\dot\vee\colon [A],[B] \mapsto [A \vee B] \qquad \dot\exists\colon [x],[A] \mapsto [\exists x A] \qquad \dot\neg\colon \ [A] \mapsto [\neg A]$$
$$sub\colon [A(x)],[x],m \mapsto [A(\bar m)]$$

are representable in EA by elementary formulæ, where A and B range over $\mathscr{L}_*$ formulæ, x is a variable symbol and $\bar m$ denotes the m-th numeral, that is, $\bar 0 = 0$ and $\overline{m+1} = s\bar m$. For example, the function sub is represented in EA by an elementary formula $\mathrm{Sub}(w,x,y,z)$ such that

$$\mathsf{EA} \vdash \forall w \forall x \forall y (\exists z\, \mathrm{Sub}(w,x,y,z) \wedge \forall z \forall z' (\mathrm{Sub}(w,x,y,z) \wedge \mathrm{Sub}(w,x,y,z') \to z = z'))$$
$$\mathsf{EA} \vdash \mathrm{Sub}(\bar k, \bar l, \bar m, \bar n) \quad \text{iff} \quad sub(k,l,m) = n$$
$$\mathsf{EA} \vdash \neg\mathrm{Sub}(\bar k, \bar l, \bar m, \bar n) \quad \text{iff} \quad sub(k,l,m) \neq n.$$

The open terms featured in the formulation of the uniform biconditionals in (1) are defined via the substitution function: for a formula $A(x)$ with at most x free, we set $[A\dot x] = sub([A],[x],x)$ and this naturally extends to multiple free variables by iteration, for $k > 1$ set $[A\dot x_1 \cdots \dot x_k] = sub([A\dot x_1 \cdots \dot x_{k-1}],[x_k],x_k)$. The formula $T[A\dot x]$ is formally represented by $\forall y(\mathrm{Sub}([A],[x],x,y) \to Ty)$, and similarly for the other syntactic functions above.

In addition to the above functions, for a (Δ^0_0-definable) collection $\mathscr{L}$ of $\mathscr{L}_*$-formulæ let $\mathrm{Sent}_{\mathscr{L}}(x)$ be a Δ^0_0 formulæ of $\mathscr{L}_0$ expressing that x is the code of an $\mathscr{L}$-sentence, and for a $\mathscr{L}_*$-theory S with an elementary set of axioms fix a Σ^0_1 formula $\mathrm{Bew}_{\mathsf{S}}(x)$ expressing that x is the code of an S-theorem.

2.1 Ordinals and notation systems

In this paper we will be concerned only with so-called *predicative ordinals*. These are the ordinals that can be constructed from 0 by repeated application of two ordinal functions, addition $\alpha, \beta \mapsto \alpha + \beta$ and the binary Veblen function $\alpha, \beta \mapsto \varphi\alpha\beta$, the latter of which we define starting from binary exponentiation:

- $\varphi 0\alpha = 2^\alpha = \sup(\{1\} \cup \{\varphi 0\beta + \varphi 0\beta \mid \beta < \alpha\})$;
- for $\beta > 0$, $\varphi\beta\alpha$ is the α-th element of the set $\{\delta \mid \forall\gamma < \beta(\varphi\gamma\delta = \delta)\}$.

Let $\epsilon_\alpha = \varphi 1\alpha$. The function $\alpha \mapsto \epsilon_\alpha$ enumerates the fixed points of $\alpha \mapsto 2^\alpha$ so in particular we have $\omega = \epsilon_0$. A *limit ordinal* is any ordinal $\beta = \sup\{\delta \mid \delta < \beta\}$. Γ_0 is the first non-zero ordinal closed under the binary Veblen function, that is, $\Gamma_0 = \inf\{\delta \mid \delta = \varphi\delta 0\} = \varphi\Gamma_0 0$. The operation of ω-exponentiation ($\alpha \mapsto \omega^\alpha$) and the standard ε-function ($\alpha \mapsto \varepsilon_\alpha$) that enumerates its fixed points can be defined by $\omega^\alpha = 2^{\omega.\alpha}$ and $\varepsilon_\alpha = \epsilon_{1+\alpha} = \varphi 1(1 + \alpha)$.

Lemma 2.3. *For all ordinals $\alpha < \Gamma_0$,*

1. $\alpha = 2^\alpha$ *iff* $\alpha = \omega$ *or* $\alpha = \omega^\alpha$;
2. $\varepsilon_\alpha = \omega^{\varepsilon_\alpha}$;
3. $\alpha = \epsilon_\alpha$ *iff* $\alpha = \varepsilon_\alpha$.

Proof. 1. Since $\alpha \leq 2^\alpha \leq \omega^\alpha$ the 'if' direction follows directly from the definition. For the other direction assume $\omega < \alpha = 2^\alpha$. Then α must be a limit ordinal and it suffices to prove that $\omega \leq \beta < \alpha$ implies $\omega^\beta < \alpha$. But if $\omega \leq \beta < \alpha$ then

$$\omega^\beta = 2^{\omega.\beta} \leq 2^{2^\beta.2^\beta} \leq 2^{2^{2^{\beta+1}}} < 2^{2^{2^\alpha}} = \alpha$$

and we are done. Cases 2 and 3 are straightforward. □

Recall Cantor's normal form theorem.

Lemma 2.4. *Every non-zero ordinal α has a unique decomposition of the form $\omega^{\alpha_0}.n_0 + \omega^{\alpha_1}.n_1 + \cdots + \omega^{\alpha_k}.n_k$ where $0 < n_i < \omega$ for each $i \leq k$ and $\alpha \geq \alpha_0 > \alpha_1 > \cdots > \alpha_k$.*

For our purposes a normal form based on binary exponentiation is more useful.

Lemma 2.5. *Every non-zero ordinal α has a unique decomposition of the form $2^{\alpha_0} + 2^{\alpha_1} + \cdots + 2^{\alpha_k}$ where $\alpha \geq \alpha_0 > \alpha_1 > \cdots > \alpha_k$.*

Proof. Let $0 < \alpha = \omega^{\alpha_0}.n_0 + \omega^{\alpha_1}.n_1 + \cdots + \omega^{\alpha_k}.n_k$ be the unique decomposition given by Lemma 2.4. For each $i \leq k$, let $0 \leq m_{i,l_i} < \cdots < m_{i,0} < \omega$ be such that $n_i = 2^{m_{i,0}} + \cdots + 2^{m_{i,l_i}}$. Using Lemma 2.3(2) we thus have

$$\alpha = \sum_{i=0}^{k} \sum_{j=0}^{l_i} 2^{\omega.\alpha_i + m_{i,j}},$$

and $\omega.\alpha_0 + m_{0,0} > \cdots > \omega.\alpha_0 + m_{0,l_0} > \cdots > \omega.\alpha_k + m_{k,0} > \cdots > \omega.\alpha_k + m_{k,l_k}$. Moreover, this decomposition must be unique. □

An ordinal presented in the form of Lemma 2.5 is considered to be in *normal form.* Using this presentation it is possible to define a commutative version of addition on ordinals, known as the *natural sum* that will prove useful in Section 6.

Definition 2.6. Let $\alpha = 2^{\alpha_1} + 2^{\alpha_2} + \cdots + 2^{\alpha_k}$ and $\beta = 2^{\beta_1} + 2^{\beta_2} + \cdots + 2^{\beta_l}$ be in normal form. The *natural sum* of α and β, written $\alpha \# \beta$ is the ordinal $\gamma = 2^{\gamma_1} + 2^{\gamma_2} + \cdots + 2^{\gamma_m}$ where $m = k + l$ and $\gamma_1 \geq \gamma_2 \geq \cdots \geq \gamma_m$ enumerates the ordinals $\alpha_1, \ldots, \alpha_k, \beta_1, \ldots, \beta_l$ (with repetitions).

Lemma 2.7. *For all α and β, $\alpha \# \beta = \beta \# \alpha$, and $\alpha \# \beta = \beta$ iff $\alpha = 0$.*

Fixed points of the function $x \mapsto 2^x$ below Γ_0 also have a unique decomposition based on the Veblen function.

Lemma 2.8. *If $\alpha = 2^\alpha < \Gamma_0$ then there exist unique ordinals $\beta, \gamma < \alpha$ such that $\alpha = \varphi\beta\gamma$.*

Proof. As $\alpha < \Gamma_0$ we have $\alpha < \varphi\alpha 0$ so $\beta = \sup\{\eta < \Gamma_0 \mid \exists\gamma(\alpha = \varphi\eta\gamma)\}$ exists, $\beta < \alpha$ and indeed $\alpha = \varphi\beta\gamma$ for some γ. It follows that $\gamma < \alpha$ as otherwise $\alpha = \varphi\beta\alpha$ which contradicts the definition of β. Since $\varphi\alpha\eta < \varphi\alpha\zeta$ whenever $\eta < \zeta$, it follows that β and γ are uniquely determined. □

2.1.1 Representation

In order to define transfinite iterations of reflection principles it is necessary to provide a representation of ordinal numbers within arithmetic. For the purposes of this paper, it suffices to consider a fixed representation of ordinals below Γ_0, which can be presented as an *elementary ordinal notation system* as defined, for example, in [24].

Such a representation system $\mathcal{O}$ is given by a set $O \subseteq \omega$ and elementary definable operations $\hat{+} \colon O^2 \to O$, $\hat{\varphi} \colon O^2 \to O$ and relation $\prec \subseteq O^2$ such that $\mathcal{O} = \langle O, \hat{+}, \hat{\varphi}, \prec \rangle$

is isomorphic to the structure $\langle\{\gamma \mid \gamma < \Gamma_0\}, +, \varphi, <\rangle$. Fix a pairing function $\langle\cdot,\cdot\rangle\colon \omega^2 \to \omega \setminus \{0,1\}$ representable in Elementary Arithmetic by a Δ^0_0 formulæ. Define $O = \{\#\alpha \mid \alpha < \Gamma_0\}$ where $\#\alpha$ is defined by transfinite recursion:

- If $\alpha \leq 1$, $\#\alpha = \alpha$;
- If $\alpha = 2^{\alpha_0} + \cdots + 2^{\alpha_n}$ is in normal form, and
  - $n > 0$ then $\#\alpha = \langle 0, \#\alpha_0, \ldots, \#\alpha_n\rangle$;
  - $n = 0$ then $\#\alpha = \langle 1, \#\beta, \#\gamma\rangle$ where $\beta, \gamma < \alpha$ are given by Lemma 2.8(4).

For $\alpha < \Gamma_0$, we write $\bar{\alpha}$ for the $\#\alpha$-th numeral, that is the term $\overline{\#\alpha}$.

Lemma 2.9. *The operation $\hat{+}$ and the relation $\prec$ are definable by Δ^0_0 formulæ. Moreover, the analogues of Lemmas 2.3 to 2.8 for the notation system $\mathcal{O}$ are all provable in* EA.

In addition to the above functions we introduce the abbreviations $\hat{2}^x = \hat{\varphi}\bar{0}x$, $\hat{\epsilon}_x = \hat{\varphi}\bar{1}x$ and $\hat{\omega}^x = \hat{2}^{\bar{\omega}.x}$, where $\bar{\omega}.0 = 0$ and $\bar{\omega}.(\hat{2}^x \hat{+} y) = \hat{2}^{\bar{\omega}\hat{+}x} \hat{+} \bar{\omega}.y$.

2.2 Transfinite induction and reflection hierarchies

Using the ordinal notation system $\mathcal{O}$ described above, the schema of transfinite induction for ordinals $\alpha < \Gamma_0$ and the theories of α-times iterated reflection can be defined.

Definition 2.10 (Transfinite induction). Let $A(x)$ be a formula of $\mathcal{L}_*$ and $x < \omega$. *Transfinite induction for A up to x*, $\mathrm{TI}(A, x)$ is the formula

$$\operatorname{Prog} \lambda x A \to A(x),$$

where $\mathrm{Prog}\lambda x A$ states that A is *progressive along* $\prec$, namely

$$\forall x \in O(\forall y \prec x A(y) \to A).$$

For a language $\mathcal{L}$ and ordinal $\alpha < \Gamma_0$, the *schema of transfinite induction up to α* is the collection of formulæ

$$\mathrm{TI}_{\mathcal{L}}({<}\alpha) = \{\mathrm{TI}(A, \bar{\beta}) \mid A \in \mathcal{L} \wedge \beta < \alpha\}.$$

Given a theory S and ordinal $\kappa < \Gamma_0$, we define by $\mathsf{S}_\kappa = \mathsf{S} + \mathrm{TI}_{\mathcal{L}_*}({<}\kappa)$. Thus, for instance, $\mathsf{PA}^* = \mathsf{EA}_{\omega+1}$. Notice that this notation does not conflict with the convention that CT_0, KF_0, etc. are theories with restricted induction.

Definition 2.11 (Reflection hierarchy). Let S be a theory with an elementary axiomatisation. The *reflection principle for* S, denoted $\mathrm{Ref}(\mathsf{S})$, is the collection of formulæ

$$\forall x(\mathrm{Bew}_{\mathsf{S}}[A\dot{x}] \to A)$$

for $A \in \mathscr{L}_*$ with at most x free. This schema induces an operation on theories $\mathsf{R}\colon \mathsf{S} \mapsto \mathsf{EA} + \mathrm{Ref}(\mathsf{S})$ which can be iterated through the transfinite (below Γ_0):

$$\mathsf{R}_0(\mathsf{S}) = \mathsf{S}, \qquad \mathsf{R}_\alpha(\mathsf{S}) = \bigcup_{\beta<\alpha} \mathsf{R}(\mathsf{R}_\beta(\mathsf{S})), \quad (\alpha > 0).$$

Notice that these theories can be represented within EA using the notation system $\mathscr{O}$ and that the provability predicates, $\mathrm{Bew}_{\mathsf{R}_\beta(\mathsf{S})}(x)$ for $\beta < \Gamma_0$, are all Σ^0_1.

3 Typed truth

With the syntactic machinery described above, it is possible to define the compositional truth theory CT and relate it to the theories of typed biconditionals.

Definition 3.1. The *compositional theory of typed truth*, CT_0 is the extension of EA by the following axioms.

CT1 $\forall x\forall y(T(x \dot{=} y) \leftrightarrow x = y)$,

CT2 $\forall x\forall y(\mathrm{Sent}_{\mathscr{L}_0}(x \dot{\wedge} y) \to (T(x \dot{\circ} y) \leftrightarrow Tx \circ Ty))$ for $\circ \in \{\wedge, \vee\}$,

CT3 $\forall x(\mathrm{Sent}_{\mathscr{L}_0}x \to (T\dot{\neg}x \leftrightarrow \neg Tx))$,

CT4 $\forall x\forall y(\mathrm{Sent}_{\mathscr{L}_0}(\dot{\forall}yx) \to (T\dot{Q}yx \leftrightarrow Qz\,T(sub(x, y, z)))$ for $Q \in \{\forall, \exists\}$.

CT is the theory extending PA^* by the above axioms or, equivalently, the extension of CT_0 by induction for all $\mathscr{L}_*$ formulæ.

The above axioms cover all the connectives, quantifiers and atomic relations of $\mathscr{L}_0$ so the typed disquotational axioms are all derivable in CT:

Lemma 3.2. UTB_0 *is a sub-theory of* CT_0.

As remarked earlier, the compositional axioms CT2–CT4 are not even derivable in UTB [12, Theorem 7.6]. Nevertheless, without induction for formulæ containing the truth predicate the compositional axioms remain proof-theoretically weak as CT_0 is a conservative extension of EA [1, 16] (a purely proof-theoretic argument is provided by [20]). When augmented by induction, however, they easily outstrip the truth biconditionals. For example, we have the following lemma which immediately implies that the formalised consistency statement for PA (and thus also UTB) is derivable in CT.

Lemma 3.3. $\mathsf{CT} \vdash \forall x(\mathrm{Bew}_{\mathsf{PA}} x \wedge \mathrm{Sent}_{\mathscr{L}_0} x \rightarrow Tx)$.

Proof. We sketch the proof. For further details see, for example, Halbach [12, Theorem 8.39]. The proof necessarily utilises induction for the truth predicate, first showing

$$\mathsf{CT} \vdash \forall x(\mathrm{Ax}_{\mathsf{PA}} x \rightarrow T(ucl\, x)) \tag{2}$$

where $\mathrm{Ax}_{\mathsf{PA}} x$ expresses that x is (a code of) an axiom of PA or a valid first order formula, and $ucl\, x$ is the function that maps a formula to its universal closure, i.e. if $x_1, \ldots, x_n$ enumerates the free variables of A then $\mathsf{EA} \vdash ucl[A] = [\forall x_1 \cdots \forall x_n A]$.

In order to establish (2) it is necessary to show CT can derive the truth of i) all axioms of first-order classical logic (in $\mathscr{L}_0$), and ii) all instances of induction for $\mathscr{L}_0$ formulæ. The first group is straightforward and follows from the compositional axioms alone. For the second group of axioms it suffices to prove

$$\mathsf{CT} \vdash \forall y(\mathrm{Sent}_{\mathscr{L}_0}(\dot{\forall} v y) \rightarrow T(\dot{\forall} v(ind\, y)))$$

where $ind[A(x)] = [A(0) \wedge \forall x(A(x) \rightarrow A(x+1)) \rightarrow A(x)]$. This is proved by induction. Let $B(x,y)$ be the formula $T(sub(ind\, y, v, x))$ which states the truth of induction for the formula y (on the variable named by v) up to x. Using the compositional axioms one can easily establish

$$\mathsf{CT}_0 \vdash \forall y(\mathrm{Sent}_{\mathscr{L}_0}(\dot{\forall} v y) \rightarrow (B(0,y) \wedge \forall x(B(x,y) \rightarrow B(sx,y))))$$

whence induction implies $\mathsf{CT} \vdash \forall y(\mathrm{Sent}_{\mathscr{L}_0}(\dot{\forall} v y) \rightarrow \forall x B(x,y))$. The compositional axioms thereby yield (ii), establishing (2).

The lemma can now be derived by induction on the length of the formalised PA-proof witnessing $\mathrm{Bew}_{\mathsf{PA}} x$. □

We now proceed with the results that yield Theorem 1.3.

Lemma 3.4. UTB_0 *is a sub-theory of* $\mathsf{R}(\mathsf{TB}_0)$.

Proof. Let $A(x)$ be a formula of $\mathscr{L}_0$ with at most x free. For every $n < \omega$ we have

$$\mathsf{TB}_0 \vdash T[A(\bar{n})] \leftrightarrow A(\bar{n}).$$

Let $B = (T[A\dot{x}] \leftrightarrow A)$. Since $\mathsf{EA} \vdash [B(\bar{n})] = [B\dot{\bar{n}}]$, the above implies

$$\mathsf{EA} \vdash \forall x \, \mathrm{Bew}_{\mathsf{TB}_0}[B\dot{x}],$$

and thus $\mathsf{R}(\mathsf{TB}_0) \vdash \forall x \, (T[A\dot{x}] \leftrightarrow A)$. □

It is trivial to generalise the above proof to the uniform disquotation axioms for any given set of formulæ closed under term substitution.

Lemma 3.5. *Let $\mathscr{L}$ be a set of formulæ such that for all $A(x) \in \mathscr{L}$ and every $n < \omega$, $A(\bar{n}) \in \mathscr{L}$. Also, let $\mathsf{TB}_0[\mathscr{L}]$ and $\mathsf{UTB}_0[\mathscr{L}]$ denote the theories derived from the local and uniform truth biconditionals for $\mathscr{L}$-formulæ. Then $\mathsf{R}(\mathsf{TB}_0[\mathscr{L}])$ proves the uniform disquotation sentences for $\mathscr{L}$, that is $\mathsf{UTB}_0[\mathscr{L}] \subseteq \mathsf{R}(\mathsf{TB}_0[\mathscr{L}])$.*

Reflecting on the local disquotation axioms not only derives the uniform instances. For instance, $\mathsf{TB}_0 \vdash T[A \wedge B] \leftrightarrow T[A] \wedge T[B]$ for all $\mathscr{L}_0$-sentence A and B, so the compositional axiom for conjunction is also derivable in $\mathsf{R}(\mathsf{TB}_0)$. However, two acts of reflection over TB proves necessary to derive the compositional axioms for either quantifier, though only one act of reflection over the uniform biconditionals is needed:

Lemma 3.6 (Halbach [11, Lemma 4.2])**.** $\mathsf{CT} \subseteq \mathsf{R}(\mathsf{UTB}_0)$.

Proof. By the uniform biconditionals for $\mathscr{L}_0$ we have

$$\mathsf{UTB}_0 \vdash \forall x \, T[A\dot{x}] \leftrightarrow T[\forall x A]$$

for every $A \in \mathscr{L}_0$ with at most x free. Since this is verifiable in EA, reflection implies the compositional axiom for the universal quantifier. The remaining compositional axioms are trivial. Finally, the induction schema is derivable in $\mathsf{R}(\mathsf{UTB}_0)$ (indeed in $\mathsf{R}(\mathsf{EA})$) since for each formula A,

$$\mathsf{EA} \vdash \forall x \mathrm{Bew}_{\mathsf{EA}}[A(\bar{0}) \wedge \forall x(A(x) \rightarrow A(x+1)) \rightarrow A(\dot{x})]. \square$$

The question naturally arises as to what truth-theoretic principles beyond the CT axioms are derivable in $\mathsf{R}(\mathsf{UTB}_0)$. The answer is, as reported earlier, none:

Theorem 3.7 (Halbach [11, Theorem 4.3][3])**.** CT *and* $\mathsf{R}(\mathsf{UTB}_0)$ *are identical theories.*

[3] Although Theorem 4.3 in [11] states only a much weaker result, namely that CT (therein named $\mathsf{PA}(S)$) and $\mathsf{R}(\mathsf{UTB}_0)$ (named AT) have the same arithmetic consequences, the proof, however, clearly establishes $\mathsf{R}(\mathsf{UTB}_0)$ as a sub-theory of CT.

One direction of Theorem 3.7 is given by Lemma 3.6 above. The converse direction requires a delicate proof-theoretic argument and is presented in detail in Section 6.

As described in the introduction, Theorem 3.7 generalises to iterations of reflection principles. In the present section we prove the corresponding generalisation of Lemma 3.6. As with the above theorem, the converse direction is more technically involved and is presented separately in Section 6; see Theorems 6.22 and 6.24.

We begin with the following observation which establishes a lower bound on the strength of the reflection hierarchy built directly over CT.

Lemma 3.8. *Let* S *be an* $\mathscr{L}_*$*-theory. For every* κ*,* $\mathsf{S}_{\epsilon_{\kappa+1}} \subseteq \mathsf{R}(\mathsf{S}_{\epsilon_\kappa})$*. Moreover, this fact is verifiable in* EA*.*

Proof. The proof is a simple generalisation of the result that transfinite induction up to ε_0 (i.e. ϵ_1) is derivable in PA.

Fix a formula $A(x)$ in $\mathscr{L}_*$ and let $A'(x) = \forall y \in O(A(y) \to A(y \hat{+} \hat{2}^x))$. Notice

$$\mathsf{EA} \vdash \mathrm{Prog}\lambda x A \to \mathrm{Prog}\lambda x A'(x).$$

Since $\mathsf{EA} \vdash A'(x) \to A(\hat{2}^x)$ we deduce

$$\mathsf{EA} \vdash \forall x \in O(\mathrm{TI}(A', x) \to \mathrm{TI}(A, \hat{2}^x)). \tag{3}$$

Define $2_0^\alpha = \alpha$ and $2_{n+1}^\alpha = 2_n^{2^\alpha}$. Iterating (3) we find, for each $n < \omega$, an $\mathscr{L}_*$ formula A_n such that

$$\mathsf{EA} \vdash \forall x \in O(\mathrm{TI}(A_n, x) \to \mathrm{TI}(A, \hat{2}_n^x)). \tag{4}$$

The remainder of the argument is now straightforward. Fix $A(x) \in \mathscr{L}_*$ and $\gamma < \epsilon_{\kappa+1}$. Recall $\epsilon_{\kappa+1} = \sup\{2_n^{\epsilon_\kappa+1} \mid n < \omega\}$ so there exists $n < \omega$ such that $\gamma < 2_n^{\epsilon_\kappa+1}$. Thus (4) reduces the argument to proving $\mathsf{R}(\mathsf{S}_{\epsilon_\kappa}) \vdash \mathrm{TI}(A_n, \hat{\epsilon}_{\bar{\kappa}})$ which immediately follows by reflection.

As the proof uses no arguments that cannot be directly formalised within EA the second part of the lemma naturally holds. □

Lemma 3.9. *For every* κ*,* $\mathsf{CT}_{\varepsilon_\kappa} = \mathsf{CT}_{\epsilon_{1+\kappa}} \subseteq \mathsf{R}_\kappa(\mathsf{CT})$*.*

Proof. Iterating Lemma 3.8 through the transfinite. □

Combining the results obtained so far we may complete part of Theorem 1.4.

Theorem 3.10. *For all* $\kappa > 0$*,* $\mathsf{CT}_{\epsilon_\kappa} \subseteq \mathsf{R}_\kappa(\mathsf{UTB}_0)$ *and for all* $\kappa > 1$*,* $\mathsf{CT}_{\epsilon_\kappa} \subseteq \mathsf{R}_\kappa(\mathsf{TB}_0)$*.*

Proof. By Lemma 3.4 we know $\mathsf{CT} \subseteq \mathsf{R}_2(\mathsf{TB}_0)$. Iterating Lemma 3.8 for $\mathsf{S} = \mathsf{TB}_0$ also reveals $\mathsf{TB}_{\epsilon_\kappa} \subseteq \mathsf{R}_\kappa(\mathsf{TB}_0)$. Since $\mathsf{CT}_{\epsilon_\kappa} = \mathsf{CT} + \mathsf{TB}_{\epsilon_\kappa}$ we deduce $\mathsf{CT}_{\epsilon_\kappa} \subseteq \mathsf{R}_\kappa(\mathsf{TB}_0)$ for $\kappa \geq 2$. Similarly for UTB_0. □

The lower bound on the proof-theoretic strength of CT is also easy to verify.

Lemma 3.11. *For every κ, CT_κ derives transfinite induction up to ϵ_κ for $\mathscr{L}_0$ formulæ.*

Proof. This result is well-known and is derivable from, for instance, Feferman's analysis of KF [9] and Fujimoto [10]. Here we give a more direct argument which borrows techniques from the ordinal analysis of truth theories presented in [21].

Let $ti(y,x)$ be given by $ti([A],x) = [\mathrm{TI}(A,\bar{x})]$ and let $I_0(x)$ denote the formula $\forall y(\mathrm{Sent}_{\mathscr{L}_0}(\dot{\forall}\bar{v}y) \to T(ti(y,x)))$. The proof of Lemma 3.8 shows that for every $\mathscr{L}_0$ formula A there exists an $\mathscr{L}_0$-formula A' such that $\mathsf{PA} \vdash \mathrm{TI}(A',x) \to \mathrm{TI}(A,\hat{2}^x)$. By Lemma 3.3 and the compositional axioms we may deduce

$$\mathsf{CT} \vdash \forall x(I_0(x) \to I_0(\hat{2}^x)), \tag{5}$$

$$\mathsf{CT} \vdash \forall x(I_0(x) \to I_0(x \hat{+} 1)), \tag{6}$$

$$\mathsf{CT} \vdash \mathrm{Prog}\, \lambda x I_0(x). \tag{7}$$

Combining (5) and (6) with induction within CT implies $\mathsf{CT} \vdash \forall x(I_0(\hat{\epsilon}_x) \to \forall y \prec \hat{\epsilon}_{x\hat{+}1}\ I_0(y))$, and hence, by (7), also

$$\mathsf{CT} \vdash \forall x(I_0(\hat{\epsilon}_x) \to I_0(\hat{\epsilon}_{x\hat{+}1})).$$

Thus we have

$$\mathsf{CT} \vdash \mathrm{Prog}\lambda x I_0(\hat{\epsilon}_x). \tag{8}$$

But then $\mathsf{CT}_\kappa \vdash I_0(\hat{\epsilon}_{\bar{\gamma}})$ for every $\gamma < \kappa$, and $\mathsf{CT} \vdash \mathrm{TI}(A,\bar{\delta})$ for every $\mathscr{L}_0$ formula A and $\delta < \epsilon_\kappa$. □

4 Ramified truth

There are two ways in which the typed biconditionals can be generalised. One either searches for sub-languages of $\mathscr{L}_*$ for which the disquotational axioms remain consistent or one sees the introduction of a truth predicate for $\mathscr{L}_0$ as the first step in an iteratable process. Following the latter approach, on top of EA one adds a truth predicate T_0 for $\mathscr{L}_0$ truth and sentences $T_0[A] \leftrightarrow A$ for $A \in \mathscr{L}_0$ (yielding TB_0). This is followed by a second truth predicate T_1 and the sentences $T_1[A] \leftrightarrow A$ for A from

$\mathscr{L}_{T_0} = \mathscr{L} \cup \{T_0\}$, then a third predicate T_2 with $T_2[A] \leftrightarrow A$ for A from $\mathscr{L}_{\{T_0,T_1\}}$, and so on. Transfinite levels can be added by including a predicate symbol T_ω and biconditionals $T_\omega[A] \leftrightarrow A$ for each $A \in \mathscr{L}_{\{T_i|i<\omega\}}$.

To expand the hierarchy to κ levels we need, for each ordinal $\alpha < \kappa$, a truth predicate T_α and biconditionals $T_\alpha[A] \leftrightarrow A$ for every sentence A from the language $\mathscr{L}_{\{T_\beta|\beta<\alpha\}}$. Using the ordinal notation system $\mathcal{O}$ the truth predicates and biconditionals can be already represented in $\mathscr{L}_{\{T_i|i<\omega\}}$ by reading T_α as $T_{\bar{\alpha}}$.

Definition 4.1 (Ramified biconditionals). For each $i < \omega$, set $\mathscr{L}_i = \mathscr{L}_{\{T_j|j\prec i\}}$. Let $\kappa < \Gamma_0$. The theory of *local truth biconditionals up to* κ, TB_0^κ, is the theory extending EA by the axiom

$$T_{\bar{\alpha}}[A] \leftrightarrow A \tag{9}$$

for every $\alpha < \kappa$ and every $A \in \mathscr{L}_{\bar{\alpha}}$. The *uniform truth biconditionals up to* κ, UTB_0^κ, is the extension of EA by the uniform form of (9) for each $\alpha < \kappa$.

For the compositional theory of κ truth predicates we follow the definition given by Halbach [12].

Definition 4.2 (Ramified truth). Let $\kappa < \Gamma_0$. The theory of *ramified truth up to* κ, RT^κ extends PA^* by the following six axiom for each $i \prec j \prec \bar{\kappa}$:

RT1 $\forall x \forall y (T_j(x \dot{=} y) \leftrightarrow x = y)$,

RT2 $\forall x (\mathrm{Sent}_{\mathscr{L}_j} x \to (T_j(\dot{\neg} x) \leftrightarrow \neg T_j x))$m

RT3 $\forall x \forall y (\mathrm{Sent}_{\mathscr{L}_j}(x \dot{\wedge} y) \to (T_j(x \dot{\circ} y) \leftrightarrow T_j x \circ T_j y))$ for $\circ \in \{\wedge, \vee\}$,

RT4 $\forall x \forall y (\mathrm{Sent}_{\mathscr{L}_j}(\dot{\forall} y x) \to (T_j(\dot{Q} y x) \leftrightarrow Q z\, T_j(sub(x, y, z)))$ for $Q \in \{\forall, \exists\}$,

RT5 $\forall x (\mathrm{Sent}_{\mathscr{L}_i} x \to (T_j[T_i(\dot{x})] \leftrightarrow T_i x))$,

RT6 $\forall x \forall y \prec \bar{\jmath} (\mathrm{Sent}_{\mathscr{L}_y} x \to (T_j[T_{\dot{y}}(\dot{x})]) \leftrightarrow T_j x))$,

where the term $[T_{\dot{y}}(\dot{x})]$ represents the function $m, \alpha \mapsto [T_{\bar{\alpha}}(\bar{m})]$.

Continuing the analogy between the disquotational and compositional theories we have:

Theorem 4.3. *For every* $\kappa < \Gamma_0$, $\mathsf{RT}^\kappa = \mathsf{R}(\mathsf{UTB}_0^\kappa)$.

Proof. For the direction $\mathsf{RT}^\kappa \subseteq \mathsf{R}(\mathsf{UTB}^\kappa)$, the only non-trivial case is axiom RT6 which follows from the observation that $\mathsf{UTB}_0^\kappa \vdash T_j[T_i[A]] \leftrightarrow T_j[A]$ for each $i < j < \bar{\kappa}$ and each sentence A of $\mathcal{L}_i$.

The converse direction is a straightforward generalisation of the argument that $\mathsf{R}(\mathsf{UTB}_0) \subseteq \mathsf{CT}$ which is proved in Section 6. □

As a consequence we can immediately deduce the proof-theoretic power of reflection hierarchy over UTB^κ.

Theorem 4.4. *Suppose* $\kappa = \omega^\lambda \geq \omega$. *The theory* $\mathsf{R}(\mathsf{UTB}_0^\kappa)$ *proves the same* $\mathcal{L}_0$ *formulæ as* $\mathsf{PA} + \mathrm{TI}(<\varphi(1+\lambda)0)$.

Proof. The theory of ramified truth with κ truth predicates, RT^κ, is known to be inter-translatable with the second-order system of *ramified analysis for* κ *levels* (often labelled $\mathsf{RA}_{<\kappa}$) [9], one of the central systems of proof-theory. These theories were examined by Feferman [7] and Schütte [25] and shown to prove the same $\mathcal{L}_0$ statements as $\mathsf{PA} + \mathrm{TI}(<\varphi(1+\lambda)0)$ where $\kappa = \omega^\lambda$. Combined with Theorem 4.3 this yields the desired result.

Alternatively, a direct argument generalising the proof of Lemma 3.11 is also possible though we will not present all the details here as it bears little difference to the proof-theoretic analysis of ramified analysis presented in, for instance, [25]. The upper bound, namely that all $\mathcal{L}_0$ theorems of RT^κ are derivable in $\mathsf{PA} + \mathrm{TI}(<\varphi(1+\lambda)0)$ is obtained by expanding the cut-elimination argument for CT (Corollary 6.16) to the analogous system for RT^κ. The lower bound, $\mathsf{RT}^\kappa \vdash \mathrm{TI}_{\mathcal{L}_0}(<\varphi(1+\lambda)0)$, can be proved by expanding Lemma 3.11 to multiple truth predicates. Let $I_\alpha(x)$ be the formula $\forall y(\mathrm{Sent}_{\mathcal{L}_\alpha}(\dot{\forall}\bar{v}y) \to T_{\bar{\alpha}}(ti(y,x)))$where ti is the function defined in the proof of Lemma 3.11. In analogy to (8) we have, for example,

$$\mathsf{RT}^\omega \vdash \mathrm{Prog}\lambda x I_n(\hat{\epsilon}_x)$$

for each $n < \omega$ and in particular

$$\mathsf{RT}^\omega \vdash \forall y(I_{n+1}(y) \to I_n(\hat{\epsilon}_y)).$$

So $\mathsf{RT}^\omega \vdash \mathrm{TI}_{\mathcal{L}_0}(<\varphi 20)$. It is not difficult to see, moreover, that

$$\mathsf{RT}^{\omega+1} \vdash \forall n T_\omega[\forall y(I_{n+1}(y) \to I_n(\hat{\epsilon}_y))],$$

i.e. $\mathsf{RT}^{\omega+1} \vdash \mathrm{Prog}\lambda x I_\omega(\hat{\varphi}\bar{2}x)$. Iterating the argument through the transfinite yields, more generally,

$$\begin{aligned} \mathsf{RT}^{\omega^\lambda} &\vdash \quad \bar{\alpha} \prec \hat{\omega}^{\bar{\lambda}} \wedge \bar{\beta} \prec \bar{\lambda} \to \forall z(I_{\alpha+\omega^\beta}(z) \to I_\alpha(\hat{\varphi}(\overline{1+\beta})z)), \\ \mathsf{RT}^{\omega^\lambda} &\vdash \forall x \forall y(x \prec \omega^\lambda \wedge y \prec \bar{\lambda} \to T_\lambda[\forall z(I_{x \hat{+} \hat{\omega}^y}(z) \to I_x(\hat{\varphi}(\bar{1} \hat{+} \dot{y})z))]), \end{aligned}$$

which suffices to derive the required result. □

5 Positive self-applicable truth

CT is a theory of *typed* truth, wherein the truth predicate only (meaningfully) applies to sentences not themselves featuring the truth predicate. This restriction can be relaxed somewhat and many 'self-referential' axioms can be consistently added. One way to achieve this is given by the Kripke–Feferman theory KF, introduced by Feferman in [9] as part of his analysis on the limits of predicativity. KF axiomatises a self-referential version of truth that avoids the inconsistency of the liar paradox by separating truth from its dual, *falsity*, which are axiomatised independently but symmetrically.

Recall the language $\mathscr{L}_{T,F}$ that extends $\mathscr{L}_0$ by two fresh unary predicate symbols, T and F. The *positive* fragment of $\mathscr{L}_{T,F}$, labelled $\mathscr{L}^+_{T,F}$, comprises the $\mathscr{L}_{T,F}$-formulæ in which the negation symbol only occurs in front of atomic formulæ from $\mathscr{L}_0$.

Definition 5.1. The *Kripke–Feferman* theory of truth KF extends PA* by the compositional axioms for the language $\mathscr{L}^+_{T,F}$:

KF1 $\forall x \forall y((T(x \dot{=} y) \leftrightarrow x = y) \wedge (T(\dot{\neg}(x \dot{=} y)) \leftrightarrow \neg x = y))$,

KF2 $\forall x \forall y((F(x \dot{=} y) \leftrightarrow \neg x = y) \wedge (F(\dot{\neg}(x \dot{=} y)) \leftrightarrow x = y))$,

KF3 $\forall x \forall y(\mathrm{Sent}_{\mathscr{L}^+_{T,F}}(x \dot{\wedge} y) \rightarrow (T(x \dot{\circ} y) \leftrightarrow Tx \circ Ty))$ for $\circ \in \{\wedge, \vee\}$,

KF4 $\forall x \forall y(\mathrm{Sent}_{\mathscr{L}^+_{T,F}}(x \dot{\wedge} y) \rightarrow (F(x \dot{\circ} y) \leftrightarrow Fx \mathbin{\bar{\circ}} Fy))$ for $\circ \in \{\wedge, \vee\}$,

KF5 $\forall x \forall y(\mathrm{Sent}_{\mathscr{L}^+_{T,F}}(\dot{\forall} yx) \rightarrow (T\dot{Q}yx \leftrightarrow Qz\, T(sub(x, y, z))))$ for $Q \in \{\forall, \exists\}$,

KF6 $\forall x \forall y(\mathrm{Sent}_{\mathscr{L}^+_{T,F}}(\dot{\forall} yx) \rightarrow (F\dot{Q}yx \leftrightarrow \bar{Q}z\, F(sub(x, y, z))))$ for $Q \in \{\forall, \exists\}$,

KF7 $\forall x((T[T\dot{x}] \leftrightarrow Tx) \wedge (T[F\dot{x}] \leftrightarrow Fx))$,

KF8 $\forall x((F[T\dot{x}] \leftrightarrow Fx) \wedge (F[F\dot{x}] \leftrightarrow Tx))$,

where $\bar{\wedge} = \vee$, $\bar{\vee} = \wedge$, $\bar{\forall} = \exists$ and $\bar{\exists} = \forall$. KF_0 is the sub-theory of KF with EA as the base theory.

The above axiomatisation does not force truth and falsity to be disjoint concepts. Indeed, Proposition 1.2 shows there are models of KF in which truth and falsity coincide, namely the greatest fixed point of Γ in which the liar sentence $L \leftrightarrow \neg T[L]$ is denoted both true and false.

The proof theory of KF has been thoroughly studied (see, e.g., Cantini [2] and Feferman [9]) and the system is significantly stronger than CT. Nevertheless, as we shall see, KF bears the same relation to UTFB_0 and TFB_0 as CT bears to UTB_0 and TB_0. Since these theories were not precisely defined in the introduction we begin with their definition.

Given a formula $A \in \mathscr{L}_{T,F}$ we define the *dual* of A, denoted $\overline{A}$, by recursion:

$$\begin{array}{lll} \overline{A \wedge B} = \overline{A} \vee \overline{B} & \overline{\forall x A} = \exists x \overline{A} & \overline{A} = \neg A \quad \text{if } A \in \mathscr{L}_0 \text{ is atomic} \\ \overline{A \vee B} = \overline{A} \wedge \overline{B} & \overline{\exists x A} = \forall x \overline{A} & \overline{\neg A} = A \\ \overline{Ts} = Fs & \overline{Fs} = Ts & \end{array}$$

TFB_0 and UTFB_0 are the theories extending EA by, respectively, the local and uniform truth and falsity biconditionals for $\mathscr{L}^+_{T,F}$:

$$\mathsf{TFB}_0 = \mathsf{EA} + \{T[A] \leftrightarrow A, F[A] \leftrightarrow \overline{A} \mid A \in \mathrm{Sent}_{\mathscr{L}^+_{T,F}}\},$$
$$\mathsf{UTFB}_0 = \mathsf{EA} + \{\forall x(T[A\dot{x}] \leftrightarrow A), \forall x(F[A\dot{x}] \leftrightarrow \overline{A}) \mid (\forall x A) \in \mathrm{Sent}_{\mathscr{L}^+_{T,F}}\}.$$

Arguing by (meta-)induction on the complexity of $\mathscr{L}^+_{T,F}$ formulæ we see that the uniform truth biconditionals are derivable in KF [2, Lemma 3.2 (ii)], whence:

Lemma 5.2. UTFB_0 *is a sub-theory of* KF_0.

UTFB_0 and TFB_0 are extensions of the theories PTB_0 and PUTB_0 of *positive disquotation* over EA, axiomatised by the local and uniform version of the biconditionals $T[A] \leftrightarrow A$ for A in $\mathscr{L}^+_T = \mathscr{L}^+_{T,F} \cap \mathscr{L}_T$ (positive formulæ not containing the predicate F). These two theories have been extensively studied (see, e.g. [4, 5, 12]), particularly in relation to subsystems of KF, and it is here that the hitherto innocuous distinction between the local and uniform biconditionals becomes relevant. Lemma 5.3 below is a simple generalisation of Cieśliński's result that $\mathsf{PTB}_0+\mathrm{Ind}(\mathscr{L}_T)$ conservatively extends PA [4, Theorem 2.2];[4] Lemma 5.4 was proved by Halbach [11, Theorem 5.2] utilising Cantini's interpretation of the theory $\widehat{\mathsf{ID}}_1$ of fixed points for arithmetical operators in $\mathsf{PUTB}_0 + \mathrm{Ind}(\mathscr{L}_T)$ [2, Corollary 3.11].

Lemma 5.3. $\mathsf{TFB}_0 + \mathrm{Ind}(\mathscr{L}_{T,F})$ *conservatively extends* PA.

Lemma 5.4. KF *is interpretable in* $\mathsf{PUTB}_0 + \mathrm{Ind}(\mathscr{L}_T)$ *and hence interpretable in* $\mathsf{UTFB}_0 + \mathrm{Ind}(\mathscr{L}_{T,F})$.

Lemma 5.4 does not mean KF is implicit in PUTB_0. Even considering a formulation of KF with only one predicate (in which falsity is defined, say by $F(x) \leftrightarrow T(\dot{\neg}x)$)

[4] Cieśliński's proof in fact establishes a stronger result than stated in Lemma 5.3. For a direct proof see [14, Theorem 13].

none of the compositional axioms of KF are derivable in PUTB_0 [12, Lemma 19.20], and $\mathsf{R}(\mathsf{PUTB}_0)$ also fails to derive analogues of $F[T\dot{x}] \leftrightarrow Fx$.

It turns out that the falsity biconditionals suffice to 'derive' the remaining compositional axioms of KF, so the theory *is* implicit in the acceptance of both truth and falsity biconditionals:

Lemma 5.5. $\mathsf{KF} \subseteq \mathsf{R}(\mathsf{UTFB}_0) \subseteq \mathsf{R}_2(\mathsf{TFB}_0)$.

Proof. The proof follows the same argument as Lemma 3.6, noticing that the self-referential axioms (KF7) and (KF8) of KF are instances of the uniform positive disquotation axioms. □

Halbach [11, Lemma 6.5] observes that, in analogy to CT, uniform reflection over the positive truth biconditionals is derivable in KF (see also Theorem 6.24 below). It is not difficult to see that this also extends to reflection over UTFB_0. In fact, the analogy goes much further, as the following theorem shows.

Theorem 5.6. *For every ordinal κ, $\mathsf{KF}_{\epsilon_\kappa} \subseteq \mathsf{R}_\kappa(\mathsf{KF}) \subseteq \mathsf{R}_{1+\kappa}(\mathsf{UTFB}_0)$.*

Proof. Apply Lemma 3.8. □

The proof-theoretic strength of the theories can also be easily computed by generalising Feferman's analysis of KF in [9].

Theorem 5.7. *For each ordinal κ, $\mathsf{KF}_{\varepsilon_\kappa}$ derives transfinite induction for $\mathscr{L}_0$ formulæ up to $\varphi\varepsilon_\kappa 0$.*

Proof. Utilising the type-free truth predicate of KF, it is possible to define not only the typed predicate of CT, but also the typed truth predicates of $\mathsf{RT}^{\varepsilon_0}$ [12, Theorem 15.25]. In fact, it follows that KF_α defines all the truth predicates of RT^α, so $\mathsf{KF}_{\varepsilon_\kappa}$ derives all $\mathscr{L}_0$ theorems of $\mathsf{RT}^{\varepsilon_\kappa}$, which, by Theorem 4.4 includes the schema of transfinite induction up to $\varphi\varepsilon_\kappa 0$ for all $\mathscr{L}_0$ formulæ. □

6 Proofs of upper bounds

In this section we present the 'upper bound' proofs of Theorem 1.4, showing that $\mathsf{R}_{1+\kappa}(\mathsf{UTB}_0)$ and $\mathsf{R}_{1+\kappa}(\mathsf{UTFB}_0)$ are sub-theories of, respectively, $\mathsf{CT} + \mathrm{TI}(<\epsilon_\kappa)$ and $\mathsf{KF} + \mathrm{TI}(<\epsilon_\kappa)$. The proofs can be seen as a generalisation of Kreisel and Lévy's argument that $\mathsf{EA} + \mathrm{Ref}(\mathsf{EA})$ is a sub-theory of PA, which we begin by sketching.

Theorem 6.1 (Kreisel and Lévy [17]). $\mathsf{PA} \vdash \mathrm{Ref}(\mathsf{EA})$.

Proof (sketch). It is well-known (see, for example, [13, Theorem 1.75]) that for each non-zero $l < \omega$ there exists a formula $\mathrm{Tr}_l(x)$ such that for every Π^0_l formula $B(x)$ (with at most x free),

$$\mathsf{PA} \vdash \forall x(\mathrm{Tr}_l[B\dot{x}] \leftrightarrow B).$$

Let U be the conjunction of the non-induction axioms of EA (EA1–5) and the sentence

$$\forall z(\mathrm{Sent}_{\Delta^0_0}(\underset{\cdot}{\forall} vz) \to \mathrm{Tr}_2(\underset{\cdot}{\forall} v(\mathit{ind}\, z)))$$

where $\mathit{ind}[A(v)] = [A(0) \wedge \forall x(A(x) \to A(sx)) \to A]$. Suppose U is a Π^0_k formula.

Fix $l < \omega$ and a Π^0_l formula $A(x)$. We prove $\mathsf{PA} \vdash \forall x(\mathrm{Bew}_{\mathsf{EA}}[A\dot{x}] \to A)$. Arguing within PA, suppose $n < \omega$ and $\mathsf{EA} \vdash A(\bar{n})$. The derivation can be easily transformed (uniformly in n) into a derivation of $\vdash U \to A(\bar{n})$ in Gentzen's sequent calculus for first-order logic (for example in the form of Definition 6.2 below), and by cut-elimination (provable in PA) into a finite cut-free derivation of $\vdash U \to A(\bar{n})$, again uniformly in n. By PA-induction we can conclude $\forall x\, \mathrm{Tr}_m[U \to A(\dot{x})]$ where $m = \max\{l, k+1\}$, from which we deduce $A(x)$ as required. □

6.1 Sequent calculi for compositional truth

To generalise the above argument to acts of reflection over truth theories we must first introduce a sequent calculus for theories of truth. In this section we will focus on the reflection hierarchies over CT and KF, though RT^κ can also be treated in this way.

We will define two sequent calculi, named T_∞ and T^+_∞, corresponding to the theories CT and KF respectively. Both calculi permit infinitary derivations utilising the ω-rule (if $\vdash A(\bar{n})$ for every n then $\vdash \forall x A$) and the elimination of all non-atomic cuts (see Lemma 6.14 below). Formally, we implement T_∞ and T^+_∞ as Gentzen-style sequent calculi that derive expressions of the form $\mathrm{T}^{(+)}_\infty \vdash^\alpha_\kappa \Gamma \Rightarrow \Delta$ (henceforth called *sequents*) where α and κ are predicative ordinals and $\Gamma \cup \Delta$ is a finite set of $\mathscr{L}_{T,F}$ sentences.[5] Numerous infinitary proof systems have been constructed for theories of truth (e.g. [21, 19, 3, 2]); the present work most closely corresponds to [21, 2].

The *rank* of a formula A of $\mathscr{L}_{T,F}$, denoted $|A|$, is defined as follows: $|A| = 0$ if A an atomic formula of $\mathscr{L}_0$; $|A| = \omega$ if $A = T(s)$ or $A = F(s)$ for some term s; $|A \wedge B| = |A \vee B| = \max\{|A|, |B|\} + 1$; $|\forall x A| = |\exists x A| = |\neg A| = |A| + 1$. The *reduced*

[5] To simplify presentation we assume both calculi may derive $\mathscr{L}_{T,F}$ formulæ but in T_∞ the falsity predicate F has no special meaning.

rank of A, $|A|^-$, is given by the same recursive rules but with $|A|^- = 0$ if A is any atomic formula. Thus either $|A| = |A|^-$ or $\omega \leq |A| \leq \omega + |A|^-$.

For the remainder of this section, unless otherwise stated, upper-case Roman letters, A, B, etc., range over sentences from $\mathscr{L}_*$ and upper-case Greek letters Γ, Δ, etc., range over finite sets of $\mathscr{L}_*$ sentences. Commas in sequents are shorthand for set union: $\Gamma, \Delta = \Gamma \cup \Delta$ and $\Gamma, A = \Gamma \cup \{A\}$.

6.1.1 The calculi T_∞ and T^+_∞

Definition 6.2 (Sequent calculi for typed and type-free truth)**.** Let $\alpha, \kappa < \Gamma_0$ be ordinals and Γ, Δ finite sets of $\mathscr{L}_*$ sentences. The relations $\mathrm{T}_\infty \vdash^\alpha_\kappa \Gamma \Rightarrow \Delta$ and $\mathrm{T}^+_\infty \vdash^\alpha_\kappa \Gamma \Rightarrow \Delta$ are defined by transfinite induction on α according to the following rules.

1. *Axioms.* For $\mathrm{S} \in \{\mathrm{T}_\infty, \mathrm{T}^+_\infty\}$ and any $\alpha, \kappa < \Gamma_0$,

   I) If Γ contains a false $\mathscr{L}_0$-atom or Δ contains a true $\mathscr{L}_0$-atom then $\mathrm{S} \vdash^\alpha_\kappa \Gamma \Rightarrow \Delta$;

   II) $\mathrm{S} \vdash^\alpha_\kappa \Gamma, A \Rightarrow \Delta, A$ if A is an atomic formula $\mathscr{L}_*$;

   III) $\mathrm{S} \vdash^\alpha_\kappa \Gamma \Rightarrow \Delta, \forall x A$ if A is an instance of Δ^0_0 induction for an $\mathscr{L}_0$ formula with at most x free;

2. *Logical rules.* For $\mathrm{S} \in \{\mathrm{T}_\infty, \mathrm{T}^+_\infty\}$, $\alpha < \beta < \Gamma_0$ and $\kappa < \Gamma_0$,

   IV) $\mathrm{S} \vdash^\alpha_\kappa \Gamma \Rightarrow \Delta, A_0$, $\mathrm{S} \vdash^\alpha_\kappa \Gamma \Rightarrow \Delta, A_1$ implies $\mathrm{S} \vdash^\beta_\kappa \Gamma \Rightarrow \Delta, A_0 \wedge A_1$;

   V) $\mathrm{S} \vdash^\alpha_\kappa \Gamma, A_i \Rightarrow \Delta$ and $i \in \{0,1\}$ implies $\mathrm{S} \vdash^\beta_\kappa \Gamma, A_0 \wedge A_1 \Rightarrow \Delta$;

   VI) $\mathrm{S} \vdash^\alpha_\kappa \Gamma, A \Rightarrow \Delta$ implies $\mathrm{S} \vdash^\beta_\kappa \Gamma \Rightarrow \Delta, \neg A$;

   VII) $\mathrm{S} \vdash^\alpha_\kappa \Gamma \Rightarrow \Delta, A$ implies $\mathrm{S} \vdash^\beta_\kappa \Gamma, \neg A \Rightarrow \Delta$;

   VIII) $\mathrm{S} \vdash^\alpha_\kappa \Gamma, A(\bar{n}) \Rightarrow \Delta$ implies $\mathrm{S} \vdash^\beta_\kappa \Gamma, \forall x A(x) \Rightarrow \Delta$;

   IX) $\mathrm{S} \vdash^\alpha_\kappa \Gamma \Rightarrow \Delta, A(\bar{n})$ for every $n < \omega$ implies $\mathrm{S} \vdash^\beta_\kappa \Gamma \Rightarrow \Delta, \forall x A(x)$;

   X) $\mathrm{S} \vdash^\alpha_\kappa \Gamma \Rightarrow \Delta, A$ and $\mathrm{S} \vdash^\alpha_\kappa \Gamma, A \Rightarrow \Delta$ implies $\mathrm{S} \vdash^\beta_\kappa \Gamma \Rightarrow \Delta$ whenever $|A| < \kappa$.

3. *Typed truth rules.* T_∞ is additionally closed under the following rules for $\alpha < \beta < \Gamma_0$ and $\kappa < \Gamma_0$.

   XI) $\mathrm{T}_\infty \vdash^\alpha_\kappa \Gamma \Rightarrow \Delta, A$ and $A \in \mathscr{L}_0$ implies $\mathrm{T}_\infty \vdash^\beta_\kappa \Gamma \Rightarrow \Delta, T[A]$;

   XII) $\mathrm{T}_\infty \vdash^\alpha_\kappa \Gamma, A \Rightarrow \Delta$ and $A \in \mathscr{L}_0$ implies $\mathrm{T}_\infty \vdash^\beta_\kappa \Gamma, T[A] \Rightarrow \Delta$.

4. *Type-free truth rules.* T_∞^+ is additionally closed under the following rules for $\alpha < \beta < \Gamma_0$ and $\kappa < \Gamma_0$.

   XIII) $\mathrm{T}_\infty^+ \vdash_\kappa^\alpha \Gamma \Rightarrow \Delta, A$ and $A \in \mathcal{L}_{T,F}^+$ implies $\mathrm{T}_\infty^+ \vdash_\kappa^\beta \Gamma \Rightarrow \Delta, T[A]$ and $\mathrm{T}_\infty^+ \vdash_\kappa^\beta \Gamma, F[\bar{A}] \Rightarrow \Delta$;

   XIV) $\mathrm{T}_\infty^+ \vdash_\kappa^\alpha \Gamma, A \Rightarrow \Delta$ and $A \in \mathcal{L}_{T,F}^+$ implies $\mathrm{T}_\infty^+ \vdash_\kappa^\beta \Gamma, T[A] \Rightarrow \Delta$ and $\mathrm{T}_\infty^+ \vdash_\kappa^\beta \Gamma \Rightarrow \Delta, F[\bar{A}]$.

Sequents in group 1 are collectively called *axioms* and the rules under 2 are *logical rules*. Rule X is referred to as the *cut rule* and the distinguished formula A therein is the *cut-formula*. Rule IX is the *ω-rule*. The rules in 3 and 4 are collectively called *truth rules*. In each rule except cut the distinguished formula in the derived sequent is called the *principal* formula.

In a sequent $\mathrm{S} \vdash_\kappa^\alpha \Gamma \Rightarrow \Delta$, α is called the *height* of the derivation and κ the *cut rank*.

We begin by observing two important lemmas that are proved by transfinite induction.

Lemma 6.3 (Consistency). *If $\mathrm{T}_\infty^{(+)} \vdash_0^\alpha \Gamma \Rightarrow \Delta$ then $\Gamma \cup \Delta \neq \emptyset$.*

Lemma 6.4. *If $\mathrm{T}_\infty^{(+)} \vdash_\kappa^\alpha \Gamma \Rightarrow \Delta$ then every cut rule used in the derivation of this sequent has cut formula with rank strictly less than κ. If, moreover, $|A| < \kappa$ for every $A \in \Gamma \cup \Delta$ then the sequent has a derivation using only sequents from formulæ with rank $< \kappa$.*

The next two lemmas list further simple properties of the calculi T_∞ and T_∞^+ that we will utilise later.

Lemma 6.5.

1. *If $\mathrm{T}_\infty \vdash_\kappa^\alpha \Gamma \Rightarrow \Delta$ then $\mathrm{T}_\infty^+ \vdash_\kappa^\alpha \Gamma \Rightarrow \Delta$.*

2. *$\mathrm{T}_\infty \vdash_0^{2|A|^-} A \Rightarrow A$.*

3. *If A is a true elementary sentence then $\mathrm{T}_\infty \vdash_0^n \emptyset \Rightarrow A$ for some n.*

Proof. 1 is proved by transfinite induction on α, noticing that the truth rules for T_∞ are are all instances of truth rules for T_∞^+.

For 2, we argue by induction on $n = |A|^-$. If $n = 0$ then A is an atom and the sequent is an instance of axiom II. The other cases are equally simple. If $A = \forall x B(x)$ for instance, $|A|^- = m + 1$ and $|B(\bar{k})|^- = m$ for every $k < \omega$ whereby the induction hypothesis and rule VIII imply $\mathrm{T}_\infty \vdash_0^{2m+1} A \Rightarrow B(\bar{k})$ for every k and so also $\mathrm{T}_\infty \vdash_0^{2n} A \Rightarrow A$.

3 is proved by induction on $|A|$. □

Lemma 6.6. *Suppose* $\mathrm{S} \in \{\mathrm{T}_\infty, \mathrm{T}_\infty^+\}$.

1. *For all finite sets* $\Gamma \subseteq \Theta$, $\Delta \subseteq \Lambda$ *and ordinals* $\beta \leq \alpha$, $\kappa \leq \lambda$, *if* $\mathrm{S} \vdash_\kappa^\beta \Gamma \Rightarrow \Delta$ *then* $\mathrm{S} \vdash_\lambda^\alpha \Theta \Rightarrow \Lambda$.
2. $\mathrm{S} \vdash_\kappa^n \Gamma \Rightarrow \Delta$ *for some* $n < \omega$ *implies* $\mathrm{S} \vdash_{\omega+m}^n \Gamma \Rightarrow \Delta$ *for some* $m < \omega$.

Proof. Both cases are proved by transfinite induction on the height of the sequent. We show 2. Suppose $\mathrm{S} \vdash_\kappa^n \Gamma \Rightarrow \Delta$. If $n = 0$ the sequent is an axiom and thus $\mathrm{S} \vdash_0^n \Gamma \Rightarrow \Delta$ follows. Otherwise $n > 0$ and we need only consider the case that the sequent $\mathrm{S} \vdash_\kappa^n \Gamma \Rightarrow \Delta$ arises through an application of the cut rule, say from $\mathrm{S} \vdash_\kappa^k \Gamma, A \Rightarrow \Delta$ and $\mathrm{S} \vdash_\kappa^k \Gamma \Rightarrow \Delta, A$, where $|A| < \kappa$ and $k < n$. By the induction hypothesis there are $m_0, m_1 < \omega$ such that $\mathrm{S} \vdash_{\omega+m_0}^k \Gamma, A \Rightarrow \Delta$ and $\mathrm{S} \vdash_{\omega+m_1}^k \Gamma \Rightarrow \Delta, A$, whence choosing $m = \max\{m_0, m_1, |A|^- + 1\}$ and applying part 1, we deduce $\mathrm{S} \vdash_{\omega+m}^n \Gamma \Rightarrow \Delta$ by cut. □

The following lemma will be important in Section 6.2 when we analyse the cut rule further.

Lemma 6.7. *Let* $\mathrm{S} \in \{\mathrm{T}_\infty, \mathrm{T}_\infty^+\}$. *For every* $\alpha, \kappa < \Gamma_0$ *and every* Γ *and* Δ,

1. *If* $\mathrm{S} \vdash_\kappa^\alpha \Gamma \Rightarrow \Delta, A$ *and* A *is a false atomic formula from* $\mathscr{L}_0$ *then* $\mathrm{S} \vdash_\kappa^\alpha \Gamma \Rightarrow \Delta$,
2. *If* $\mathrm{S} \vdash_\kappa^\alpha \Gamma, A \Rightarrow \Delta$ *and* A *is a true atomic formula from* $\mathscr{L}_0$ *then* $\mathrm{S} \vdash_\kappa^\alpha \Gamma \Rightarrow \Delta$,
3. $\mathrm{S} \vdash_\kappa^\alpha \Gamma \Rightarrow \Delta, A_0 \wedge A_1$ *implies for each* $i \in \{0,1\}$, $\mathrm{S} \vdash_\kappa^\alpha \Gamma \Rightarrow \Delta, A_i$,
4. $\mathrm{S} \vdash_\kappa^\alpha \Gamma, A_0 \wedge A_1 \Rightarrow \Delta$ *implies* $\mathrm{S} \vdash_\kappa^\alpha \Gamma, A_0, A_1 \Rightarrow \Delta$,
5. $\mathrm{S} \vdash_\kappa^\alpha \Gamma \Rightarrow \Delta, \neg A$ *implies* $\mathrm{S} \vdash_\kappa^\alpha \Gamma, A \Rightarrow \Delta$,
6. $\mathrm{S} \vdash_\kappa^\alpha \Gamma, \neg A \Rightarrow \Delta$ *implies* $\mathrm{S} \vdash_\kappa^\alpha \Gamma \Rightarrow \Delta, A$,
7. $\mathrm{S} \vdash_\kappa^\alpha \Gamma \Rightarrow \Delta, \forall x A$ *and* $|A| \geq \omega$ *implies* $\mathrm{S} \vdash_\kappa^\alpha \Gamma \Rightarrow \Delta, A(\bar{n})$ *for every* $n < \omega$.

Proof. Each case is proved by transfinite induction on α. We show, for example, 7. Suppose

$$\mathrm{S} \vdash_\kappa^\alpha \Gamma \Rightarrow \Delta, \forall x A \tag{10}$$

and argue by a case distinction on the last rule applied to deduce this sequent. If the sequent is an axiom then this is not an instance of Δ^0_0 induction since $|\forall x A| \geq \omega$, so the sequent $\mathrm{S} \vdash^\alpha_\kappa \Gamma \Rightarrow \Delta, A(\bar{n})$ is also an axiom for every n. Also, if $\forall x A$ is not principal in the rule used to deduce (10) then result follows from the induction hypothesis. We may assume, therefore, that (10) derives by ω-rule from the sequents

$$\mathrm{S} \vdash^\beta_\kappa \Gamma \Rightarrow \Delta', A(\bar{n})$$

for $\beta < \alpha$. If $\Delta' = \Delta$ apply Lemma 6.6(1) and we are done. The only other possibility is $\Delta' = \Delta \cup \{\forall x A\}$, whence an application of the induction hypothesis and Lemma 6.6 yields $\mathrm{S} \vdash^\alpha_\kappa \Gamma \Rightarrow \Delta, A(\bar{n})$ for every n. □

6.1.2 Embedding

Lemma 6.8. *For every formula $A(x)$ with at most x free there exists $n < \omega$ such that*

1. *if $A \in \mathcal{L}_0$ then $\mathrm{T}_\infty \vdash^n_\omega \emptyset \Rightarrow \forall x(A \leftrightarrow T[A\dot{x}])$,*
2. *if $A \in \mathcal{L}^+_{T,F}$ then $\mathrm{T}^+_\infty \vdash^n_\omega \emptyset \Rightarrow \forall x(A \leftrightarrow T[A\dot{x}])$ and $\mathrm{T}^+_\infty \vdash^n_0 \emptyset \Rightarrow \forall x(\overline{A} \leftrightarrow F[A\dot{x}])$.*

Proof. Both cases follow similar arguments so we present only 1. Fix $A \in \mathcal{L}_0$ with at most x free and let $m = |A|^-$. Formally, the formula $T[A\dot{x}]$ is shorthand for $\forall y(\mathrm{Sub}([A], [x], x, y) \rightarrow Ty)$ where Sub is the elementary formula representing the syntactic function *sub*. Thus to show 1, we need to derive

$$\mathrm{T}_\infty \vdash^n_\omega \emptyset \Rightarrow \forall x(A \leftrightarrow \forall y(\mathrm{Sub}([A], [x], x, y) \rightarrow Ty))$$

for some n. By Lemma 6.5(2) we have

$$\mathrm{T}_\infty \vdash^{2m}_0 A(\bar{k}) \Rightarrow A(\bar{k}) \tag{11}$$

for every $k < \omega$. By Lemma 6.5(3) there also exists $m' < \omega$ such that for every $k < \omega$ and $l \neq sub([A], [x], k)$,

$$\mathrm{T}_\infty \vdash^{m'}_0 \mathrm{Sub}([A], [x], \bar{k}, \bar{l}) \Rightarrow \emptyset.$$

Combining this with (11) and Lemma 6.6 we conclude

$$\mathrm{T}_\infty \vdash^n_\omega \emptyset \Rightarrow \forall x(A \leftrightarrow T[A\dot{x}])$$

for some n depending only on $|A|$. □

Given Lemma 6.8 we can readily transfer derivations in the truth theories UTB_0 and UTFB_0 into derivations in T_∞ or T^+_∞.

Lemma 6.9 (First embedding lemma).

1. *If* $\mathsf{UTB}_0 \vdash A(x_0, \dots, x_k)$ *then there exists* $n < \omega$ *s.t. for all* $m_0, \dots, m_k < \omega$, $\mathrm{T}_\infty \vdash^n_{\omega+n} \emptyset \Rightarrow A(\bar{m}_0, \dots, \bar{m}_k)$.
2. *If* $\mathsf{UTFB}_0 \vdash A(x_0, \dots, x_k)$ *then there exists* $n < \omega$ *s.t. for all* $m_0, \dots, m_k < \omega$, $\mathrm{T}^+_\infty \vdash^n_{\omega+n} \emptyset \Rightarrow A(\bar{m}_0, \dots, \bar{m}_k)$.

Proof. By induction on the length of the UTB_0 or UTFB_0. Lemmas 6.5(3) and 6.8 cover the axioms and applications of modus ponens are replaced by the cut rule. □

By making use of infinite derivations it is possible to derive not only instances of transfinite induction, but also the compositional axioms for truth.

Lemma 6.10 (Second embedding lemma). *Let* $\omega < \kappa < \Gamma_0$.

1. *If* $\mathsf{CT}_\kappa \vdash A$ *then there exists* $\gamma < \kappa$ *and* $n < \omega$ *such that* $\mathrm{T}_\infty \vdash^{\gamma.n}_{\omega+n} \emptyset \Rightarrow A$.
2. *If* $\mathsf{KF}_\kappa \vdash A$ *then there exists* $\gamma < \kappa$ *and* $n < \omega$ *such that* $\mathrm{T}^+_\infty \vdash^{\gamma.n}_{\omega+n} \emptyset \Rightarrow A$.

Proof. Recall $\mathsf{S}_\kappa = \mathsf{S} + \mathrm{TI}_{\mathscr{L}_*}(<\kappa)$. We begin by showing that transfinite induction up to κ is derivable in T_∞. Fix $A(x) \in \mathscr{L}_*$ and an ordinal $\alpha < \kappa$. We prove

$$\mathrm{T}_\infty \vdash^{\omega+\alpha.8+3}_0 \mathrm{Prog}\lambda x A \Rightarrow A\bar{\alpha} \tag{12}$$

by transfinite induction on α, so in particular

$$\mathrm{T}^{(+)}_\infty \vdash^{\max\{\omega,\alpha\}.9}_0 \mathrm{Prog}\lambda x A \Rightarrow A\bar{\alpha}.$$

If $\alpha = 0$ (12) this is trivial. Suppose $\beta > 0$ and (12) holds for all $\alpha < \beta$. We have

$$\mathrm{T}_\infty \vdash^{\omega+\alpha.8+4}_0 \mathrm{Prog}\lambda x A \Rightarrow \bar{\alpha} \prec \bar{\beta} \to A\bar{\alpha} \tag{13}$$

for every $\alpha < \beta$. Combining (13) with Lemma 6.5(3) we deduce

$$\mathrm{T}_\infty \vdash^{\omega+\delta}_0 \mathrm{Prog}\lambda x A \Rightarrow \forall x(x \prec \bar{\beta} \to A),$$

where $\delta = \sup\{\alpha.8 + 5 \mid \alpha < \beta\}$. Further applications of the logical rules combined with Lemma 6.5(2) yields

$$\mathrm{T}_\infty \vdash^{\omega+\delta+2}_0 \mathrm{Prog}\lambda x A, \forall x(x \prec \bar{\beta} \to A) \to A\bar{\beta} \Rightarrow A\bar{\beta},$$

from which we deduce $\mathrm{T}_\infty \vdash_0^{\omega+\delta+3} \mathrm{Prog}\lambda x A \Rightarrow A\bar{\beta}$. Notice that $\delta \leq \beta$.8.

The second part consists in deriving the compositional axioms for the two truth theories within the infinitary calculus. These all follow similar arguments so we present only one case, the compositional axiom for the universal quantifier over $\mathscr{L}^+_{T,F}$ formulæ. Fix $n < \omega$. Suppose $n = \#A$ for some formula $A(x) \in \mathscr{L}^+_{T,F}$. From the proof of Lemma 6.8 there therefore exists $k_n < \omega$ such that for every $m < \omega$

$$\mathrm{T}_\infty \vdash_0^{k_n} T sub(\bar{n}, [x], \bar{m}) \Rightarrow A(\bar{m}).$$

Thus we may successively deduce

$$\begin{aligned}
&\mathrm{T}_\infty \vdash_0^{k_n+1} \mathrm{Sent}(\underset{.}{\forall} x\bar{n}), \forall x T sub(\bar{n}, [x], x) \Rightarrow A\bar{m}\\
&\mathrm{T}_\infty \vdash_0^{k_n+2} \mathrm{Sent}(\underset{.}{\forall} x\bar{n}), \forall x T sub(\bar{n}, [x], x) \Rightarrow \forall x A\\
&\mathrm{T}_\infty \vdash_0^{k_n+3} \mathrm{Sent}(\underset{.}{\forall} x\bar{n}), \forall x T sub(\bar{n}, [x], x) \Rightarrow T(\underset{.}{\forall} x\bar{n})\\
&\mathrm{T}_\infty \vdash_0^{k_n+5} \mathrm{Sent}(\underset{.}{\forall} x\bar{n}) \Rightarrow \forall x T sub(\bar{n}, [x], x) \rightarrow T(\underset{.}{\forall} x\bar{n}).
\end{aligned}$$

Likewise we may also deduce

$$\mathrm{T}_\infty \vdash_0^{k_n+5} \mathrm{Sent}(\underset{.}{\forall} x\bar{n}) \Rightarrow T(\underset{.}{\forall} x\bar{n}) \rightarrow \forall x T sub(\bar{n}, [x], x),$$

and so

$$\mathrm{T}_\infty \vdash_0^{\omega} \emptyset \Rightarrow \mathrm{Sent}(\forall_x \bar{n}) \rightarrow (T(\underset{.}{\forall} x\bar{n}) \leftrightarrow \forall x T sub(\bar{n}, [x], x)). \tag{14}$$

For any $n \notin \{[A] \mid A(x) \in \mathscr{L}^+_{T,F}\}$ we can also derive (14) on account of $\mathrm{Sent}(\underset{.}{\forall} x\bar{n})$ being a false Δ^0_0 statement (and so $\mathrm{T}_\infty \vdash_0^{\omega} \mathrm{Sent}(\underset{.}{\forall} x\bar{n}) \Rightarrow \emptyset$ is derivable). Thus we have

$$\mathrm{T}_\infty \vdash_0^{\omega+1} \emptyset \Rightarrow \forall z(\mathrm{Sent}(\underset{.}{\forall} xz) \rightarrow (T(\underset{.}{\forall} xz) \leftrightarrow \forall x T sub(z, [x], x))).$$

The remainder of the proof proceeds by (finite) induction on the length of the CT_κ or KF_κ derivation following the proof of the first embedding lemma. □

6.2 Cut elimination

In this section we prove that it is possible to reduce the cut rank of certain sequents at the expense of an increase in the ordinal height.

Theorem 6.11 (Partial cut elimination).

1. $\mathrm{T}_\infty \vdash^{\alpha}_{\omega+m} \Gamma \Rightarrow \Delta$ *implies* $\mathrm{T}_\infty \vdash^{2^{\alpha}_m}_{\omega} \Gamma \Rightarrow \Delta$,

2. $\mathrm{T}^+_\infty \vdash^\alpha_{\omega+m} \Gamma \Rightarrow \Delta$ *implies* $\mathrm{T}^+_\infty \vdash^{2^\alpha_m}_{\omega+1} \Gamma \Rightarrow \Delta$.

The key ingredient for this result is the following lemma which shows how to avoid the 'border-line' cuts in deriving a sequent. Iterating this process permits the elimination of the most complex cuts involving the truth predicate throughout a derivation (Lemma 6.14 below) from which Theorem 6.11 follows.

Lemma 6.12 (Reduction lemma). *If* $|A| \leq \omega + k + 1$, $\mathrm{T}^{(+)}_\infty \vdash^\alpha_{\omega+k+1} \Gamma \Rightarrow \Delta, A$ *and* $\mathrm{T}^{(+)}_\infty \vdash^\beta_{\omega+k+1} \Gamma, A \Rightarrow \Delta$ *then* $\mathrm{T}^{(+)}_\infty \vdash^{\alpha\#\beta}_{\omega+k+1} \Gamma \Rightarrow \Delta$, *where* $\alpha \# \beta$ *is the natural sum of* α *and* β *(see Definition 2.6).*

Proof. Let S be either T_∞ or T^+_∞. Fix $k < \omega$ and suppose $|A| = \omega + k + 1$ (if $|A| \leq \omega + k$ the result is immediate). The proof proceeds by a case distinction on the form of A.

We consider the case $A = \forall x B(x)$ for which we argue by induction on β; the other cases are similar. Suppose

$$\mathrm{S} \vdash^\beta_{\omega+k+1} \Gamma, \forall x B \Rightarrow \Delta. \tag{15}$$

If the sequent is an axiom then so is $\mathrm{S} \vdash^\beta_{\omega+k+1} \Gamma \Rightarrow \Delta$ and we are done. Otherwise (15) derives from a sequent

$$\mathrm{S} \vdash^{\beta_0}_{\omega+k+1} \Gamma' \Rightarrow \Delta'$$

for some $\beta_0 < \beta$. If $\forall x B$ is not principal in this rule the desired sequent can be derived via the induction hypothesis and reapplying the rule. If, however, A is principal then we may assume $\Delta' = \Delta$ and $\Gamma' = \Gamma \cup \{A, B(s)\}$ for some term s. The induction hypothesis therefore implies

$$\mathrm{S} \vdash^{\alpha\#\beta_0}_{\omega+k+1} \Gamma, B(s) \Rightarrow \Delta.$$

By Lemma 6.7 we have also $\mathrm{S} \vdash^\alpha_{\omega+k+1} \Gamma \Rightarrow \Delta, B(s)$, whence $\mathrm{S} \vdash^{\alpha\#\beta_0+1}_{\omega+k+1} \Gamma \Rightarrow \Delta$. Since $\alpha \# \beta_0 + 1 \leq \alpha \# \beta$ we are done. □

If we focus attention on derivations in T_∞ only it is possible to also remove cuts on the truth predicate:

Lemma 6.13 (Truth reduction lemma). *If* $|A| = \omega$, $\mathrm{T}_\infty \vdash^\alpha_\omega \Gamma \Rightarrow \Delta, A$ *and* $\mathrm{T}_\infty \vdash^\beta_\omega \Gamma, A \Rightarrow \Delta$ *then* $\mathrm{T}_\infty \vdash^{\alpha\#\beta}_\omega \Gamma \Rightarrow \Delta$.

Proof. The proof proceeds by induction on $\alpha \# \beta$. We consider only the case $A = T\bar{n}$ and A is principal in the derivation of the two assumed sequents, so $n = \# B$ for some $B \in \mathscr{L}_0$ and we have

$$\mathrm{T}_\infty \vdash^{\alpha_0}_{\omega} \Gamma \Rightarrow \Delta, T\bar{n}, B \qquad\qquad \mathrm{T}_\infty \vdash^{\beta_0}_{\omega} \Gamma, T\bar{n}, B \Rightarrow \Delta$$

for some $\alpha_0 < \alpha$ and $\beta_0 < \beta$. Notice $\alpha_0 \# \beta < \alpha \# \beta > \alpha \# \beta_0$, so the induction hypothesis may be applied, yielding

$$\mathrm{T}_\infty \vdash^{\alpha_0 \# \beta}_{\omega} \Gamma \Rightarrow \Delta, B \qquad\qquad \mathrm{T}_\infty \vdash^{\alpha \# \beta_0}_{\omega} \Gamma, B \Rightarrow \Delta.$$

Since $|B| < \omega$, we conclude $\mathrm{T}_\infty \vdash^{\alpha \# \beta}_{\omega} \Gamma \Rightarrow \Delta$ by cut. □

Lemma 6.14. *For every $k < \omega$, $\mathrm{T}_\infty \vdash^{\alpha}_{\omega+k+1} \Gamma \Rightarrow \Delta$ implies $\mathrm{T}_\infty \vdash^{2^\alpha}_{\omega+k} \Gamma \Rightarrow \Delta$, and for every $k > 0$, $\mathrm{T}^+_\infty \vdash^{\alpha}_{\omega+k+1} \Gamma \Rightarrow \Delta$ implies $\mathrm{T}^+_\infty \vdash^{2^\alpha}_{\omega+k} \Gamma \Rightarrow \Delta$.*

Proof. By induction on α. Suppose

$$\mathrm{T}^{(+)}_\infty \vdash^{\alpha}_{\omega+k+1} \Gamma \Rightarrow \Delta. \tag{16}$$

If this sequent is an axiom then $\mathrm{T}^{(+)}_\infty \vdash^{2^\alpha}_{\omega+k} \Gamma \Rightarrow \Delta$ is immediate. Suppose, therefore, (16) follows by the cut rule from

$$\mathrm{T}^{(+)}_\infty \vdash^{\alpha_0}_{\omega+k+1} \Gamma, A \Rightarrow \Delta \qquad\qquad \mathrm{T}^{(+)}_\infty \vdash^{\alpha_1}_{\omega+k+1} \Gamma \Rightarrow \Delta, A$$

where $\alpha_0, \alpha_1 < \alpha$ and $|A| \leq \omega + k$. Applying the induction hypothesis yields

$$\mathrm{T}^{(+)}_\infty \vdash^{2^{\alpha_0}}_{\omega+k} \Gamma, A \Rightarrow \Delta \qquad\qquad \mathrm{T}^{(+)}_\infty \vdash^{2^{\alpha_1}}_{\omega+k} \Gamma \Rightarrow \Delta, A.$$

Since $2^{\alpha_0} \# 2^{\alpha_1} \leq 2^\alpha$, Lemma 6.12 (or Lemma 6.13 if $k = 0$) yields $\mathrm{T}^{(+)}_\infty \vdash^{2^\alpha}_{\omega+k} \Gamma \Rightarrow \Delta$ as required. The remaining cases are simple. □

Combining partial cut elimination with the results of the previous section we obtain

Theorem 6.15. *Suppose $\kappa > 0$.*

1. *If $\mathsf{CT}_{\epsilon_\kappa} \vdash A$ then there exists $\gamma < \epsilon_\kappa$ such that $\mathrm{T}_\infty \vdash^{\gamma}_{\omega} \emptyset \Rightarrow A$.*
2. *If $\mathsf{KF}_{\epsilon_\kappa} \vdash A$ then there exists $\gamma < \epsilon_\kappa$ such that $\mathrm{T}^+_\infty \vdash^{\gamma}_{\omega+1} \emptyset \Rightarrow A$.*

Proof. To show 1, suppose $\mathsf{CT}_{\epsilon_\kappa} \vdash A$. By Lemma 6.10 there exists $\gamma < \epsilon_\kappa$ and $n < \omega$ such that $\mathrm{T}_\infty \vdash^{\gamma.n}_{\omega+n} \emptyset \Rightarrow A$. Theorem 6.11 implies $\mathrm{T}_\infty \vdash^{2^{\gamma.n}_n}_{\omega} \emptyset \Rightarrow A$, but $2^{\gamma.n}_n < \epsilon_\kappa$ so we are done. The other case is analogous. □

Before proceeding with the embedding of $T_\infty^{(+)}$ derivations in CT (KF) we observe that part 1 of Theorem 6.11 (and so part 1 from Theorem 6.15) can be expanded to eliminate *all* cuts:

Corollary 6.16. *If* $\mathsf{CT}_{\varepsilon_\kappa} \vdash A$ *then* $T_\infty \vdash_0^{\varepsilon_\gamma} \emptyset \Rightarrow A$ *for some* $\gamma < \varepsilon_\kappa$.

Proof. We prove, by induction on $\alpha \geq \omega^\omega$, that $T_\infty \vdash_\omega^\alpha \Gamma \Rightarrow \Delta$ implies $T_\infty \vdash_0^{\epsilon_\alpha} \Gamma \Rightarrow \Delta$, which combined with the previous theorem yields the result.

The only non-trivial case is if $T_\infty \vdash_\omega^\alpha \Gamma \Rightarrow \Delta$ arises from a cut on

$$T_\infty \vdash_\omega^{\alpha_0} \Gamma \Rightarrow \Delta, A \qquad\qquad T_\infty \vdash_\omega^{\alpha_1} \Gamma, A \Rightarrow \Delta$$

with $|A| < \omega$. Then the induction hypothesis implies

$$T_\infty \vdash_0^{\epsilon_{\alpha_0}} \Gamma \Rightarrow \Delta, A \qquad\qquad T_\infty \vdash_0^{\epsilon_{\alpha_1}} \Gamma, A \Rightarrow \Delta$$

and so

$$T_\infty \vdash_{|A|}^{\max\{\epsilon_{\alpha_0}, \epsilon_{\alpha_1}\}+1} \Gamma \Rightarrow \Delta. \tag{17}$$

We claim $T_\infty \vdash_0^{\epsilon_\alpha} \Gamma \Rightarrow \Delta$.

Let T_∞^- denote the subsystem of T_∞ without the induction axiom (axiom III). If $B(x)$ is an instance of elementary induction we have $T_\infty^- \vdash_0^{\omega+1} \emptyset \Rightarrow \forall x B$ by the ω-rule and the argument in Lemma 6.10. Thus $T_\infty \vdash_\kappa^\beta \Gamma \Rightarrow \Delta$ implies $T_\infty^- \vdash_\kappa^{\omega+1+\beta} \Gamma \Rightarrow \Delta$ for all β, κ, Γ and Δ. Since the principal formulæ of the T_∞^--axioms are all atomic, the reduction lemma and partial cut elimination theorem for T_∞^- can be generalised to derivations with both finite and transfinite cut rank:

1. For all α and β, $T_\infty^- \vdash_{|A|}^\alpha \Gamma, A \Rightarrow \Delta$ and $T_\infty^- \vdash_{|A|}^\beta \Gamma \Rightarrow \Delta, A$ imply $T_\infty^- \vdash_{|A|}^{\alpha \# \beta} \Gamma \Rightarrow \Delta$;

2. For all κ, if $T_\infty^- \vdash_{\kappa+1}^\alpha \Gamma \Rightarrow \Delta$ then $T_\infty^- \vdash_\kappa^{2^\alpha} \Gamma \Rightarrow \Delta$.

Only the proof of 1 in the case $|A| = 0$ is not already covered by the earlier argument. In this case, however, one may assume both sequents $T_\infty^- \vdash_{|A|}^\alpha \Gamma, A \Rightarrow \Delta$ and $T_\infty^- \vdash_{|A|}^\beta \Gamma \Rightarrow \Delta, A$ are instances of either axiom I or II and so the desired sequent is either an axiom or derivable from Lemma 6.7(1/2). Combining these observations with (17) yields the desired result. □

6.3 Deriving reflection

The aim is to prove that reflection for T_∞ sequents of height ϵ_κ is derivable in $\mathsf{CT}_{\epsilon_{\kappa+1}}$ since this, when combined with Theorem 6.15, implies $\mathsf{CT}_{\epsilon_{\kappa+1}} \vdash \mathrm{Ref}(\mathsf{CT}_{\epsilon_\kappa})$. Informally speaking, this will achieved by defining a predicate $\mathrm{Bew}'_\infty(a, y)$ that expresses 'a encodes an ordinal α, y is a formula A of $\mathscr{L}_*$ and $\mathrm{T}_\infty \vdash^\alpha_\omega \emptyset \Rightarrow A$' and proving

$$\mathsf{CT}_{\epsilon_{\kappa+1}} \vdash \mathrm{Bew}_{\mathsf{CT}_{\epsilon_\kappa}}(x) \wedge \mathrm{Sent}_{\mathscr{L}_T}(x) \to \exists a \prec \hat{\epsilon}_{\bar{\kappa}}\, \mathrm{Bew}'_\infty(a, x), \tag{18}$$

which formalises Theorem 6.15, and

$$\mathsf{CT}_{\epsilon_{\kappa+1}} \vdash \mathrm{Bew}'_\infty(\bar{\alpha}, [A]) \to A \tag{19}$$

for each $\alpha < \epsilon_\kappa$ and $A \in \mathscr{L}_T$. It is not possible to completely formalise T_∞ derivations within arithmetic due to the infinitary nature of the ω-rule: a witness to a sequent $\mathrm{T}_\infty \vdash^\alpha_\kappa \Gamma \Rightarrow \Delta$ is essentially a well-founded tree (with order-type α) labelled by sequents locally satisfying the rules in Definition 6.1. It is well-known, however, that for showing (18) and (19) it suffices to consider a recursive counterpart of the ω-rule which can be given a purely finitary representation:

- a witness of $\mathrm{T}_\infty \vdash^\alpha_\kappa \Gamma \Rightarrow \Delta, \forall x A$ is (the code of) a recursive function $f\colon \omega \to \omega \times \omega \times \omega$ such that for every $n < \omega$, $f(n)$ has the form $(p, \#\beta, \#\lambda)$ where p is a witness of $\mathrm{T}_\infty \vdash^\beta_\lambda \Gamma \Rightarrow \Delta, A(\bar{n})$.

Let $\mathrm{Bew}^{(+)}_\infty(a, k, x, y)$ be the formula that expresses 'a and k encode ordinals α and κ, x and y are finite sets Γ and Δ of $\mathscr{L}_T$ formulæ and there exists p witnessing $\mathrm{T}^{(+)}_\infty \vdash^\alpha_\kappa \Gamma \Rightarrow \Delta$.'

Using the formalisation above we can prove the following.

Lemma 6.17. *For each $\alpha < \kappa$, we have*

$$\mathsf{PA}_\kappa \vdash \forall x(\mathrm{Bew}_{\mathsf{CT}_{\bar{\alpha}}} x \wedge \mathrm{Sent}_{\mathscr{L}_0} x \to \exists b \prec \bar{\alpha} \exists k \mathrm{Bew}_\infty(b.k, \bar{\omega} \hat{+} k, \bar{\emptyset}, \{x\})),$$
$$\mathsf{PA}_\kappa \vdash \forall x(\mathrm{Bew}_{\mathsf{KF}_{\bar{\alpha}}} x \wedge \mathrm{Sent}_{\mathscr{L}_0} x \to \exists b \prec \bar{\alpha} \exists k \mathrm{Bew}^+_\infty(b.k, \bar{\omega} \hat{+} k, \bar{\emptyset}, \{x\})),$$
$$\mathsf{PA}_\kappa \vdash \forall x \forall b \prec \bar{\alpha} \forall g, d \subset \mathscr{L}_{T,F}(\mathrm{Bew}^{(+)}_\infty(b, \hat{\omega} \hat{+} x, g, d) \to \mathrm{Bew}^{(+)}_\infty(\hat{2}^b_x, \bar{\omega}, g, d)),$$

where $\bar{\emptyset}$ represents the code of the empty set and $\{x\}$ the singleton set containing the formula represented by x.

To derive (19) we formalise the obvious soundness argument for derivations in $\mathrm{T}^{(+)}_\infty$, proving an arithmetical approximation to the statement '$\mathrm{T}^{(+)}_\infty \vdash^\alpha_\kappa \Gamma \Rightarrow \Delta$ implies the formula $\bigwedge \Gamma \to \bigvee \Delta$ is true'. Focusing on CT, if $\kappa > \omega$, or $\Gamma \cup \Delta \not\subseteq \mathscr{L}_0$

the derivation of this sequent may utilise formulæ outside $\mathscr{L}_0$, and the phrase 'A is true' cannot be formalised simply as $T[A]$. To overcome this we use a combination of the cut elimination theorem and partial truth predicates for $\mathscr{L}_{T,F}$ formulæ. For $n > 0$, let $\mathrm{Tr}_n(x, P)$ (where P is a free predicate symbol) be a partial truth predicate for $\mathscr{L}_{T,F}$-formulæ over P given by

$$\begin{aligned}\mathrm{Tr}_{n+1}(x, P) \leftrightarrow P(x) \vee \exists yz < x((x = [T\dot{y}] \wedge Ty) \vee (x = [F\dot{y}] \wedge Fy)\\ \vee (x = y \dot{\wedge} z \wedge \mathrm{Tr}_n(y, P) \wedge \mathrm{Tr}_n(z, P))\\ \vee (x = \dot{\neg} y \wedge \neg \mathrm{Tr}_n(y, P))\\ \vee (x = \dot{\forall} yz \wedge \forall v \mathrm{Tr}_n(sub(z, y, v)))).\end{aligned}$$

Using the predicates Tr_k we can define partial truth predicate of arithmetic as well as truth predicates over the theories CT and KF:

$$\begin{aligned}&\mathrm{Tr}_0^{\mathsf{PA}}(x) \leftrightarrow x \neq x & &\mathrm{Tr}_{n+1}^{\mathsf{PA}}(x) \leftrightarrow \mathrm{Tr}_{n+1}(x, \mathrm{Tr}_0^{\mathsf{PA}})\\ &\mathrm{Tr}_0^{\mathsf{CT}}(x) \leftrightarrow \mathrm{Sent}_{\mathscr{L}_0}(x) \wedge T(x) & &\mathrm{Tr}_{n+1}^{\mathsf{CT}}(x) \leftrightarrow \mathrm{Tr}_{n+1}(x, \mathrm{Tr}_0^{\mathsf{CT}})\\ &\mathrm{Tr}_0^{\mathsf{KF}}(x) \leftrightarrow \mathrm{Sent}_{\mathscr{L}_{T,F}^+}(x) \wedge T(x) & &\mathrm{Tr}_{n+1}^{\mathsf{KF}}(x) \leftrightarrow \mathrm{Tr}_{n+1}(x, \mathrm{Tr}_0^{\mathsf{KF}})\end{aligned}$$

Lemma 6.18. *For each choice of* $\mathsf{S} \in \{\mathsf{PA}, \mathsf{CT}, \mathsf{KF}\}$, *The formula* $\mathrm{Tr}_{2n}^{\mathsf{S}}$ *is* $\Delta_1^0(\mathscr{L}_{T,F})$ *if* $n = 0$, *and equivalent to a* $\Sigma_{2n}^0(\mathscr{L}_{T,F})$ *formula otherwise, while* $\mathrm{Tr}_{2n+1}^{\mathsf{S}}$ *is equivalent to a* $\Pi_{2n}^0(\mathscr{L}_{T,F})$ *formula for every* n.

Proof. Since $\mathrm{Sent}_{\mathscr{L}}(x)$ is a Δ_1^0 formula, $\mathrm{Tr}_0^{\mathsf{CT}}(x)$ and $\mathrm{Tr}_0^{\mathsf{KF}}$ are $\Delta_1^0(\mathscr{L}_{T,F})$. Therefore $\mathrm{Tr}_1^{\mathsf{S}}(x)$ is equivalent to a $\Pi_1^0(\mathscr{L}_{T,F})$ formula, $\mathrm{Tr}_2^{\mathsf{S}}(x)$ to a $\Sigma_2^0(\mathscr{L}_{T,F})$ formula, and so on. □

In addition to the rank functions $|\cdot|$ and $|\cdot|^-$, define $|\cdot|^+$ according to the same recursive rules but with $|A|^+ = \omega$ whenever $A \in \mathscr{L}_{T,F}^+$ and A is not arithmetical.

Lemma 6.19. *For each formula* $A \in \mathscr{L}_{T,F}$ *with only* x *free and each* $k < \omega$,

$$\begin{aligned}|A|^- &< k & &\textit{implies} & &\mathsf{PA} \vdash \forall x(\mathrm{Tr}_k^{\mathsf{PA}}[A\dot{x}] \leftrightarrow A),\\ |A| &< \omega + k & &\textit{implies} & &\mathsf{CT} \vdash \forall x(\mathrm{Tr}_k^{\mathsf{CT}}[A\dot{x}] \leftrightarrow A),\\ |A|^+ &< \omega + k & &\textit{implies} & &\mathsf{KF} \vdash \forall x(\mathrm{Tr}_k^{\mathsf{KF}}[A\dot{x}] \leftrightarrow A).\end{aligned}$$

Proof. All cases proceed by induction on k. If $k > 1$ (or in the PA case for any k), the result is deduced directly from the definition of the respective truth predicate and, if applicable, the induction hypothesis. The remaining cases are $|A| \leq \omega$ for CT and $|A|^+ \leq \omega$ for KF. We show only the former.

If $|A| < \omega$ then $A \in \mathscr{L}_0$, so

$$\mathsf{CT} \vdash \mathrm{Tr}_0^{\mathsf{CT}}[A\dot{x}] \leftrightarrow T[A\dot{x}] \leftrightarrow A.$$

If $|A| = \omega$ then either $A = Tt$ or $A = Ft$ for some term t. In the former case we observe

$$\begin{aligned}\mathsf{CT} \vdash \mathrm{Tr}_1^{\mathsf{CT}}[A\dot{x}] &\leftrightarrow \mathrm{Tr}_1^{\mathsf{CT}}(sub([T(t)], [x], x)) \\ &\leftrightarrow \mathrm{Tr}_1^{\mathsf{CT}}(sub([T(x)], [x], t)) \\ &\leftrightarrow \mathrm{Tr}_1^{\mathsf{CT}}[T(\dot{t})] \\ &\leftrightarrow T(t)\end{aligned}$$

and similarly for the latter case. □

We can now formalise the soundness argument for CT. We first introduce a new quantifier $\forall g \subset \mathscr{L}$ abbreviating quantification over codes for finite subsets of $\mathscr{L}$, so $\forall g \subseteq \mathscr{L}_{T,F} A(g)$ expresses 'for all g, if g is the code of a finite set of $\mathscr{L}_{T,F}$ formulæ then $A(g)$,' a formula $x \in g$ expressing that g encodes a finite set of which x is a member, and terms $g \cup d$ and $|g|^{(\pm)}$ denoting, respectively, the union of two finite sets and maximal rank among formulæ in the set g.

Lemma 6.20. *For each $\beta < \alpha$ and each n we have,*

$$\begin{aligned}\mathsf{PA}_\alpha \vdash \forall \gamma < \bar{\beta} \forall g{,}d \subset \mathscr{L}_{T,F}(&|g \cup d|^- < \bar{n} \wedge \\ &\mathrm{Bew}_\infty(\gamma, \bar{n}, g, d) \to (\forall x \in g\, \mathrm{Tr}_n^{\mathsf{PA}}(x) \to \exists y \in d \mathrm{Tr}_n^{\mathsf{PA}}(y))),\\ \mathsf{CT}_\alpha \vdash \forall \gamma < \bar{\beta} \forall g{,}d \subset \mathscr{L}_{T,F}(&|g \cup d| < \bar{\omega} \hat{+} \bar{n} \wedge \\ &\mathrm{Bew}_\infty(\gamma, \bar{\omega} \hat{+} \bar{n}, g, d) \to (\forall x \in g\, \mathrm{Tr}_n^{\mathsf{CT}}(x) \to \exists y \in d \mathrm{Tr}_n^{\mathsf{CT}}(y))),\\ \mathsf{KF}_\alpha \vdash \forall \gamma < \bar{\beta} \forall g{,}d \subset \mathscr{L}_{T,F}(&|g \cup d|^+ < \bar{\omega} \hat{+} \bar{n} \wedge \\ &\mathrm{Bew}_\infty^+(\gamma, \bar{\omega} \hat{+} \bar{n}, g, d) \to (\forall x \in g\, \mathrm{Tr}_n^{\mathsf{KF}}(x) \to \exists y \in d \mathrm{Tr}_n^{\mathsf{KF}}(y))).\end{aligned}$$

Proof. Fix n, α and β, and let $\forall \gamma < \bar{\beta}\, A^-(\gamma)$, $\forall \gamma < \bar{\beta}\, A(\gamma)$ and $\forall \gamma < \bar{\beta}\, A^+(\gamma)$ denote respectively the three formulæ listed above. It suffices to prove $\mathsf{PA} \vdash \mathrm{Prog}_\prec \lambda x A^-$, $\mathsf{CT} \vdash \mathrm{Prog}_\prec \lambda x A$ and $\mathsf{KF} \vdash \mathrm{Prog}_\prec \lambda x A^+$, but these follow directly from the definition of the calculi T_∞ and T_∞^+. □

Combining Lemma 6.20 with the first embedding lemma and the cut reduction lemma for T_∞ we naturally obtain

Lemma 6.21. $\mathsf{CT} \vdash \mathrm{Ref}(\mathsf{UTB}_0)$.

Proof. Let A be a formula with free variable x. We argue informally within CT. Fix a and suppose $\mathsf{UTB}_0 \vdash A(\bar{a})$. By Lemma 6.9 there exists $n < \omega$ such that $\mathrm{T}_\infty \vdash^n_{\omega+n} \emptyset \Rightarrow A(\bar{a})$. Cut elimination therefore yields an $m < \omega$ for which $\mathrm{T}_\infty \vdash^m_\omega \emptyset \Rightarrow A(\bar{a})$ holds, whence Lemma 6.20 implies $\mathrm{Tr}^{\mathsf{CT}}_{|A|}[A\dot{\bar{a}}]$ and Lemma 6.19 implies $A(a)$. □

As a consequence we may deduce the remaining half of Theorem 1.4.

Theorem 6.22. *For all* $\kappa < \Gamma_0$*,*

1. $\mathsf{R}(\mathsf{CT}_{\epsilon_\kappa}) \subseteq \mathsf{CT}_{\epsilon_\kappa+1}$,
2. $\mathsf{R}_\kappa(\mathsf{UTB}) \subseteq \mathsf{CT}_{\epsilon_\kappa}$.

Moreover, these are verifiable in $\mathsf{PA}_{\kappa+1}$.

Proof. We begin with 1. Let α be such that $\alpha = 2^\alpha$. Observe that α is a limit ordinal and additive principal. We will show that reflection for CT_α is derivable in $\mathsf{CT}_{\alpha+1}$. Let A be a formula in $\mathscr{L}_T$ with x free. Arguing within $\mathsf{CT}_{\alpha+1}$, fix $a < \omega$ and suppose $\mathsf{CT}_\alpha \vdash A\bar{a}$. By the second embedding lemma there exists $\gamma < \alpha$ and n such that $\mathrm{T}_\infty \vdash^{\gamma.n}_{\omega+n} \emptyset \Rightarrow A\bar{a}$ and so $\mathrm{T}_\infty \vdash^\delta_\omega \emptyset \Rightarrow A\bar{a}$ where $\delta = 2^{\gamma.n}_n < \alpha$. By Lemma 6.20 we obtain $\mathrm{Tr}^{\mathsf{CT}}_{|A|}[A\bar{a}]$ and we are done.

2 is proved by transfinite induction on κ. The base case, $\kappa = 0$, is a consequence of Lemma 6.21. If $\kappa = \eta + 1$ then the induction hypothesis implies $\mathsf{R}_\eta(\mathsf{UTB}) \subseteq \mathsf{CT}_{\epsilon_\eta}$ and, in particular,

$$\mathsf{PA}_\kappa \vdash \forall x(\mathrm{Bew}_{\mathsf{R}_\eta(\mathsf{UTB})}x \rightarrow \mathrm{Bew}_{\mathsf{CT}_{\epsilon_\eta}}x),$$

whence $\mathsf{R}(\mathsf{R}_\eta(\mathsf{UTB})) \subseteq \mathsf{R}(\mathsf{CT}_{\epsilon_\eta})$. By 1, therefore, $\mathsf{R}_\kappa(\mathsf{UTB}) \subseteq \mathsf{CT}_{\epsilon_\eta+1} = \mathsf{CT}_{\epsilon_\kappa}$. In the case κ is a limit ordinal $\mathsf{R}_\kappa(\mathsf{UTB}) = \bigcup_{\eta<\kappa}\mathsf{R}_\eta(\mathsf{UTB}) \subseteq \bigcup_{\eta<\kappa}\mathsf{CT}_{\epsilon_\eta} = \mathsf{CT}_{\epsilon_\kappa}$ which is all verifiable in $\mathsf{PA}_{\kappa+1}$. □

Applying the same reasoning to KF we have

Theorem 6.23.

1. $\mathsf{KF} \vdash \mathrm{Ref}(\mathsf{UTFB}_0)$.
2. *If* $2^\kappa = \kappa$ *then* $\mathsf{KF}_{\kappa+1} \vdash \mathrm{Ref}(\mathsf{KF}_\kappa)$.
3. *For each* κ, $\mathsf{R}_\kappa(\mathsf{UTFB}) \subseteq \mathsf{KF}_{\epsilon_\kappa}$.

We finish this section by determining the proof-theoretic ordinal of the systems CT_κ and KF_κ, thereby also the strength of $\mathsf{R}_\kappa(\mathsf{UTB})$ and $\mathsf{R}_\kappa(\mathsf{UTB})$. It is also worth noting that this is the only result for which the proofs for the CT and KF statements differ significantly.

Theorem 6.24.

1. *If A is an $\mathscr{L}_0$ theorem of either* $\mathsf{R}_\kappa(\mathsf{UTB})$, $\mathsf{R}_\kappa(\mathsf{CT})$ *or* $\mathsf{CT}_{\varepsilon_\kappa}$ *then* $\mathsf{EA} + \mathrm{TI}(<\varepsilon_{\varepsilon_\kappa}) \vdash A$.

2. *If A is an $\mathscr{L}_0$ theorem of either* $\mathsf{R}_\kappa(\mathsf{UTFB})$, $\mathsf{R}_\kappa(\mathsf{KF})$ *or* $\mathsf{KF}_{\varepsilon_\kappa}$ *then* $\mathsf{EA} + \mathrm{TI}(<\varphi\varepsilon_\kappa 0) \vdash A$.

Proof. The first part can be deduced directly from the proof-theoretic analysis presented above. By Theorem 6.22 it suffices to assume $\mathsf{CT}_{\varepsilon_\kappa} \vdash A$ and A is arithmetical. By Corollary 6.16 $\mathrm{T}_\infty \vdash_0^\gamma \emptyset \Rightarrow A$ for some $\gamma < \varepsilon_{\varepsilon_\kappa}$. Since A is arithmetical, the first part of Lemma 6.20 can be applied to deduce $\mathsf{PA}_{\varepsilon_{\varepsilon_\kappa}} \vdash A$.

We do not have the resources present to establish 2 via an analogous argument. Nevertheless, the result can be readily obtained from Feferman's original analysis of KF. Theorem 4.1.2 of [9] establishes the proof-theoretic equivalence of KF (therein denoted Ref(PA)) and the second-order system $(\Pi^0_1\text{-}\mathsf{CA})_{<\varepsilon_0}$ formalising ε_0-times iterated Π^0_1-comprehension. The proof trivially extends to show that the theories $\mathsf{KF}_{\varepsilon_\kappa}$ and $(\Pi^0_1\text{-}\mathsf{CA})_{<\varepsilon_\kappa}$ also prove the same $\mathscr{L}_0$-theorems, the latter of which is proof-theoretically equivalent to $\mathsf{PA}_{\varphi\varepsilon_\kappa 0}$ [8]. □

References

[1] Jon Barwise. *Admissible Sets and Structures: An Approach to Definability Theory.* Springer–Verlag, 1975.

[2] Andrea Cantini. Notes on formal theories of truth. *Zeitschrift für mathematische Logik und Grundlagen der Mathematik*, 35:97–130, 1989.

[3] Andrea Cantini. A theory of formal truth arithmetically equivalent to ID1. *Journal of Symbolic Logic*, 55(1):244–259, 1990.

[4] Cezary Cieśliński. T-equivalences for positive sentences. *Review of Symbolic Logic*, 4(2):319–325, 2011.

[5] Cezary Cieśliński. Typed and untyped disquotational truth. In *Unifying the Philosophy of Truth*, Theodora Achourioti, Henri Galinon, José Martínez Fernández, and Kentaro Fujimoto, eds., pages 307–320. Springer, 2015.

[6] Solomon Feferman. Transfinite recursive progressions of axiomatic theories. *Journal of Symbolic Logic*, 27:259–316, 1962.

[7] Solomon Feferman. Systems of predicative analysis. *Journal of Symbolic Logic*, 29:1–30, 1964.

[8] Solomon Feferman. *Theories of finite type related to mathematical practice*, In *Handbook of Mathmatical Logic*, pages 913–971. North-Holland, Amsterdam, 1977.

[9] Solomon Feferman. Reflecting on incompleteness. *Journal of Symbolic Logic*, 56:1–49, 1991.

[10] Kentaro Fujimoto. Relative truth definability of axiomatic truth theories. *Bulletin of Symbolic Logic*, 16(3):305–344, 09 2010.

[11] Volker Halbach. Disquotational truth and analyticity. *Journal of Symbolic Logic*, 66:1959–1973, 2001.

[12] Volker Halbach. *Axiomatic Theories of Truth*. Cambridge University Press, 2011.

[13] Petr Hájek and Pavel Pudlák. *Metamathematics of First-Order Arithmetic*. Springer-Verlag, Berlin, 1998.

[14] Leon Horsten and Graham E. Leigh. Truth is simple. *Mind*, to appear.

[15] Paul Horwich. *Truth*. Clarendon Press, second edition edition, 1998.

[16] Henryk Kotlarski, Stanislav Krajewski, and Alistair Lachlan. Construction of satisfaction classes for nonstandard models. *Canadian Mathematical Bulletin*, 24:283–93, 1981.

[17] Georg Kreisel and Azriel Lévy. Reflection principles and their use for establishing the complexity of axiomatic systems. *Mathematical Logic Quarterly*, 14:97–142, 1968.

[18] Saul Kripke. Outline of a theory of truth. *Journal of Philosophy*, 72(19):690–716, 1975.

[19] Graham E. Leigh. A proof-theoretic account of classical principles of truth. *Annals of Pure and Applied Logic*, 164:1009–1024, 2013.

[20] Graham E. Leigh. Conservativity for theories of compositional truth via cut elimination. *Journal of Symbolic Logic*, 80(3):845–865, 2015.

[21] Graham E. Leigh and Michael Rathjen. An ordinal analysis for theories of self-referential truth. *Archive for Mathematical Logic*, 49(2):213–247, 2010.

[22] Vann McGee. Maximal consistent sets of Tarski's schema (T). *Journal of Philosophical Logic*, 21:235–241, 1992.

[23] Willard Van Orman Quine. *Philosophy of Logic*. Harvard University Press, Cambridge, Massachusetts, 1970.

[24] Michael Rathjen. The realm of ordinal analysis. In *Sets and Proofs, Proceedings of the Logic Colloquium '97*, 1997.

[25] Kurt Schütte. *Proof Theory*. Springer, Berlin, 1977.

Received 29 June 2015

A PREDICATIVE VARIANT OF A REALIZABILITY TRIPOS FOR THE MINIMALIST FOUNDATION

Maria Emilia Maietti
Dipartimento di Matematica, Università di Padova
`maietti@math.unipd.it`

Samuele Maschio
Dipartimento di Matematica, Università di Padova
`maschio@math.unipd.it`

Abstract

Here we present a predicative variant of a realizability tripos validating the intensional level of the Minimalist Foundation extended with Formal Church thesis, for short **CT**.

The original concept of tripos was introduced in the 80s by J. M. E. Hyland, P. T. Johnstone and A. M. Pitts in order to build various kinds of toposes including realizability ones.

Our categorical structure provides the key ingredient to build a predicative variant of a realizability topos satisfying **CT**, like Hyland's Effective topos, where to validate the extensional level of the Minimalist Foundation.

The adjective *predicative* refers to the fact that our categorical structure is formalized in Feferman's theory of inductive definitions $\widehat{ID_1}$.

1 Introduction

Constructive mathematics is mathematics developed with constructive proofs, that is proofs enjoying a computational method to construct witnesses of their existential statements. As a consequence constructively definable number theoretic functions are all computable. It is indeed often said that constructive mathematics is abstract mathematics which is implicitly computable.

We acknowledge many useful fruitful discussions with Laura Crosilla, Takako Nemoto, Giovanni Sambin and Thomas Streicher on topics of this paper. We are grateful to Andrew Swan very much for proof-reading our paper and to the referees for their suggestions.

To give evidence of such a claim it is convenient to call "constructive" those proofs that can be formalized in a foundation enjoying a so called "realizability model" where one may extract the computational contents of its proofs by interpreting its sets as data types and its functions as programs. The most basic example of such a constructive foundation, at least for constructive arithmetics, is Intuitionistic Arithmetic HA in [40]. Its realizability semantics is the well-known Kleene realizability interpretation (see for example [40]) which makes HA consistent with the so called Formal Church thesis, for short **CT**, expressing that from a total number-theoretic relation we can extract a computable function. Actually, most constructive foundations in the literature are consistent with **CT** and this is the case also for the Minimalist Foundation.

The Minimalist Foundation, for short **MF**, was conceived by the first author in joint work with G. Sambin in [28] as a common core among the most relevant constructive and classical foundations, introduced both in type theory, in category theory and in axiomatic set theory. In [28] **MF** is also required to be a two-level system equipped with an *intensional level* suitable for extraction of computational contents from its proofs, an *extensional level* formulated in a language as close as possible to that of ordinary mathematics and an interpretation of the latter in the former showing that the extensional level has been obtained by abstraction from the intensional one according to Sambin's forget-restore principle in [35].

A two-level formal system of this kind for **MF** was completed in [23]. Both intensional and extensional levels of **MF** consist of type systems based on versions of Martin-Löf's type theory with the addition of a primitive notion of propositions and some related constructors: the intensional one, called **mTT**, is based on [31] and the extensional one, called **emTT**, on [30]. Actually, **mTT** can be considered a *predicative* version of Coquand's Calculus of Constructions [13].

The two-level structure of **MF** has various kinds of benefits.

First of all it provides a framework for computer-aided formalization of its constructive proofs. Indeed the intensional level of **MF** has enough decidable properties to be a base for a proof-assistant in which to formalize the constructive proofs done at the extensional level via the interpretation provided in [23].

Moreover, the presence of two levels is crucial to easily show the compatibility of **MF** with the other foundations at the "right" level: the intensional level of **MF** can be easily interpreted in intensional theories such as those formulated in type theory, for example Martin-Löf's type theory [31] or Coquand's Calculus of Constructions, while its extensional level can be easily interpreted in extensional theories such as those formulated in axiomatic set theory, for example Aczel's constructive set theory [4], or those formulated in category theory, for example the internal languages of topoi or pretopoi [21, 22].

Finally, the two levels of **MF** and their link resemble a well-known construction in category theory, namely the tripos-to-topos construction of a realizability topos in [18]. This is because the interpretation of the extensional level of **MF** in [23] is done in a *quotient completion* built on the intensional level of **MF**. Such a quotient completion had been studied categorically in [26], [25] under the name of "elementary quotient completion" and related to the well known notion of exact completion on a lex or regular category in [27]. Then, an analogy between **MF** and the tripos-to-topos construction of a realizability topos can be described as follows: the categorical structure of the intensional level of **MF** plays the role of a tripos, its elementary quotient completion plays the role of the realizability topos construction, while the extensional level of **MF** plays the role of the internal language of a generic elementary topos.

In this paper we strengthen this analogy by building a realizability categorical structure for the intensional level **mTT** of **MF** in Feferman's classical predicative theory of inductive definitions $\widehat{ID_1}$ (see e. g. [15]). This is obtained by extracting the categorical structure behind the realizability interpretation in [24] for **mTT** in $\widehat{ID_1}$. As an advantage we get an easier proof of validity for **mTT** by defining a partial typed interpretation as in [38].

Our categorical semantics for **mTT** is called "effective" since it validates the formal Church thesis and constitutes the key ingredient to build a predicative variant of Hyland's "Effective Topos" [17] in $\widehat{ID_1}$, where to interpret the extensional level of **MF** extended with **CT**.

A predicative study of the Effective Topos, and more generally of realizability toposes, had been already developed in the context of algebraic set theory by B. van den Berg and I. Moerdijk, in particular in [41], by taking Aczel's Constructive Zermelo-Fraenkel set theory (for short CZF) in [4] as the predicative constructive set theory to be realized in their categorical structure.

A precise comparison between our work and that in [41] is expected to mirror the relationship between **MF** and CZF described in [23] and it is left to future work. We just recall that **mTT**, and the whole **MF**, is a much weaker theory than CZF concerning the proof-theoretic strength, because it can be interpreted in a strictly predicative theory as Feferman's $\widehat{ID_1}$ as [24] shows, while CZF is known not to be a predicative theory in Feferman's sense (see [16], [1], [11]).

2 The Minimalist Foundation

A peculiarity of constructive mathematics with respect to classical mathematics is the absence of a commonly accepted standard foundation as Zermelo-Fraenkel set theory for classical mathematics.

Various logical systems are available in the literature to formalize constructive mathematics: they range from axiomatic set theories à la Zermelo-Fraenkel, such as Aczel's CZF [4, 1, 2, 3] or Friedman's IZF [8], to the internal set theory of categorical universes such as topoi or pretopoi [21, 19, 22], to type theories such as Martin-Löf's type theory [31] or Coquand's Calculus of Inductive Constructions [13, 14]. No existing constructive foundation has yet superseded the others as the standard one.

The Minimalist Foundation, for short **MF**, was conceived in [28] to serve as a common core among the most relevant constructive and classical foundations. A key novelty of **MF** required in [28] is to be a two-level formal system equipped with an *intensional level* suitable for extraction of computational contents from its proofs, an *extensional level* formulated in a language as close as possible to that of ordinary mathematics and an interpretation of the latter in the former showing that the extensional level is obtained by abstraction from the intensional one according to Sambin's forget-restore principle in [35].

The two-level formal system of **MF** was completed in [23] with an interpretation of the extensional level into a quotient model of the intensional level analyzed categorically in [26, 25, 27].

The two-level structure of **MF** has at least two main advantages. On one hand the compatibility of **MF** with the most relevant constructive and classical foundations can be done at the most suitable level, namely the intensional level with intensional foundations, mostly designed as type theories, and the extensional one with usual extensional foundations, mostly designed as axiomatic set theories. On the other hand the two-level structure of **MF** has the advantage to meet the usual practice of developing mathematics in an extensional set theory, represented by the extensional level of **MF**, whose equalities are undecidable, with the practice of formalizing mathematical proofs in a computer-aided way by means of an interactive proof-assistant based on intensional type theory with decidable properties, including decidable type-checking of proofs and equalities, such as for example Agda [10] (on Martin-Löf's type theory [31]), or Coq [12, 9] or Matita [6, 5] (on the Calculus of Inductive Constructions).

2.1 Distinct features of MF

Here we present the main distinct conceptual features of both levels of **MF**. Their design is certainly influenced by the need of building a foundation for constructive mathematics which is compatible with the most relevant constructive and classical foundations at the appropriate level. An immediate consequence is that both levels of **MF** must be *predicative* theories to be compatible with well-known predicative theories, such as Martin-Löf's type theory or Aczel's CZF.

To meet this goal one could think of using Heyting arithmetic, possibly extended with finite types, as the extensional level for **MF**. However, in order to formalize most of constructive mathematics, and in particular constructive topology, in an extensional language close to that used in common practice, it would be good to have a theory with a more expressive language including quotient sets and the power of any set. On the other hand, it is worth noting that the power of a non-empty set inherits an impredicative nature as soon as it is considered a set and hence in a predicative set theory it must be considered an entity greater than a set, like a collection or a class. This fact led to introduce the notion of *collection* beside that of *set* at both levels of **MF**.

Concerning the intensional level of **MF**, the authors in [28] thought of designing it as an intensional dependent type theory à la Martin-Löf like that in [31]. Then, to make the extensional level interpretable in the intensional one easily and in a modular way, in [23] also the extensional level was designed as a dependent theory à la Martin-Löf like that in [30].

The final outcome in [23] was to design the intensional level of **MF**, called **mTT**, as a **predicative version** of Coquand's Calculus of Constructions in [13], for short CoC, which is essentially the basic system behind the proof-assistants Coq [12, 9] and Matita [6, 5].

The main features of CoC and of its extension in Coq that are strictly connected with the design of **mTT** are the following:

- sets include *sets in first-order intensional Martin-Löf's type theory* (i.e. the fragment of Martin-Löf's type theory in [31] corresponding to first-order logic with list types but without universes or well-founded sets) and there is a *primitive notion of propositions*, closed under intuitionistic connectives and quantifiers and equipped with proof-terms; hence propositions are though of as sets of their proofs;

- there is a *universe of propositions*, which is a set in CoC.

It is worth noting that only the second feature makes CoC impredicative. This feature allows one to represent the power of a set in a suitable model of quotients, called the setoid model (see [7]).

It is also important to recall that the CoC-universe of propositions is inconsistent with an identification of sets as propositions typical of Martin-Löf's type theory (see [13]), for short MLtt. As a consequence, the existential quantifier of CoC can not be that in MLtt and it does not yield choice principles, like the axiom of choice (see [31, 30]), as shown in [39].

In particular, a relevant consequence of the above features of CoC, which is a peculiar feature of **MF** discussed in [29, 34], is the possibility of distinguishing between the notion of a type theoretic function between sets A, B

$$f \in A \to B$$

called operation in **MF** (see [29]), and the notion of functional relation determined by a relation $R(x, y)\, prop\, [\, x \in A, y \in B\,]$ for which we can prove

$$\forall x \in A\; \exists!\, y \in B\; R(x, y)$$

Indeed in CoC, as well as in **MF**, the so called *axiom of unique choice*

$$(\ \mathbf{AC!}\ ) \qquad \forall x \in A\; \exists!\, y \in B\; R(x, y) \;\longrightarrow\; \exists f \in A \to B\; \forall x \in A\; R(x, f(x))$$

which allows one to extract a type theoretic function from a functional relation, is not valid (see [39] for a proof). This distinction between type-theoretic functions and functional relations, beside the non-validity of **AC!**, is also a property of Feferman's theories in [15].

The design of **mTT** in [23] proposes a way to turn the mentioned features of CoC in a predicative form by *extending first-order Martin-Löf's intensional type theory in [31]* with

- a notion of *collection* beside that of *set*: collections include sets but also certain types that can not be considered sets predicatively;
- a primitive notion of *proposition* closed under intuitionistic connectives and quantifiers over both sets and collections;
- a notion of *small proposition* denoting propositions closed under intuitionistic connectives and quantifiers restricted to sets;
- proof-terms for all propositions: *small propositions* are defined as *sets* of their proofs, while generic *propositions* are defined as *collections* of their proofs;
- a collection of small propositions and a collection of propositional functions on any set;

The last feature is what in the quotient model in [23] allows to define a power-collection of a set A as a suitable quotient on the collection of propositional functions on A.

Accordingly, the extensional level of **MF** in [23], called **emTT**, is an extension of the extensional version of first-order Martin-Löf's type theory in [30] with the following distinct features:

- a notion of *collection* beside that of *set* as in the intensional level of **MF**;
- a primitive notion of *proposition*, closed under intuitionistic connectives and quantifiers over collections and a notion of *small proposition* denoting propositions closed under intuitionistic connectives and quantifiers restricted to sets;
- *proof-irrelevance* of all propositions, namely all propositions are equipped with at most a canonical proof-term to denote when they are true;
- *a power-collection for each set* where subsets are equivalence classes of small propositions depending on the set and quotiented under equiprovability;
- *effective quotient sets* of equivalence relations defined by small propositions.

An important consequence of **MF**-design is *the compatibility of* **MF** *with classical predicative theories as Feferman's predicative theories [15].* Indeed it is well known that the *addition of the principle of excluded middle* can turn a predicative theory where functional relations between sets form a set, as Aczel's CZF or Martin-Löf's type theory, into an impredicative one where *power-collections become sets.*

In the next section we are going to describe in more details the type theory **mTT** of the intensional level and we refer to [23] for the description of the type theory **emTT** of the extensional level and of its interpretation in **mTT**.

2.2 The intensional level of the Minimalist Foundation

Here we describe the type theory **mTT** representing the intensional level of **MF** in [23], which extends that presented in [28]. **mTT** is a dependent type theory written in the style of Martin-Löf's type theory [31] by means of the following four kinds of judgements:

$$A\ type\ [\Gamma] \quad A = B\ type\ [\Gamma] \quad a \in A\ [\Gamma] \quad a = b \in A\ [\Gamma]$$

that is the type judgement (expressing that something is a specific type), the type equality judgement (expressing that two types are equal), the term judgement (expressing that something is a term of a certain type) and the term equality judgement (expressing the *definitional equality* between terms of the same type), respectively, all under a context Γ.

The word *type* is used as a meta-variable to indicate four kinds of entities: collections, sets, propositions and small propositions, namely

$$type \in \{col, set, prop, prop_s\}$$

Therefore, in **mTT** types are actually formed by using the following judgements:

$$A\ set\ [\Gamma] \qquad B\ col\ [\Gamma] \qquad \phi\ prop\ [\Gamma] \qquad \psi\ prop_s\ [\Gamma]$$

saying that A is a set, that B is a collection, that ϕ is a proposition and that ψ is a small proposition.

Here as in [24], and contrary to [23] where we use only capital latin letters as meta-variables for types, we use greek letters ψ, ϕ as meta-variables for propositions and capital latin letters A, B as meta-variables for sets or collections.

As in the intensional version of Martin-Löf's type theory, in **mTT** there are two kinds of equality concerning terms: one is the definitional equality of terms of the same type given by the judgement

$$a = b \in A\ [\Gamma]$$

which is decidable, and the other is the propositional equality written

$$\mathsf{Id}(A, a, b)\ prop\ [\Gamma]$$

which is not necessarily decidable.

We now proceed by briefly describing the various kinds of types in **mTT**, starting from small propositions and propositions and then passing to sets and finally collections.

Small propositions in **mTT** include all the logical constructors of intuitionistic predicate logic with equality and quantifications restricted to sets:

$$\phi\ prop_s \equiv \bot \mid \phi \wedge \psi \mid \phi \vee \psi \mid \phi \rightarrow \psi \mid (\forall x \in A)\, \phi \mid (\exists x \in A)\, \phi \mid \mathsf{Id}(A, a, b)$$

provided that A is a set.

Then, *propositions* in **mTT** include all the logical constructors of intuitionistic predicate logic with equality and quantifications on all kinds of types, i. e. sets and collections. Of course, small propositions are also propositions. Propositions can be generated as follows:

$$\phi\ prop\ \equiv \phi\ prop_s \mid \phi \wedge \psi \mid \phi \vee \psi \mid \phi \rightarrow \psi \mid (\forall x \in D)\, \phi \mid (\exists x \in D)\, \phi \mid \mathsf{Id}(D, d, b)$$

In order to close sets under comprehension, for example to include the set of positive natural numbers $\{x \in \mathsf{N} \mid x \geq 1\}$, and to define operations on such sets, we need to think of propositions as types of their proofs: small propositions are seen as sets of their proofs while generic propositions are seen as collections of their proofs. That is, we add to **mTT** the following rules

$$\textbf{prop}_s\textbf{-into-set)}\ \frac{\phi\ prop_s}{\phi\ set} \qquad\qquad \textbf{prop-into-col)}\ \frac{\phi\ prop}{\phi\ col}$$

Before explaining the differences between the notion of set and that of collection we describe their constructors in **mTT**.

Sets in **mTT** are characterized as inductively generated types and they include the following:

$$A\ set\ \equiv \phi\ prop_s \mid \mathsf{N}_0 \mid \mathsf{N}_1 \mid \mathsf{N} \mid \mathsf{List}(A) \mid (\Sigma x \in A)\, B \mid A + B \mid (\Pi x \in A)\, B$$

where the notation N_0 stands for the empty set, N_1 for the singleton set, N for the set of natural numbers, $\mathsf{List}(A)$ for the set of lists on the set A, $(\Sigma x \in A)\, B$ for the strong indexed sum, called here also dependent sum, of the family of sets $B\ set\ [x \in A]$ indexed on the set A, $A + B$ for the disjoint sum of the set A with the set B, $(\Pi x \in A)\, B$ for the dependent product set of the family of sets $B\ set\ [x \in A]$ indexed on the set A.

It is worth noting that the set N of natural numbers is not present in a primitive way in **mTT** since its rules can be derived by putting $\mathsf{N} \equiv \mathsf{List}(\mathsf{N}_1)$. Here, as in [24], we add it to the syntax of **mTT** because it plays a prominent role in the realizability interpretation in $\widehat{ID_1}$ and we want to avoid complications due to list encodings.

Finally, *collections* in **mTT** include the following types:

$$D\ col\ \equiv\ A\ set\ \mid\ \phi\ prop\ \mid\ \mathsf{prop_s}\ \mid\ A \rightarrow \mathsf{prop_s}\ \mid\ (\Sigma x \in D)\, E$$

where $\mathsf{prop_s}$ stands for the collection of small propositions and $A \rightarrow \mathsf{prop_s}$ for the collection of propositional functions of the set A, while $(\Sigma x \in D)\, E$ stands for the dependent sum of the collection family $E\ col\ [x \in D]$ indexed on the collection D. Collection constructors here are kept to a minimum in order to interpret power-collections of sets and contexts with dependent types which will be present in the extensional level of **MF**.

All sets are collections thanks to the following rule:

$$\textbf{set-into-col)}\ \frac{A\ set}{A\ col}$$

We end by mentioning the following relevant technical peculiarities of **mTT**:

- elimination rules of propositions act only toward propositions, as in CoC, to avoid the validity of choice principles contrary to what happens in Martin-Löf's type theory [1].

- in **mTT** we add explicitly substitution term equality rules of the form

$$\text{sub)}\ \frac{c \in C\,[\,x_1 \in A_1,\ \ldots,\ x_n \in A_n\,] \qquad a_1 = b_1 \in A_1\ \ldots\ a_n = b_n \in A_n[a_1/x_1, \ldots, a_{n-1}/x_{n-1}]}{c[a_1/x_1, \ldots, a_n/x_n] = c[b_1/x_1, \ldots, b_n/x_n] \in C[a_1/x_1, \ldots, a_n/x_n]}$$

in place of the usual term equality rules preserving term constructions typical of Martin-Löf's type theory MLtt in [31]. This choice yields a restriction of the valid equality rules in **mTT** with respect to those valid in MLtt. In particular in **mTT** the so called ξ-rule of lambda-terms

$$\xi\ \frac{c = c' \in C\ [x \in B]}{(\lambda x)c = (\lambda x)c' \in (\Pi x \in B)\, C}$$

is not derivable.

It is worth recalling from [23] that the term equality rules of **mTT** are enough to interpret an extensional level including extensional equality of functions, as that represented by **emTT**, by means of the quotient model described in [23] and studied abstractly in [26, 25, 27].

mTT can be essentially viewed as a fragment of CoC by identifying collections with sets.

Moreover, **mTT** can be easily interpreted in intensional Martin-Löf's type theory MLtt in [31] by interpreting sets as MLtt-sets in the first universe and collections simply as MLtt-sets, propositions as sets according to the well-known isomorphism in [30] and the universe of small propositions as the first universe of MLtt.

2.3 The auxiliary type theory $\mathbf{mTT}^a$

Here we describe an auxiliary type theory, called $\mathbf{mTT}^a$, which is essentially an extension of **mTT** which we will validate in our categorical structure. The reason for interpreting $\mathbf{mTT}^a$, instead of simply **mTT**, is that the rules of $\mathbf{mTT}^a$ enjoy an easier proof of validity in our predicative variant of a realizability tripos.

[1] If you allow an elimination of existential quantifiers towards any type, you could build a function mapping a proof of an existential quantification $p \in (\exists x \in A)\,\phi$ towards the corresponding indexed sums $(\Sigma x \in A)\,\phi$ and by means of the first indexed sum projection you can extract a *choice function* whose value $f(p) \in A$ is a witness of the existential quantification.

First of all, in $\mathbf{mTT}^a$, as well as in the version of $\mathbf{mTT}$ interpreted in [24], the collection of small propositions $\mathsf{prop_s}$ is defined with codes à la Tarski as in [31], contrary to the version in [23], to make the interpretation easier to understand. Its rules are the following.
Elements of the collection of small propositions are generated as follows:

$$\text{Pr}_1)\ \widehat{\bot} \in \mathsf{prop_s} \qquad \text{Pr}_2)\ \frac{p \in \mathsf{prop_s} \quad q \in \mathsf{prop_s}}{p \mathbin{\widehat{\vee}} q \in \mathsf{prop_s}}$$

$$\text{Pr}_3)\ \frac{p \in \mathsf{prop_s} \qquad q \in \mathsf{prop_s}}{p \mathbin{\widehat{\rightarrow}} q \in \mathsf{prop_s}} \qquad \text{Pr}_4)\ \frac{p \in \mathsf{prop_s} \qquad q \in \mathsf{prop_s}}{p \mathbin{\widehat{\wedge}} q \in \mathsf{prop_s}}$$

$$\text{Pr}_5)\ \frac{A\ set \quad a \in A \quad b \in A}{\widehat{\mathsf{Eq}}(A, a, b) \in \mathsf{prop_s}} \qquad \text{Pr}_6)\ \frac{p \in \mathsf{prop_s}\ [x \in A] \qquad A\ set}{\widehat{(\exists x \in A)}\, p \in \mathsf{prop_s}}$$

$$\text{Pr}_7)\ \frac{p \in \mathsf{prop_s}\ [x \in A] \qquad A\ set}{\widehat{(\forall x \in A)}\, p \in \mathsf{prop_s}}$$

Elements of the collection of small propositions can be decoded as small propositions via an operator as follows

$$\tau\text{-Pr})\ \frac{p \in \mathsf{prop_s}}{\tau(p)\ prop_s}$$

and this operator satisfies the following definitional equalities:

$$\text{eq-Pr}_1)\ \tau(\widehat{\bot}) = \bot\, prop_s \qquad \text{eq-Pr}_2)\ \frac{p \in \mathsf{prop_s} \quad q \in \mathsf{prop_s}}{\tau(p \mathbin{\widehat{\vee}} q) = \tau(p) \vee \tau(q)\, prop_s}$$

$$\text{eq-Pr}_3)\ \frac{p \in \mathsf{prop_s} \qquad q \in \mathsf{prop_s}}{\tau(p \mathbin{\widehat{\rightarrow}} q) = \tau(p) \rightarrow \tau(q)\, prop_s} \qquad \text{eq-Pr}_4)\ \frac{p \in \mathsf{prop_s} \quad q \in \mathsf{prop_s}}{\tau(p \mathbin{\widehat{\wedge}} q) = \tau(p) \wedge \tau(q)\, prop_s}$$

$$\text{eq-Pr}_5)\ \frac{A\ set \quad a \in A \quad b \in A}{\tau(\, \widehat{\mathsf{Eq}}(A, a, b)\,) = \mathsf{Eq}(A, a, b)\, prop_s}$$

$$\text{eq-Pr}_6)\ \frac{p \in \mathsf{prop_s}\ [x \in A] \qquad A\ set}{\tau(\widehat{(\exists x \in A)}\, p) = (\exists x \in A)\, \tau(p)\, prop_s}$$

$$\text{eq-Pr}_7)\ \frac{p \in \mathsf{prop_s}\ [x \in A] \qquad A\ set}{\tau(\widehat{(\forall x \in A)}\, p) = (\forall x \in A)\, \tau(p)\, prop_s}$$

Moreover, for the same reasons explained in [24] and essentially due to the need of interpreting the universe of small propositions in a clear way, even in $\mathbf{mTT}^a$ we add the collection Set of set codes whose related rules are the following. We do not add corresponding elimination and conversion rules as those of universes à la Tarski in [31] since they are not needed to prove the validity of $\mathbf{mTT}$-rules.

Collection of sets

F-Se) $\mathsf{Set}\ col$

Elements of the collection of sets are generated as follows:

$$\text{sp-i-p)}\ \frac{p \in \mathsf{props_s}}{\sigma(p) \in \mathsf{Set}} \qquad \text{Se}_e)\ \widehat{\mathsf{N_0}} \in \mathsf{Set}$$

$$\text{Se}_s)\ \widehat{\mathsf{N_1}} \in \mathsf{Set} \qquad \text{Se}_n)\ \widehat{\mathsf{N}} \in \mathsf{Set}$$

$$\text{Se}_l)\ \frac{a \in \mathsf{Set}}{\widehat{\mathsf{List}}(a) \in \mathsf{Set}} \qquad \text{Se}_u)\ \frac{a \in \mathsf{Set} \quad b \in \mathsf{Set}}{a \mathbin{\widehat{+}} b \in \mathsf{Set}}$$

$$\text{Se}_\Sigma)\ \frac{b \in \mathsf{Set}\ [x \in A] \qquad A\ set}{(\widehat{\Sigma x \in A})\, b \in \mathsf{Set}} \qquad \text{Se}_\Pi)\ \frac{b \in \mathsf{Set}\ [x \in A] \qquad A\ set}{(\widehat{\Pi x \in A})\, b \in \mathsf{Set}}$$

Set codes will be used to easily interpret the code of quantified small propositions.

Finally to further simplify the definition of the realizability interpretation, in $\mathbf{mTT}^a$ the elimination rules of some types, including disjoint sums, lists and natural numbers, are restricted to act toward *non-dependent types* and they are equipped with an extra equality rule expressing the uniqueness of the eliminator constructor as follows

2.3.1 Rules of disjoint sum

$$\text{+-f)}\ \frac{A\ set \qquad B\ set}{A + B\ set}$$

$$\text{+-i}_1)\ \frac{a \in A \qquad A\ set \qquad B\ set}{\mathsf{inl}(a) \in A + B} \qquad \text{+-i}_2)\ \frac{b \in B \qquad A\ set \qquad B\ set}{\mathsf{inr}(b) \in A + B}$$

$$\text{+-e)}\ \frac{c \in A + B \qquad C\ col \qquad d \in C\,[x \in A] \qquad e \in C\,[y \in B]}{\mathsf{El}_+(c, (x)\,d, (y)\,e) \in C}$$

$$\text{+-c}_1)\ \frac{a \in A \qquad C\ col \qquad d \in C\,[x \in A] \qquad e \in C\,[y \in B]}{\mathsf{El}_+(\mathsf{inl}(a), (x)\,d, (y)\,e) = d[a/x] \in C}$$

$$\text{+-c}_2)\ \frac{b \in B \qquad C\ col \qquad d \in C\,[x \in A] \qquad e \in C\,[y \in B]}{\mathsf{El}_+(\mathsf{inr}(b), (x)\,d, (y)\,e) = e[b/y] \in C}$$

$$\text{+-}\eta)\ \frac{p \in C + D \qquad t \in A\ [z \in C + D]}{\mathsf{El}_+(\,p, (x)\,t[\mathsf{inl}(x)/z]\,, (y)\,t[\mathsf{inr}(y)/z]\,) = t[p/z] \in A}$$

2.3.2 Rules of lists

List-f) $$\frac{A\ set}{\mathsf{List}(A)\ set}$$

List-i$_1$) $$\frac{A\ set}{\epsilon \in \mathsf{List}(A)}$$ List-i$_2$) $$\frac{A\ set \qquad b \in \mathsf{List}(A) \qquad a \in A}{\mathsf{cons}(b,a) \in \mathsf{List}(A)}$$

List-e) $$\frac{c \in \mathsf{List}(A) \qquad B\ col \qquad d \in B \qquad e \in B\,[x \in B, y \in A]}{\mathsf{El}_{\mathsf{List}}(c, d, (x,y)\,e) \in C}$$

List-c$_1$) $$\frac{B\ col \qquad d \in B \qquad e \in B\,[x \in B, y \in A]}{\mathsf{El}_{\mathsf{List}}(\epsilon, d, (x,y)\,e) = d \in C}$$

List-c$_2$) $$\frac{b \in \mathsf{List}(A) \qquad a \in A \qquad B\ col \qquad d \in B \qquad e \in B\,[x \in B, y \in A]}{\mathsf{El}_{\mathsf{List}}(\mathsf{cons}(b,a), d, (x,y)\,e) = e[\mathsf{El}_{\mathsf{List}}(b, d, (x,y)\,e)/x, a/y] \in C}$$

List-η) $$\frac{\begin{array}{l} B\ col \quad d \in B \quad e \in B\ [x \in B, y \in A] \quad t \in B\ [z \in \mathsf{List}(A)] \\ c \in \mathsf{List}(A) \qquad t[\epsilon/z] = a \in B \\ t[\mathsf{cons}(u,y)/z] = e[t[u/z]/x] \in B\ [u \in \mathsf{List}(A), y \in A] \end{array}}{\mathsf{El}_{\mathsf{List}}(c, d, (x,y)e) = t[c/z] \in L}$$

2.3.3 Rules of natural numbers set

N-f) $\mathsf{N}\ set$ N-i$_1$) $0 \in \mathsf{N}$ N-i$_2$) $$\frac{a \in \mathsf{N}}{\mathsf{succ}(a) \in \mathsf{N}}$$

N-e) $$\frac{a \in \mathsf{N} \quad A\ col \quad d \in A \quad e \in A\,[x \in A]}{\mathsf{El}_{\mathsf{N}}(a, d, (x)\,e) \in A}$$ N-c$_1$) $$\frac{A\ col \quad d \in A \quad e \in A\,[x \in A]}{\mathsf{El}_{\mathsf{N}}(0, d, (x)\,e) = d \in A}$$

N-c$_2$) $$\frac{a \in \mathsf{N} \qquad A\ col \qquad d \in A \qquad e \in A\,[x \in A]}{\mathsf{El}_{\mathsf{N}}(\mathsf{succ}(a), d, (x)\,e) = e[\mathsf{El}_{\mathsf{N}}(a, d, (x)\,e)/x] \in A}$$

N-η) $$\frac{\begin{array}{l} c \in \mathsf{N} \qquad A\ col \qquad t \in A\,[z \in \mathsf{N}] \qquad d \in A \qquad e \in A\,[x \in A] \\ t[0/z] = d \in A \qquad t[\mathsf{succ}(y)/z] = e[t[y/z]/x] \in A\,[y \in \mathsf{N}] \end{array}}{\mathsf{El}_{\mathsf{N}}(c, d, (x)e) = t[c/z] \in A}$$

These rules do not change the expressive power of disjoint sums, lists and natural numbers. The reason is that, as first shown in [22], the above kinds of elimination rules with related equality rules are equivalent to the original ones of **mTT** provided that we add to $\mathbf{mTT}^a$ the following rules of extensional propositional equality of Martin-Löf's type theory in [30], which we also adopt in the extensional level of **MF** instead of those of the propositional identity Id:

$$\text{Eq-f)}\ \frac{A\,col \qquad a \in A \qquad b \in A}{\mathsf{Eq}(A,a,b)\,prop} \qquad \text{Eq-f}_s)\ \frac{A\,set \qquad a \in A \qquad b \in A}{\mathsf{Eq}(A,a,b)\,prop_s}$$

$$\text{Eq-i)}\ \frac{a \in A}{\mathsf{eq}(a) \in \mathsf{Eq}(A,a,a)} \qquad \text{Eq-e)}\ \frac{p \in \mathsf{Eq}(A,a,b)}{a = b \in A} \qquad \text{Eq-}\eta)\ \frac{d \in \mathsf{Eq}(A,a,b)}{d = \mathsf{eq}(a) \in \mathsf{Eq}(A,a,b)}$$

and we add the usual equality rules preserving each type constructor as in [31, 30] or as those present in the extensional level of **MF** in [23].

Then we can equivalently define (see [31]) the strong indexed sums with the following rules

2.3.4 Rules of strong indexed sums

$$\Sigma\text{-f)}\ \frac{A\,set \qquad B\,set\,[x \in A]}{(\Sigma x \in A)\,B\,set} \qquad \Sigma\text{-f}_{col})\ \frac{A\,col \qquad B\,col\,[x \in A]}{(\Sigma x \in A)\,B\,col}$$

$$\Sigma\text{-i)}\ \frac{B\,col\,[x \in A] \qquad a \in A \qquad b \in B[a/x]}{\langle a,b\rangle \in (\Sigma x \in A)\,B}$$

$$\Sigma\text{-e}_1)\ \frac{c \in (\Sigma x \in A)\,B}{\pi_1(c) \in A} \qquad \Sigma\text{-e}_2)\ \frac{c \in (\Sigma x \in A)\,B}{\pi_2(c) \in B[\pi_1(c)/x]}$$

$$\Sigma\text{-c}_1)\ \frac{B\,col\,[x \in A] \qquad a \in A \qquad b \in B[a/x]}{\pi_1(\langle a,b\rangle) = a \in A};$$

$$\Sigma\text{-c}_2)\ \frac{B\,col\,[x \in A] \qquad a \in A \qquad b \in B[a/x]}{\pi_2(\langle a,b\rangle) = b \in B[a/x]}$$

$$\Sigma\text{-}\eta)\ \frac{c \in (\Sigma x \in A)\,B}{\langle \pi_1(c), \pi_2(c)\rangle = c \in (\Sigma x \in A)\,B}$$

Therefore we can easily show:

Proposition 2.1. *We can interpret* **mTT** *into* $\mathbf{mTT}^a$ *as the identity on all constructors except for those of the propositional equality* Id *which are interpreted as those of the extensional one* Eq, *and except for the strong indexed sum elimination constructor which is interpreted via projections.*

Proof. We briefly describe how to interpret the rules of **mTT**-strong indexed sums. Given $d \in (\Sigma x \in B)\, C$, $M\, col\, [\, z \in (\Sigma x \in B)\, C\,]$ and $m \in M[\langle x, y\rangle/z]\, [\, x \in B, y \in C\,]$ then

$$El_{\Sigma}(d, m) \;\equiv^{def}\; m[\pi_1(d)/x, \pi_2(d)/y]$$

is of type $M[\langle \pi_1(d), \pi_2(d)\rangle/z]$ by definition. But by the substitution rules and the rule conv) [2] (see the rules of **mTT** in [23]) and the above Σ-η of $\mathbf{mTT}^a$ we conclude that it is of type $M(d)$ as well, as required.

Concerning the propositional equality: the constructor $\mathsf{id}_A(a)$ of **mTT** is interpreted as $\mathsf{eq}(a)$ of $\mathbf{mTT}^a$ and the elimination constructor $El_{\mathsf{Id}}(p, (x)c)$ as $c[a/x]$, given that its type $C(a, a, \mathsf{eq}(a))$ happens to be equal to $C(a, b, p)$ by the rules subT) and conv) in [23] since from $p \in \mathsf{Eq}(A, a, b)$ we get $a = b \in A$ and also $p = \mathsf{eq}(a) \in \mathsf{Eq}(A, a, b)$ by the rules of Eq. □

3 Feferman's theory of inductive definitions $\widehat{ID_1}$

The system $\widehat{ID_1}$ is a predicative fragment of second-order arithmetic, more precisely it is the predicative fragment of second-order arithmetic extending Peano arithmetic with some (not necessarily least) fixpoints for each positive arithmetical operator. Its number terms are number variables (or simply variables) $\xi_1, ..., \xi_n ...$, the constant 0 and the terms built by applying the unary successor functional symbol *succ* and the binary sum and product functional symbols $+$ and $*$ to number terms. Set terms are only set variables $X, Y, Z....$ The *arithmetical* formulas are obtained starting from $t = s$ and $t \,\varepsilon\, X$ with t, s number terms and X a set variable, by applying the connectives $\wedge, \vee, \neg, \rightarrow$ and the number quantifiers $\forall x$, $\exists x$. Moreover let us give the following two definitions.

Definition 3.1. *An occurrence of a set variable X in an arithmetical formula φ is* positive *or* negative *according to the following conditions.*

1. *the occurrence of X in $t \,\varepsilon\, X$, where t is a number term, is positive;*

2. *a positive (negative) occurrence of X in ψ, is positive (negative) in $\psi \wedge \phi$, $\phi \wedge \psi$, $\phi \vee \psi$, $\psi \vee \phi$, $\phi \rightarrow \psi$, $\exists x\, \psi$ and $\forall x\, \psi$;*

[2] We just recall that this rule says that from $a \in A$ and $A = B$ *type* we get $a \in B$.

3. *a positive (negative) occurrence of* X *in* ψ, *is negative (positive) in* $\psi \to \phi$ *and* $\neg\psi$.

Definition 3.2. *An arithmetical formula* φ *with exactly one free number variable* x *and one free set variable* X *which occurs only positively is called an* admissible *formula.*

In order to define the system $\widehat{ID_1}$ we add to the language of second-order arithmetic a unary predicate symbol P_φ for every admissible formula φ . The atomic formulas of $\widehat{ID_1}$ are

1. $t = s$ with t and s number terms;

2. $t \,\varepsilon\, X$ with t a number term and X a set variable;

3. $P_\varphi(t)$ with t a number term and φ an admissible formula.

All formulas of $\widehat{ID_1}$ are obtained from atomic formulas by applying connectives, number quantifiers and set quantifiers.

The axioms of $\widehat{ID_1}$ are the axioms of Peano Arithmetic plus the following three axiom schemata:

1. *Comprehension schema*: for all formulas $\varphi(x)$ of $\widehat{ID_1}$ without set quantifiers

$$\exists X\, \forall x\, (x \,\varepsilon\, X \leftrightarrow \varphi(x))$$

provided that X is not free in $\varphi(x)$

2. *Induction schema*: for all formulas $\varphi(x)$ of $\widehat{ID_1}$ without set quantifiers

$$(\varphi(0) \wedge \forall x\, (\varphi(x) \to \varphi(succ(x)))) \to \forall x\, \varphi(x)$$

3. *Fixpoint schema*: for all admissible formulas φ with x and X free

$$\varphi[P_\varphi/X] \leftrightarrow P_\varphi(x)$$

where $\varphi[P_\varphi/X]$ is the result of substituting in φ every atomic subformula $t \,\varepsilon\, X$ with $P_\varphi(t)$.

The system $\widehat{ID_1}$ allows us to define predicates as fixpoints, by using axiom schema 3, if they are presented in an appropriate way (i. e. using admissible formulas).

3.1 Notations of recursive functions in $\widehat{ID}_1$

A *numeral* is a term of the form $succ(succ...succ(0))$. As usual we denote numerals with boldface lower case letters $\mathbf{n}$.

In $\widehat{ID}_1$ one can certainly represent a Gödelian coding of recursive functions by means of the Kleene predicate $T(x, y, z)$ and the primitive recursive (meta)function U. First of all we define *applicative terms* as follows (notice that these terms are not part of the syntax of $\widehat{ID}_1$, but are auxiliary terms):

1. every number variable is an applicative term;
2. every numeral is an applicative term;
3. if t and s are applicative terms, then $\{t\}(s)$ is an applicative term.

We use the abbreviation $\{s\}(t_1, ..., t_n)$ for applicative terms $s, t_1, ..., t_n, ...$ as follows

1. $\{s\}()$ is s;
2. $\{s\}(t_1, ..., t_{n+1})$ is $\{\{s\}(t_1, ..., t_n)\}(t_{n+1})$.

If $\varphi(\overline{x}, x)$ is a formula of $\widehat{ID}_1$ and t is an applicative term, then we define $\varphi(\overline{x}, t)$ by induction on the definition of applicative terms t for all formulas as follows:

1. $\varphi(\overline{x}, y)$ is itself;
2. if $\mathbf{n}$ is a numeral, $\varphi(\overline{x}, \mathbf{n})$ is itself;
3. $\varphi(\overline{x}, \{t\}(s))$ is $\exists x(T(t, s, x) \wedge \varphi(\overline{x}, U(x)))$.

Notice that if $\{t\}(s)$ is an applicative term, the formula $\{t\}(s) = \{t\}(s)$ turns out to be equivalent to what is usually denoted with $\{t\}(s)\downarrow$ i. e. the formula $\exists x\, T(t, s, x)$. In particular, for a *generic* applicative term t it can be proved that the formula $t = t$ is provable when *the applicative term t converges*. Hence it makes sense to introduce the formula $t \simeq s$ as an abbreviation for $t = t \vee s = s \rightarrow t = s$ for every pair of *generic* applicative terms t and s.

If t is an applicative term with all variables among $x_1, ..., x_n$, then there is a numeral $\Lambda x_1 ... \Lambda x_n.t$ for which

$$\widehat{ID}_1 \vdash \forall x_1 ... \forall x_n(\{\Lambda x_1 ... \Lambda x_n.t\}(x_1, ..., x_n) \simeq t)$$

For $1 \leq j \leq n$ we define a numeral π_j^n as $\Lambda x_1 .... \Lambda x_n.x_j$. These numerals obviously satisfy the following

$$\widehat{ID}_1 \vdash \{\pi_j^n\}(x_1, ..., x_n) = x_j$$

Any n-ary primitive recursive (meta)function f can be represented by a numeral $\mathbf{f}$ through the Gödelian coding in such a way that

$$\widehat{ID_1} \vdash \{\mathbf{f}\}(x_1, ..., x_n) = f(x_1, ..., x_n)$$

In particular there exist numerals $\mathbf{p}, \mathbf{p}_1, \mathbf{p}_2$ and $\mathbf{s}$ representing a primitive recursive pairing function p with primitive recursive projections p_1, p_2 and the successor function.

We define for $1 \leq j \leq n$, numerals $\mathbf{p}^n$ and $\mathbf{p}_j^n$, representing the encoding of n-tuples of natural numbers and the relative jth projections as follows:

1. $\mathbf{p}^1$ and $\mathbf{p}_1^1$ are both π_1^1;
2. $\mathbf{p}^{n+1}$ is $\Lambda x_1 ... \Lambda x_{n+1}.\{\mathbf{p}\}(\{\mathbf{p}^n\}(x_1, ..., x_n), x_{n+1})$;
3. $\mathbf{p}_j^{n+1}$ is $\Lambda x.\{\mathbf{p}_j^n\}(\{\mathbf{p}_1\}(x))$ if $1 \leq j \leq n$;
4. $\mathbf{p}_{n+1}^{n+1}$ is $\mathbf{p}_2$.

We have that for $n \geq 1$

1. $\widehat{ID_1} \vdash \{\mathbf{p}^n\}(\{\mathbf{p}_1^n\}(x), ..., \{\mathbf{p}_n^n\}(x)) = x$
2. $\widehat{ID_1} \vdash \{\mathbf{p}_j^n\}(\{\mathbf{p}^n\}(x_1, ..., x_n)) = x_j$ for every $1 \leq j \leq n$.

We can bijectively encode finite lists of natural numbers $[n_0, ..., n_k]$ with natural numbers in such a way that the component functions $(\ )_j$, the length function $lh(\ )$ and the concatenation function cnc of lists with natural numbers are primitive recursive and that the empty list is coded by 0. In particular there exists a numeral $\mathbf{cnc}$ for which $\widehat{ID_1} \vdash \{\mathbf{cnc}\}(x, y) = cnc(x, y)$.

Moreover there exists a list recursor, i.e. a numeral **listrec** for which

1. $\widehat{ID_1} \vdash \{\mathbf{listrec}\}(0, y, z) \simeq y$
2. $\widehat{ID_1} \vdash \{\mathbf{listrec}\}(\{\mathbf{cnc}\}(x, x'), y, z) \simeq \{z\}(\{\mathbf{listrec}\}(x, y, z), x')$

4 The effective pretripos for mTT

In this section we are going to define in $\widehat{ID_1}$ a predicative categorical structure, called *effective pretripos for* **mTT**, which represents a predicative variant of a realizability tripos validating **mTT**. In a broad sense it can be considered a predicative variant of the effective tripos giving rise to Hyland's effective topos Eff in [17]. Indeed, our ultimate goal is to use our effective pretripos to build a predicative variant of a realizability topos like Eff.

Recall from [18, 33] that a tripos is an indexed category

$$P : \mathcal{C}^{op} \longrightarrow \mathbf{Cat}$$

which is a Lawvere-first order hyperdoctrine in the category of Heyting algebras enriched with a weak subobject classifier, called a generic predicate in [18], capable of producing power-sets in the category obtained by applying the so called tripos-to-topos construction. This weak classifier is of an impredicative nature and it must be necessarily so.

Here we are going to define a predicative variant of a tripos with the idea of getting just power-collections and not power-sets in the corresponding predicative variant of the tripos-to-topos construction. These will be structured in a fully analogous way to the two-level structure of **MF** where the universes of small propositions and of propositional functions on any set at the intensional level of **MF** are enough to model power-collections of sets at the extensional level of **MF** by means of a quotient model (see [23]).

We now briefly outline the categorical structure of our predicative variant of a realizability tripos by describing what we are going to include in it:

- We define an indexed category of "realized" propositions

$$\mathbf{Prop} : \mathbf{Cont}^{op} \to \mathbf{Cat}$$

  on a category **Cont** of "realized contexts" and realized morphisms between them, equipped with the structure of a Lawvere's first order hyperdoctrine but in the category of Heyting prealgebras [3]. The category **Cont** will host a realizability interpretation of $\mathbf{mTT}^a$-contexts as that in [24]. This category is also equivalent to its full subcategory $\mathcal{C}$ of realized collections, which are defined as subsets of natural numbers in $\widehat{ID_1}$ equipped with an equivalence relation, whose morphisms turn out to be suitable recursive operations. Each fibre of **Prop** represents the category of realized propositions defined in a proof-irrelevant way as subsets of a singleton.

  We use the category of contexts **Cont** instead of $\mathcal{C}$ as the base of our categorical structure, because the realizability interpretation of $\mathbf{mTT}^a$-contexts and generic $\mathbf{mTT}^a$-judgements becomes simpler.

[3] A Heyting prealgebra is a preorder whose posetal reflection is a Heyting algebra.

It is worth noting that the category **Cont** has also an indexed structure of families of realized collections

$$\mathbf{Col} : \mathbf{Cont}^{op} \to \mathbf{Cat}$$

whose fibre on the empty context [ ] is equivalent to **Cont**. Moreover, it contains **Prop** as a sub-indexed category

$$\mathbf{Prop} \hookrightarrow \mathbf{Col}$$

- We define a realized collection US via a fixpoint formula of $\widehat{ID_1}$, which will host the realizability interpretation of the collection of $\mathbf{mTT}^a$-sets. This is defined as in [24] following a technique due to Beeson [8].

  US is crucial to define the (indexed) category of families of realized sets

  $$\mathbf{Set} : \mathbf{Cont}^{op} \to \mathbf{Cat}$$

  which is a sub-indexed structure of **Col**

  $$\mathbf{Set} \hookrightarrow \mathbf{Col}$$

  Namely families of realized sets are families of realized collections classified by the non-dependent realized collection US, in the sense that US represents the indexed functor **Set** via a natural bijection

  $$\mathbf{Set}(\Gamma) \simeq \mathbf{Cont}(\Gamma, \mathsf{US})$$

  for objects Γ in **Cont**.

- We define a realized collection USP as a sub-collection of US, which will host the realizability interpretation of the collection of $\mathbf{mTT}^a$-small propositions. This is also defined as in [24].

  The construction of USP is crucial to define the first-order hyperdoctrine of realized small propositions

  $$\mathbf{Prop}_s : \mathbf{Cont}^{op} \to \mathbf{Cat}$$

  which is a subindexed category both of **Prop** and of **Set**

  $$\begin{array}{ccc} \mathbf{Set} & \hookrightarrow & \mathbf{Col} \\ \uparrow & & \uparrow \\ \mathbf{Prop}_s & \hookrightarrow & \mathbf{Prop} \end{array}$$

and is classified by USP in the sense that USP represents the indexed functor $\mathbf{Prop}_s$ via a natural bijection for objects Γ in **Cont**

$$\mathbf{Prop}_s(\Gamma) \simeq \mathbf{Cont}(\Gamma, \mathsf{USP})$$

This classification property provides an *intensional predicative* version of the original weak subobject classifier property of a tripos.

In the next sections we will often include lemmas and theorems without proofs because their proofs just involve straightforward verifications.

4.1 The category of realized collections in $\widehat{ID_1}$

Here we are going to define the category of *realized collections*. We will denote such a category as $\mathcal{C}$.

A realized collection will denote a quotient of a subset of natural numbers acting as realizers. It is represented in $\widehat{ID_1}$ by a first-order formula defining the realizers together with an equivalence relation $x \sim y$. Morphisms between realized collections will be defined as recursive functions between them preserving the corresponding equivalence relations and called *recursive operations.*

We start by giving the notion of dependent realized collection, namely a family of realized collections depending on a finite number of variables. From this notion we will deduce that of realized collection.

Definition 4.1. *Let $\overline{x}$ be a (possibly empty) list of distinct variables of the language of $\widehat{ID_1}$. A* realized collection of $\widehat{ID_1}$ depending on $\overline{x}$ *(or simply a dependent realized collection) is a pair $A(\overline{x}) := (|A(\overline{x})|, x \sim_{A(\overline{x})} y)$ where*

1. *$|A(\overline{x})|$ is a first-order definable class of $\widehat{ID_1}$, i. e. it is a formal expression*

   $$\{x \mid \phi_A(\overline{x}, x)\}$$

   where x is a variable different from those in $\overline{x}$ and $\phi_A(\overline{x}, x)$ is a first-order formula of $\widehat{ID_1}$, namely a formula without set variables and set quantifiers, but possibly with fixpoint predicates P_φ, with all free variables among those in $\overline{x}$ and x. We will write $x \,\varepsilon\, A(\overline{x})$ as an abbreviation for $\phi_A(\overline{x}, x)$, since we may think of $A(\overline{x})$ as a subset $|A(\overline{x})|$ of natural numbers, called realizers, *equipped with a relation $\sim_{A(\overline{x})}$.*

2. *$x \sim_{A(\overline{x})} y$ is a first-order definable equivalence relation on $|A(\overline{x})|$, i. e. it is a first-order formula of $\widehat{ID_1}$, where x and y are distinct variables and they are different from those in $\overline{x}$, with all free variables among those in $\overline{x}$, x or y for which:*

(a) $x \sim_{A(\overline{x})} y \vdash_{\widehat{ID_1}} x \,\varepsilon\, A(\overline{x}) \wedge y \,\varepsilon\, A(\overline{x})$

(b) $x \,\varepsilon\, A(\overline{x}) \vdash_{\widehat{ID_1}} x \sim_{A(\overline{x})} x$

(c) $x \sim_{A(\overline{x})} y \vdash_{\widehat{ID_1}} y \sim_{A(\overline{x})} x$

(d) $x \sim_{A(\overline{x})} y \wedge y \sim_{A(\overline{x})} z \vdash_{\widehat{ID_1}} x \sim_{A(\overline{x})} z$

We identify dependent realized collections $A(\overline{x})$ and $B(\overline{x})$ for which

$$\widehat{ID_1} \vdash x \sim_{A(\overline{x})} y \leftrightarrow x \sim_{B(\overline{x})} y$$

(this automatically ensures that $\widehat{ID_1} \vdash x \,\varepsilon\, A(\overline{x}) \leftrightarrow x \,\varepsilon\, B(\overline{x})$, namely the validity of subset extensional equality).

Definition 4.2. *A* realized collection *of $\widehat{ID_1}$ is a realized collection depending on the empty list.*

Definition 4.3. *Given two realized collections A and B, a* recursive operation *(or simply an operation) from A to B is an equivalence class $[\mathbf{n}]_{\approx_{A,B}}$ of numerals for which*

$$x \sim_A y \vdash_{\widehat{ID_1}} \{\mathbf{n}\}(x) \sim_B \{\mathbf{n}\}(y)$$

with respect to the equivalence relation given by

$$\mathbf{n} \approx_{A,B} \mathbf{m} \text{ if and only if } x \,\varepsilon\, A \vdash_{\widehat{ID_1}} \{\mathbf{n}\}(x) \sim_B \{\mathbf{m}\}(x)$$

Definition 4.4. *We call $\mathcal{C}$ the category of realized collections of $\widehat{ID_1}$ and recursive operations between them where the composition of morphisms and identities are defined as follows.*

If $[\mathbf{n}]_{\approx_{A,B}}$ is an operation from a realized collection A to a realized collection B and $[\mathbf{m}]_{\approx_{B,C}}$ is an operation from a realized collection B to a realized collection C, then their composition is the operation

$$[\mathbf{m}]_{\approx_{B,C}} \circ [\mathbf{n}]_{\approx_{A,B}} := [\,\Lambda x.\{\mathbf{m}\}(\{\mathbf{n}\}(x))\,]_{\approx_{A,C}}$$

If A is a realized collection, then its identity id_A is defined as $[\pi_1^1]_{\approx_{A,A}}$.

4.2 The category of realized contexts in $\widehat{ID_1}$

Here we are going to define the category **Cont** of *realized contexts* and *realized morphisms* between them. This category will be used to interpret the telescopic contexts of dependent types of $\mathbf{mTT}^a$. We will deduce the categorical properties which are necessary to validate $\mathbf{mTT}^a$ from those of the category of realized collections $\mathcal{C}$, being **Cont** equivalent to $\mathcal{C}$. Indeed the categorical structure of $\mathcal{C}$ will be easier to describe.

We start by giving some abbreviations on list of variables.

Fix two countable sequences of variables $x_1, ..., x_n...$ and $y_1, ..., y_n...$ in such a way that all these variables are distinct. We denote by $\overline{x}_{|_j}$ the empty list if $j = 0$ or the list $x_1, ..., x_j$ otherwise. Similarly we define $\overline{y}_{|_j}$.

We use the abbreviation $\Lambda\overline{x}_{|_j}$ for $\Lambda x_1...\Lambda x_j$ if $j > 0$, while $\Lambda\overline{x}_{|_0}$ means no Λ-quantification. In case of an empty list of variables $A(\,)$ means A.

If $\overline{\mathbf{k}}$ is a finite list of numerals with length n, then for $j \leq n$, we use the abbreviation $\{\overline{\mathbf{k}}_{|_j}\}(\overline{t})$ for the empty list if $j = 0$, while $\{\overline{\mathbf{k}}_{|_j}\}(\overline{t})$ is the list $\{\mathbf{k}_1\}(\overline{t}), ..., \{\mathbf{k}_j\}(\overline{t})$ otherwise; we write $\{\overline{\mathbf{k}}\}(\overline{t})$ as an abbreviation for $\{\overline{\mathbf{k}}_{|_n}\}(\overline{t})$.

Definition 4.5. *A* realized context *(or simply a context) of $\widehat{ID_1}$ is a (possibly empty) finite list*

$$\Gamma = [A_1, ..., A_j(\overline{x}_{|_{j-1}}), ..., A_n(\overline{x}_{|_{n-1}})]$$

where $A_j(\overline{x}_{|_{j-1}})$ is a collection of $\widehat{ID_1}$ depending on $\overline{x}_{|_{j-1}}$ for $1 \leq j \leq n$, which satisfies the following conditions:

1. $x_{j+1} \sim_{A_{j+1}(\overline{x}_{|_j})} y_{j+1} \vdash_{\widehat{ID_1}} x_1 \,\varepsilon\, A_1 \wedge ... \wedge x_j \,\varepsilon\, A_j(\overline{x}_{|_{j-1}})$

2. $x_1 \sim_{A_1} y_1 \wedge ... \wedge x_j \sim_{A_j(\overline{x}_{|_{j-1}})} y_j \vdash_{\widehat{ID_1}}$

$$x_{j+1} \sim_{A_{j+1}(\overline{x}_{|_j})} y_{j+1} \leftrightarrow x_{j+1} \sim_{A_{j+1}(\overline{y}_{|_j})} y_{j+1}$$

for every $1 \leq j \leq n-1$.

Moreover, for a realized context Γ of $\widehat{ID_1}$, the length $\ell(\Gamma)$ of Γ is the length of Γ as a list.

Finally, if $\Gamma = [A_1, ..., A_n(\overline{x}_{|_{n-1}})]$ is a realized context of $\widehat{ID_1}$ with positive length n, then

1. $\overline{x}_{|_n} \,\varepsilon\, \Gamma$ *is an abbreviation for* $x_1 \,\varepsilon\, A_1 \wedge ... \wedge x_n \,\varepsilon\, A_n(\overline{x}_{|_{n-1}})$

2. $\overline{x}_{|_n} \sim_\Gamma \overline{y}_{|_n}$ *is an abbreviation for* $x_1 \sim_{A_1} y_1 \wedge ... \wedge x_n \sim_{A_n(\overline{x}_{|_{n-1}})} y_n$

If Γ *is the empty list, then* $\overline{x}_{|0}\,\varepsilon\,\Gamma$ *and* $\overline{y}_{|0} \sim_\Gamma \overline{y}_{|0}$ *are both the true constant* $\top$.

Definition 4.6. *If* Γ *and* Γ' *are contexts of* $\widehat{ID_1}$, *then a* realized morphism *from* Γ *to* Γ' *is an equivalence class* $[\overline{\mathbf{k}}]_{\approx_{\Gamma,\Gamma'}}$ *of lists of numerals with length equal to the length of* Γ' *satisfying the following requirements: if* $\Gamma' = [B_1, ..., B_n(\overline{x}_{|n-1})]$ *with* $n > 0$, *then for all* $1 \leq j \leq n$:

$$\overline{x}_{|\ell(\Gamma)} \sim_\Gamma \overline{y}_{|\ell(\Gamma)} \vdash_{\widehat{ID_1}} \{\mathbf{k}_j\}(\overline{x}_{|\ell(\Gamma)}) \sim_{B_j(\{\overline{\mathbf{k}}_{|j-1}\}(\overline{x}_{|\ell(\Gamma)}))} \{\mathbf{k}_j\}(\overline{y}_{|\ell(\Gamma)})$$

with respect to the equivalence relation $\approx_{\Gamma,\Gamma'}$ *defined by* $\overline{\mathbf{k}} \approx_{\Gamma,\Gamma'} \overline{\mathbf{k}'}$ *if and only if*

$$\overline{x}_{|\ell(\Gamma)}\,\varepsilon\,\Gamma \vdash_{\widehat{ID_1}} \{\mathbf{k}_j\}(\overline{x}_{|\ell(\Gamma)}) \sim_{B_j(\{\overline{\mathbf{k}}_{|j-1}\}(\overline{x}_{|\ell(\Gamma)}))} \{\mathbf{k}'_j\}(\overline{x}_{|\ell(\Gamma)})$$

for every $1 \leq j \leq n$.

In the case in which $\Gamma' = [\,]$, *then the unique realized morphism is the class* $!_{\Gamma,[\,]} := [\,]_{\approx_{\Gamma,[\,]}}$ *containing only the empty list.*

Definition 4.7. *If* $\overline{\mathbf{k}}$ *and* $\overline{\mathbf{h}}$ *are lists of numerals and* n *is a natural (meta)number, then*

$$\overline{\mathbf{h}} \circ^n \overline{\mathbf{k}} := [\,\Lambda\overline{x}_{|n}.\{\mathbf{h}_1\}(\{\overline{\mathbf{k}}\}(\overline{x}_{|n})), ..., \Lambda\overline{x}_{|n}.\{\mathbf{h}_{\ell(\overline{\mathbf{h}})}\}(\{\overline{\mathbf{k}}\}(\overline{x}_{|n}))\,]$$

Definition 4.8. *If* $[\overline{\mathbf{k}}]_{\approx_{\Gamma,\Gamma'}} : \Gamma \to \Gamma'$ *and* $[\overline{\mathbf{h}}]_{\approx_{\Gamma',\Gamma''}} : \Gamma' \to \Gamma''$ *are realized morphisms between contexts of* $\widehat{ID_1}$, *then we define their composition as the realized morphism*

$$[\overline{\mathbf{h}}]_{\approx_{\Gamma',\Gamma''}} \circ [\overline{\mathbf{k}}]_{\approx_{\Gamma,\Gamma'}} := [\,\overline{\mathbf{h}} \circ^{\ell(\Gamma)} \overline{\mathbf{k}}\,]_{\approx_{\Gamma,\Gamma''}} : \Gamma \to \Gamma''$$

If Γ *is a context of* $\widehat{ID_1}$, *then its identity is defined as the realized morphism*

$$[\,\pi_1^{\ell(\Gamma)}, ..., \pi_{\ell(\Gamma)}^{\ell(\Gamma)}\,]_{\approx_{\Gamma,\Gamma}} : \Gamma \to \Gamma$$

if $\ell(\Gamma) > 0$, *while it is the realized morphism*

$$[\,]_{\approx_{[\,],[\,]}} : [\,] \to [\,]$$

if $\Gamma = [\,]$.

Theorem 4.9. *Realized contexts of* $\widehat{ID_1}$ *and realized morphisms between them with their compositions and identities form a category denoted by* **Cont**.

As it happens in dependent type theory contexts can be equivalently represented as the indexed sums of their components. To this purpose we define the following realized morphisms which will act as projections to extract the components of a context:

Definition 4.10. *If Γ is a context of $\widehat{ID_1}$ and n is a natural (meta)number, we define the realized morphisms* pr_Γ *and* $\mathsf{pr}_\Gamma^{(n)}$ *in* **Cont** *as follows:*

- $\mathsf{pr}_{[\,]}$ *is* $id_{[\,]}$ *and* $\mathsf{pr}_{[A]}$ *is* $[\,]_\approx : [A] \to [\,]$;[4]

  $\mathsf{pr}_{[\Gamma,A]}$ *is* $[\pi_1^{\ell(\Gamma)+1}, ..., \pi_{\ell(\Gamma)}^{\ell(\Gamma)+1}]_\approx : [\Gamma, A] \to \Gamma$ *if* $\ell(\Gamma) > 0$;

- pr_Γ^0 *is* id_Γ *and* $\mathsf{pr}_\Gamma^{(i+1)}$ *is* $\mathsf{pr}_{\mathsf{cod}(\mathsf{pr}_\Gamma^{(i)})} \circ \mathsf{pr}_\Gamma^{(i)}$

where $\mathsf{cod}(\mathsf{pr}_\Gamma^{(i)})$ *denotes the codomain of* $\mathsf{pr}_\Gamma^{(i)}$.

Now we define the indexed sum of the last two components of a context:

Definition 4.11. *Suppose* $[\Gamma, A(\overline{x}_{|\ell(\Gamma)}), B(\overline{x}_{|\ell(\Gamma)+1})]$ *is a realized context of* $\widehat{ID_1}$. *We define the* indexed sum collection

$$\Sigma^\Gamma(\, A(\overline{x}_{|\ell(\Gamma)}),\, B(\overline{x}_{|\ell(\Gamma)+1})\,)$$

as a collection depending on $\overline{x}_{|\ell(\Gamma)}$ *determined by the following conditions:*[5]

$$x \,\varepsilon\, \Sigma^\Gamma(\, A(\overline{x}_{|\ell(\Gamma)}),\, B(\overline{x}_{|\ell(\Gamma)+1})\,) \equiv^{def} p_1(x) \,\varepsilon\, A(\overline{x}_{|\ell(\Gamma)}) \,\wedge\, p_2(x) \,\varepsilon\, B(\overline{x}_{|\ell(\Gamma)}, p_1(x))$$

$$x \sim_{\Sigma^\Gamma(\, A(\overline{x}_{|\ell(\Gamma)}), B(\overline{x}_{|\ell(\Gamma)+1})\,)} y \equiv^{def} p_1(x) \sim_{A(\overline{x}_{|\ell(\Gamma)})} p_1(y) \,\wedge\, p_2(x) \sim_{B(\overline{x}_{|\ell(\Gamma)}, p_1(x))} p_2(y)$$

Clearly, the indexed sum collection allows to represent a context in an equivalent way as follows:

Lemma 4.12. *Suppose* $[\Gamma, A(\overline{x}_{|\ell(\Gamma)}), B(\overline{x}_{|\ell(\Gamma)+1})]$ *is a realized context of* $\widehat{ID_1}$. *Then,* $[\Gamma, A(\overline{x}_{|\ell(\Gamma)}), B(\overline{x}_{|\ell(\Gamma)+1})]$ *is isomorphic to* $[\Gamma, \Sigma^\Gamma(A(\overline{x}_{|\ell(\Gamma)}), B(\overline{x}_{|\ell(\Gamma)+1}))]$ *in* **Cont**.

Proof. If Γ is not empty, just take the realized morphism from $[\Gamma, A(\overline{x}_{|\ell(\Gamma)}), B(\overline{x}_{|\ell(\Gamma)+1})]$ to $[\Gamma, \Sigma^\Gamma(\, A(\overline{x}_{|\ell(\Gamma)}),\, B(\overline{x}_{|\ell(\Gamma)+1})\,)]$ determined by the list

$$[\,\pi_1^{\ell(\Gamma)+2}, ..., \pi_{\ell(\Gamma)}^{\ell(\Gamma)+2}, \Lambda\overline{x}_{|\ell(\Gamma)}.\Lambda x.\Lambda y.\{\mathbf{p}\}(x,y)\,]$$

Its inverse from $[\Gamma, \Sigma^\Gamma(\, A(\overline{x}_{|\ell(\Gamma)}),\, B(\overline{x}_{|\ell(\Gamma)+1})\,)]$ to $[\Gamma, A(\overline{x}_{|\ell(\Gamma)}), B(\overline{x}_{|\ell(\Gamma)+1})]$ is the realized morphism determined by the list

$$[\,\pi_1^{\ell(\Gamma)+1}, ..., \pi_{\ell(\Gamma)}^{\ell(\Gamma)+1}, \Lambda\overline{x}_{|\ell(\Gamma)}.\Lambda x.\{\mathbf{p_1}\}(x), \Lambda\overline{x}_{|\ell(\Gamma)}.\Lambda x.\{\mathbf{p_2}\}(x)\,]$$

Instead, if Γ is empty, we can consider the realized isomorphism determined by $[\mathbf{p}]$ and $[\mathbf{p_1}, \mathbf{p_2}]$. □

[4]The subscripts on $\approx$ will be omitted when they will be clear from the context

[5]Here as usual x and y are fresh distinct variables

Theorem 4.13. **Cont** *is equivalent to* $\mathcal{C}$.

Proof. From $\mathcal{C}$ to **Cont** take the functor **F** sending any collection A of $\mathcal{C}$ to $[A]$ and every realized morphism $[\mathbf{n}]_{\approx_{A,B}}$ to $[\mathbf{n}]_{\approx_{[A],[B]}}$. Then, define a functor **E** from **Cont** to $\mathcal{C}$ as follows:

1. $\mathbf{E}([\,])$ is $(\ \{x \,|\, x = 0\}\ ,\ \ x = y\ \wedge\ x = 0\ )$,

   $\mathbf{E}([A])$ is A,

   $\mathbf{E}([\Gamma, A])$ is $\Sigma^{[\,]}(\ \mathbf{E}(\Gamma)\ ,\ A(\{\mathbf{p}_1^{\ell(\Gamma)}\}(x_1), ..., \{\mathbf{p}_{\ell(\Gamma)}^{\ell(\Gamma)}\}(x_1))\ )$;

2. if $\ell(\Gamma) > 0$ and $\ell(\Gamma') > 0$, then $[\overline{\mathbf{k}}]_{\approx_{\Gamma,\Gamma'}} : \Gamma \to \Gamma'$ is sent to

   $$\left[\Lambda x.\{\mathbf{p}^{\ell(\Gamma')}\}(\{\overline{\mathbf{k}}\}(\{\mathbf{p}_1^{\ell(\Gamma)}\}(x), ..., \{\mathbf{p}_{\ell(\Gamma)}^{\ell(\Gamma)}\}(x)))\right]_{\approx_{\mathbf{E}(\Gamma),\mathbf{E}(\Gamma')}}$$

   if $\Gamma = [\,]$ and $\ell(\Gamma') > 0$, then $[\overline{\mathbf{k}}]_{\approx_{[\,],\Gamma'}} : [\,] \to \Gamma'$ is sent to

   $$[\ \Lambda x.\{\mathbf{p}^{\ell(\Gamma')}\}(\overline{\mathbf{k}})\ ]_{\approx_{\mathbf{E}([\,]),\mathbf{E}(\Gamma')}}$$

   and $!_{\Gamma,[\,]} : \Gamma \to [\,]$ is sent to $[\,\Lambda x.0\,]_{\approx_{\mathbf{E}(\Gamma),\mathbf{E}([\,])}}$.

□

4.3 Families of realized collections as an indexed category

Here we are going to define an indexed category on the category of realized contexts

$$\mathbf{Col} : \mathbf{Cont}^{op} \to \mathbf{Cat}$$

whose fibre on a context Γ will be defined as a presentation of the slice category $\mathbf{Cont}/\Gamma$ in terms of *families of realized collections depending on the context* Γ.

Definition 4.14. *If* Γ *is a context of* $\widehat{ID_1}$ *(i. e. an object of* **Cont***), then* A *is a* family of realized collections *on* Γ *(or a* realized collection depending on Γ*) if and only if* $[\Gamma, A]$ *is a context of* $\widehat{ID_1}$.

Definition 4.15. *Let* $\mathbf{Col}(\Gamma)$ *be the category whose objects are families of realized collections on* Γ *and a morphism from a family of realized collections* A *to another family* B *is a realized morphism from* $\mathsf{pr}_{[\Gamma,A]}$ *to* $\mathsf{pr}_{[\Gamma,B]}$ *in the slice category* $\mathbf{Cont}/\Gamma$. *Composition and identities are inherited from those of* $\mathbf{Cont}/\Gamma$.

Lemma 4.16. *Let* Γ *be an object of* **Cont** *with* $\ell(\Gamma) > 0$ *and* A, B *be objects of* $\mathbf{Col}(\Gamma)$. *If* $\mathbf{n}$ *is a numeral for which*

$$\overline{x}_{|\ell(\Gamma)+1} \sim_{[\Gamma,A]} \overline{y}_{|\ell(\Gamma)+1} \vdash_{\widehat{ID_1}} \{\mathbf{n}\}(\overline{x}_{|\ell(\Gamma)+1}) \sim_B \{\mathbf{n}\}(\overline{y}_{|\ell(\Gamma)+1})$$

then

$$\gamma^{A,B}_{\mathbf{n},\Gamma} := [\pi_1^{\ell(\Gamma)+1}, ..., \pi_{\ell(\Gamma)}^{\ell(\Gamma)+1}, \mathbf{n}]_{\approx_{[\Gamma,A],[\Gamma,B]}}$$

is a well defined realized morphism from A *to* B *in* $\mathbf{Col}(\Gamma)$.

Conversely, for every $f : A \to B$ *in* $\mathbf{Col}(\Gamma)$ *there exists a numeral* $\mathbf{n}$ *for which* $f = \gamma^{A,B}_{\mathbf{n},\Gamma}$, *and in this case we say that* f is represented by $\mathbf{n}$.

If $[\mathbf{n}]_{\approx}$ is an arrow from A to B in $\mathbf{Col}([\,])$, then we will denote it also by $\gamma^{A,B}_{\mathbf{n},[\,]}$.

We will omit A, B and Γ in the notation $\gamma^{A,B}_{\mathbf{n},\Gamma}$, when they will be clear from the context.

Lemma 4.17. *Suppose* $f = [\overline{\mathbf{k}}]_{\approx_{\Gamma',\Gamma}} : \Gamma' \to \Gamma$ *in* **Cont**.

1. *If* $A(\overline{x}_{|\ell(\Gamma)})$ *is an object of* $\mathbf{Col}(\Gamma)$, *then the conditions*

   (a) $x \,\varepsilon\, \mathbf{Col}_f(A(\overline{x}_{|\ell(\Gamma)})) \equiv^{def} \overline{x}_{|\ell(\Gamma')} \,\varepsilon\, \Gamma' \wedge x \,\varepsilon\, A(\{\overline{\mathbf{k}}\}(\overline{x}_{|\ell(\Gamma')}))$

   (b) $x \sim_{\mathbf{Col}_f(A(\overline{x}_{|\ell(\Gamma)}))} y \equiv^{def} \overline{x}_{|\ell(\Gamma')} \,\varepsilon\, \Gamma' \wedge x \sim_{A(\{\overline{\mathbf{k}}\}(\overline{x}_{|\ell(\Gamma')}))} y$

   determine an object $\mathbf{Col}_f(A(\overline{x}_{|\ell(\Gamma)}))$ *of* $\mathbf{Col}(\Gamma')$.

2. *If* $g = \gamma_{\mathbf{n}}$ *is an arrow in* $\mathbf{Col}(\Gamma)$ *from* A *to* B, *then the numeral*

   $$\mathbf{n}' := \Lambda \overline{x}_{|\ell(\Gamma')+1}.\{\mathbf{n}\}(\{\overline{\mathbf{k}}\}(\overline{x}_{|\ell(\Gamma')}), x_{\ell(\Gamma')+1})$$

   determines an arrow $\mathbf{Col}_f(g) := \gamma_{\mathbf{n}'}$ *in* $\mathbf{Col}(\Gamma')$ *from* $\mathbf{Col}_f(A)$ *to* $\mathbf{Col}_f(B)$.

Moreover, $\mathbf{Col}_f(h \circ g) = \mathbf{Col}_f(h) \circ \mathbf{Col}_f(g)$ *if* $g : A \to B$ *and* $h : B \to C$ *are arrows in* $\mathbf{Col}(\Gamma)$ *and* $\mathbf{Col}_f(id_A) = id_{\mathbf{Col}_f(A)}$ *if* A *is an object of* $\mathbf{Col}(\Gamma)$, *i. e.* $\mathbf{Col}_f$ *is a functor from* $\mathbf{Col}(\Gamma)$ *to* $\mathbf{Col}(\Gamma')$.

Moreover we have the following property.

Lemma 4.18. *If* $f : \Gamma' \to \Gamma''$ *and* $g : \Gamma'' \to \Gamma'''$ *are arrows in* **Cont** *and* Γ *is an object of* **Cont**, *then*

1. $\mathbf{Col}_{g \circ f} = \mathbf{Col}_f \circ \mathbf{Col}_g$

2. $\mathbf{Col}_{id_\Gamma} = \mathbf{id}_{\mathbf{Col}_f(\Gamma)}$

Now we are ready to define the indexed category of families of realized collections as follows:

Definition 4.19. *Let* $\mathbf{Col} : \mathbf{Cont}^{op} \to \mathbf{Cat}$ *be the functor defined by the pair of assignment* $\Gamma \mapsto \mathbf{Col}(\Gamma)$, $f \mapsto \mathbf{Col}_f$.

$\mathbf{Col}_f$ *is called* the substitution functor along f.

In the following lemma we introduce the notation of pullback projections which will be used later to characterize the interpretation of the substitution of terms in types and the interpretation of the context operation of weakening:

Lemma 4.20. *If* Γ *and* Γ' *are objects of* $\mathbf{Cont}$, $f : \Gamma' \to \Gamma$ *in* $\mathbf{Cont}$ *and* A *is an object in* $\mathbf{Col}(\Gamma)$, *then* $\mathbf{Col}_f(A)$ *fits into a pullback in* $\mathbf{Cont}$ *as follows*

$$\begin{array}{ccc} [\Gamma', \mathbf{Col}_f(A)] & \xrightarrow{q(f,[\Gamma,A])} & [\Gamma, A] \\ \downarrow{\scriptstyle \mathsf{pr}} & & \downarrow{\scriptstyle \mathsf{pr}} \\ \Gamma' & \xrightarrow{\quad f \quad} & \Gamma \end{array}$$

where, if f *is represented by the list* $[\mathbf{k}_1, ..., \mathbf{k}_{\ell(\Gamma)}]$, *then* $q(f, [\Gamma, A])$ *is represented by the list*

$$[\, \Lambda\overline{x}_{|_{\ell(\Gamma')+1}}.\{\mathbf{k}_1\}(\overline{x}_{|_{\ell(\Gamma')}}), ..., \Lambda\overline{x}_{|_{\ell(\Gamma')+1}}.\{\mathbf{k}_{\ell(\Gamma)}\}(\overline{x}_{|_{\ell(\Gamma')}}), \pi^{\ell(\Gamma')+1}_{\ell(\Gamma')+1}\,]$$

Now, we are going to describe the categorical structure of each fibre $\mathbf{Col}(\Gamma)$ for a fixed context Γ.

Hence, *in all the following lemmas* Γ *is an object of* $\mathbf{Cont}$ *with* $\ell(\Gamma) = n$.

We start by showing that each fibre $\mathbf{Col}(\Gamma)$ is closed under finite products.

Lemma 4.21. *The object*

$$\mathbf{1}^\Gamma := (\,\{x|\,\overline{x}_{|_n}\,\varepsilon\,\Gamma \wedge x = 0\},\ \overline{x}_{|_n}\,\varepsilon\,\Gamma \wedge x = 0 \wedge x = y\,)$$

is a terminal object in $\mathbf{Col}(\Gamma)$, *i. e. for every* A *in* $\mathbf{Col}(\Gamma)$, *there exists a unique arrow* $!_{A,\mathbf{1}^\Gamma} : A \to \mathbf{1}^\Gamma$ *in* $\mathbf{Col}(\Gamma)$.

Lemma 4.22. *If* A *and* B *are objects of* $\mathbf{Col}(\Gamma)$, *then the object* $A \times^\Gamma B$ *defined by the following conditions:*

1. $x\,\varepsilon\,A \times^\Gamma B \equiv^{def} \mathsf{p}_1(x)\,\varepsilon\,A \wedge \mathsf{p}_2(x)\,\varepsilon\,B$

2. $x \sim_{A\times^\Gamma B} y \equiv^{def} \mathsf{p}_1(x) \sim_A \mathsf{p}_1(y) \wedge \mathsf{p}_2(x) \sim_B \mathsf{p}_2(y)$

with $\mathbf{p}_i^\Gamma := \Lambda\overline{x}_{|n+1}.\{\mathbf{p}_i\}(x_{n+1})$ *for* $i = 1, 2$ *yields the following* binary product *diagram in* $\mathbf{Col}(\Gamma)$

$$A \xleftarrow[\pi_1^{A,B} := \gamma_{\mathbf{p}_1^\Gamma}]{\qquad\qquad} A \times^\Gamma B \xrightarrow[\pi_2^{A,B} := \gamma_{\mathbf{p}_2^\Gamma}]{\qquad\qquad} B$$

i. e. for every $f : C \to A$ *and* $g : C \to B$ *in* $\mathbf{Col}(\Gamma)$*, there exists a unique arrow* $\langle f, g\rangle : C \to A \times^\Gamma B$ *in* $\mathbf{Col}(\Gamma)$ *for which the following diagram commutes*

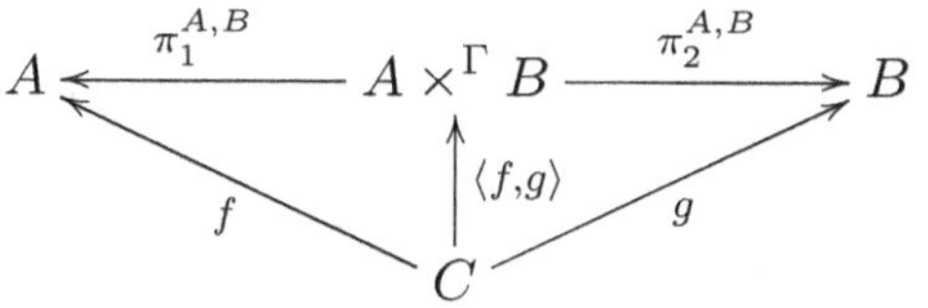

Now we are going to show how to form equalizers in $\mathbf{Col}(\Gamma)$.

Lemma 4.23. *If* A *is an object of* $\mathbf{Col}(\Gamma)$ *and* $f, g : \mathbf{1}^\Gamma \to A$ *are arrows in* $\mathbf{Col}(\Gamma)$ *represented by numerals* $\mathbf{n}_f$ *and* $\mathbf{n}_g$ *respectively, then* $\mathbf{Eq}^\Gamma(A, f, g)$ *given by the following conditions*

1. $x \,\varepsilon\, \mathbf{Eq}^\Gamma(A, f, g) \equiv^{def} \{\mathbf{n}_f\}(\overline{x}_{|n}, 0) \sim_A \{\mathbf{n}_g\}(\overline{x}_{|n}, 0)$
2. $x \sim_{\mathbf{Eq}^\Gamma(A,f,g)} y \equiv^{def} x \,\varepsilon\, \mathbf{Eq}^\Gamma(A, f, g) \wedge y \,\varepsilon\, \mathbf{Eq}^\Gamma(A, f, g)$

is a well defined object of $\mathbf{Col}(\Gamma)$.

Lemma 4.24. *Suppose* $f_1, f_2 : A \to B$ *in* $\mathbf{Col}(\Gamma)$ *and* $f_i = \gamma_{\mathbf{n}_i}$ *for* $i = 1, 2$*. If for* $i = 1, 2$ *we define* f_i' *to be* $\gamma_{\mathbf{n}'_i} : \mathbf{1}^{[\Gamma,A]} \to \mathbf{Col}_{\mathsf{pr}_{[\Gamma,A]}}(B)$ *in* $\mathbf{Col}([\Gamma, A])$ *with* $\mathbf{n}_i' := \Lambda\overline{x}_{|n+2}.\{\mathbf{n}_i\}(\overline{x}_{|n+1})$*, then*

$$E(f_1, f_2) := \Sigma^\Gamma(\,A,\, \mathbf{Eq}^{[\Gamma,A]}(\mathbf{Col}_{\mathsf{pr}_{[\Gamma,A]}}(B), f_1', f_2')\,)$$

$$e(f_1, f_2) := \gamma_{\mathbf{p}_1^\Gamma} : E(f_1, f_2) \to A$$

define an equalizer for f_1 and f_2 in $\mathbf{Col}(\Gamma)$*, i. e. for every* $e' : E' \to A$ *for which* $f_1 \circ e' = f_2 \circ e'$*, there exists a unique arrow* $g : E' \to E(f_1, f_2)$ *in* $\mathbf{Col}(\Gamma)$ *for which the following diagram commutes.*

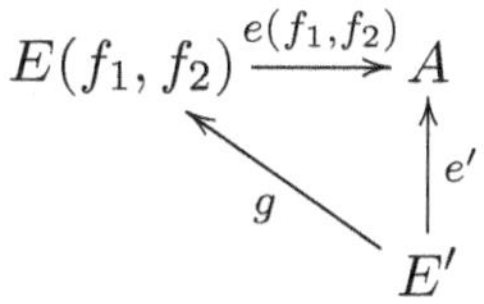

Here we are going to show that each fibre $\mathbf{Col}(\Gamma)$ is closed under finite coproducts.

Lemma 4.25. *The object* $\mathbf{0}^\Gamma := (\{x|\perp\}, \perp)$ *is an initial object of* $\mathbf{Col}(\Gamma)$*, i. e. for every* A *in* $\mathbf{Col}(\Gamma)$ *there exists a unique arrow* $!_{\mathbf{0}^\Gamma, A} : \mathbf{0}^\Gamma \to A$ *in* $\mathbf{Col}(\Gamma)$*.*

Lemma 4.26. *If* A *and* B *are objects of* $\mathbf{Col}(\Gamma)$*, then the object* $A +^\Gamma B$ *of* $\mathbf{Col}(\Gamma)$ *defined by the following conditions:*

1. $x \,\varepsilon\, A +^\Gamma B \equiv^{def} (p_1(x) = 0 \wedge p_2(x) \,\varepsilon\, A) \vee (p_1(x) = 1 \wedge p_2(x) \,\varepsilon\, B)$
2. $x \sim_{A+^\Gamma B} y \equiv^{def} p_1(x) = p_1(y) \wedge$
   $$((p_1(x) = 0 \wedge p_2(x) \sim_A p_2(y)) \vee (p_1(x) = 1 \wedge p_2(x) \sim_B p_2(y)))$$

with $\mathbf{j}_1^\Gamma := \Lambda\overline{x}_{|n+1}.\{\mathbf{p}\}(0, x_{n+1})$ *and* $\mathbf{j}_2^\Gamma := \Lambda\overline{x}_{|n+1}.\{\mathbf{p}\}(1, x_{n+1})$ *yields the following* binary coproduct *diagram in* $\mathbf{Col}(\Gamma)$

$$A \xrightarrow{\mathsf{j}_1^{A,B} := \gamma_{\mathbf{j}_1^\Gamma}} A +^\Gamma B \xleftarrow{\mathsf{j}_2^{A,B} := \gamma_{\mathbf{j}_2^\Gamma}} B$$

i. e. for every object C *of* $\mathbf{Col}(\Gamma)$ *and every pair of arrows* $f : A \to C$ *and* $g : B \to C$*, there is a unique arrow* $\mathsf{case}(f, g) : A +^\Gamma B \to C$ *in* $\mathbf{Col}(\Gamma)$ *for which the following diagram commutes*

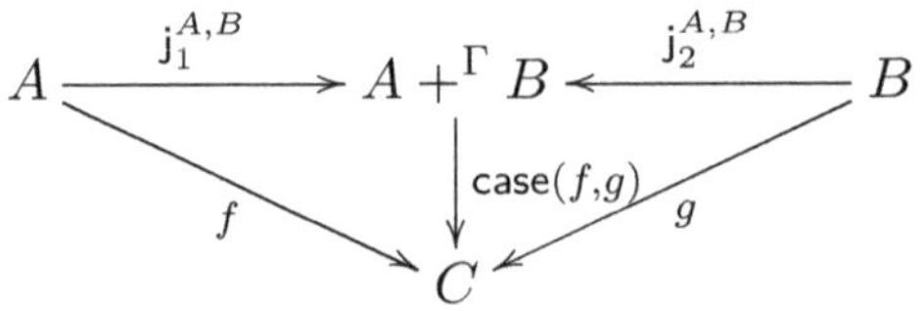

Now we are going to show how $\mathbf{Col}(\Gamma)$ is closed under exponential objects, namely under function spaces:

Lemma 4.27. *If* A *and* B *are objects of* $\mathbf{Col}(\Gamma)$*, then the object* $A \Rightarrow^\Gamma B$ *defined by*

1. $x \,\varepsilon\, A \Rightarrow^\Gamma B \equiv^{def} \overline{x}_{|n} \,\varepsilon\, \Gamma \wedge \forall t \forall s\, (t \sim_A s \to \{x\}(t) \sim_B \{x\}(s))$
2. $x \sim_{A \Rightarrow^\Gamma B} y \equiv^{def} x \,\varepsilon\, A \Rightarrow^\Gamma B \wedge y \,\varepsilon\, A \Rightarrow^\Gamma B \wedge \forall t\, (t \,\varepsilon\, A \to \{x\}(t) \sim_B \{y\}(t))$

together with the arrow

$$\mathsf{ev}^{A,B} := \gamma_{\mathbf{ev}^\Gamma} : (A \Rightarrow^\Gamma B) \times^\Gamma A \to B$$

where $\mathbf{ev}^\Gamma$ *is* $\Lambda\overline{x}_{|n+1}.\{\{\mathbf{p}_1\}(x_{n+1})\}(\{\mathbf{p}_2\}(x_{n+1}))$ *defines an* exponential of A and B in $\mathbf{Col}(\Gamma)$ *i. e. for every object* C *of* $\mathbf{Col}(\Gamma)$ *and every arrow* $f : C \times^\Gamma A \to B$ *in*

$\mathbf{Col}(\Gamma)$, *there exists a unique arrow* $\mathsf{Cur}(f) : C \to A \Rightarrow^\Gamma B$ *for which the following diagram commutes in* $\mathbf{Col}(\Gamma)^6$

$$\begin{array}{ccc}
C \times^\Gamma A & \xrightarrow{\quad f \quad} & B \\
{\scriptstyle \mathsf{Cur}(f) \times^\Gamma id_A} \downarrow & \nearrow {\scriptstyle \mathsf{ev}^{A,B}} & \\
(A \Rightarrow^\Gamma B) \times^\Gamma A & &
\end{array}$$

$\mathbf{Col}(\Gamma)$ *has also list objects (see for instance [22] for a categorical definition)*

Lemma 4.28. *If A is an object of $\mathbf{Col}(\Gamma)$, then the object $\mathbf{List}^\Gamma(A)$ defined by*

1. $x \,\varepsilon\, \mathbf{List}^\Gamma(A) \equiv^{def} \overline{x}_{|n} \,\varepsilon\, \Gamma \wedge \forall j\, (\, j < lh(x) \to (x)_j \,\varepsilon\, A\,)$

2. $x \sim_{\mathbf{List}^\Gamma(A)} y \equiv^{def} \overline{x}_{|n} \,\varepsilon\, \Gamma \wedge lh(x) = lh(y) \wedge \forall j\, (\, j < lh(x) \to (x)_j \sim_A (y)_j\,)$

together with the arrows

$$\epsilon^A := \gamma_{\Lambda \overline{x}_{|n+1}.0} : \mathbf{1}^\Gamma \to \mathbf{List}^\Gamma(A)$$

$$\mathsf{cons}^A := \gamma_{\mathbf{cnc}^\Gamma} : \mathbf{List}^\Gamma(A) \times^\Gamma A \to \mathbf{List}^\Gamma(A)$$

where $\mathbf{cnc}^\Gamma$ is $\Lambda \overline{x}_{|n+1}.\{\mathbf{cnc}\}(\{\mathbf{p_1}\}(x_{n+1}), \{\mathbf{p_2}\}(x_{n+1}))$, defines a *list object on A in* $\mathbf{Col}(\Gamma)$, *i. e. for every object B of $\mathbf{Col}(\Gamma)$ and every pair of arrows $f : \mathbf{1}^\Gamma \to B$ and $g : B \times^\Gamma A \to B$ in $\mathbf{Col}(\Gamma)$, there exists a unique arrow*

$$listrec(f, g) : \mathbf{List}^\Gamma(A) \to B$$

for which the following diagram commutes in $\mathbf{Col}(\Gamma)$

$$\begin{array}{ccccc}
\mathbf{1}^\Gamma & \xrightarrow{\epsilon^A} & \mathbf{List}^\Gamma(A) & \xleftarrow{\mathsf{cons}^A} & \mathbf{List}^\Gamma(A) \times^\Gamma A \\
& {\scriptstyle f} \searrow & \downarrow {\scriptstyle listrec(f,g)} & & \downarrow {\scriptstyle listrec(f,g) \times^\Gamma id_A} \\
& & B & \xleftarrow{\quad g \quad} & B \times^\Gamma A
\end{array}$$

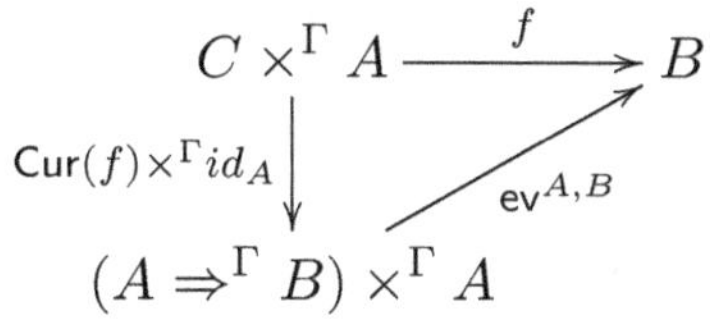

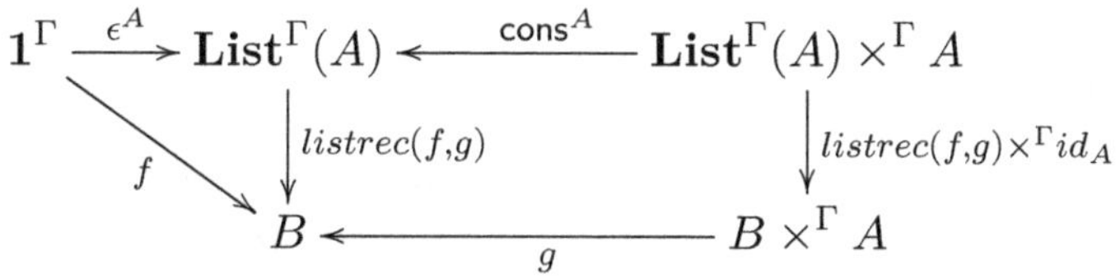

Theorem 4.29. *For every Γ in $\mathbf{Cont}$, $\mathbf{Col}(\Gamma)$ is a finitely complete cartesian closed category with finite coproducts and list objects and for every morphism f in $\mathbf{Cont}$ the functors $\mathbf{Col}_f$ preserve this structure.*

---

[6]For $f : A \to C$ and $g : B \to D$ in $\mathbf{Col}(\Gamma)$, we use the notation $f \times^\Gamma g$ for the arrow $\langle f \circ \pi_1^{A,B}, g \circ \pi_2^{A,B} \rangle : A \times^\Gamma B \to C \times^\Gamma D$.

Proof. This is a consequence of the previous lemmas (see [20]) and it is an immediate verification to see that all these structures are preserved by the functors $\mathbf{Col}_f$. $\square$

Remark 4.30. *The object* $\mathbf{N}^\Gamma$ *of* $\mathbf{Col}(\Gamma)$ *defined by the following:*

1. $x \,\varepsilon\, \mathbf{N}^\Gamma \equiv^{def} \overline{x}_{|_{\ell(\Gamma)}} \,\varepsilon\, \Gamma \wedge x = x$
2. $x \sim_{\mathbf{N}^\Gamma} y \equiv^{def} \overline{x}_{|_{\ell(\Gamma)}} \,\varepsilon\, \Gamma \wedge x = y$

together with the arrows

$$\mathsf{z}^\Gamma := \gamma_{\Lambda \overline{x}_{|_{\ell(\Gamma)+1}}.0} : \mathbf{1}^\Gamma \to \mathbf{N}^\Gamma \qquad \mathsf{s}^\Gamma := \gamma_{\Lambda \overline{x}_{|_{\ell(\Gamma)+1}}.\{\mathsf{s}\}(x_{\ell(\Gamma)+1})} : \mathbf{N}^\Gamma \to \mathbf{N}^\Gamma.$$

defines a natural numbers object in $\mathbf{Col}(\Gamma)$, *i. e. for every A in* $\mathbf{Col}(\Gamma)$ *and for every pair of arrows* $f : \mathbf{1}^\Gamma \to A$ *and* $g : A \to A$ *in* $\mathbf{Col}(\Gamma)$*, there exists a unique arrow* $rec(f,g) : \mathbf{N}^\Gamma \to A$ *for which the following diagram commutes.*

$$\begin{array}{ccccc}
\mathbf{1}^\Gamma & \xrightarrow{\mathsf{z}^\Gamma} & \mathbf{N}^\Gamma & \xrightarrow{\mathsf{s}^\Gamma} & \mathbf{N}^\Gamma \\
& \searrow^{f} & \downarrow{\scriptstyle rec(f,g)} & & \downarrow{\scriptstyle rec(f,g)} \\
& & A & \xrightarrow[g]{} & A
\end{array}$$

It is immediate to see that this natural numbers object is preserved by the substitution functors $\mathbf{Col}_f$. *A natural numbers object can be defined also as* $\mathbf{List}^\Gamma(\mathbf{1}^\Gamma)$, *but it is convenient to consider the representation* $\mathbf{N}^\Gamma$ *to simplify the realizability interpretation of* $\mathbf{mTT}^a$.

Corollary 4.31. **Cont** *is a finitely complete cartesian closed category with finite coproducts and list objects.*

Proof. This is an immediate consequence of theorem 4.29 and 4.13 as $\mathcal{C}$ is clearly isomorphic to $\mathbf{Col}([\,])$. $\square$

Definition 4.32. *If* $f : [\Gamma, \mathbf{1}^\Gamma] \to [\Gamma, A]$ *is an arrow in* **Cont***, then we define* $\widetilde{f} : \Gamma \to [\Gamma, A]$ *as* $f \circ j$ *where* $j : \Gamma \to [\Gamma, \mathbf{1}^\Gamma]$ *is the isomorphism in* **Cont** *defined by the list*

$$[\,\pi_1^{\ell(\Gamma)}, ..., \pi_{\ell(\Gamma)}^{\ell(\Gamma)}, \Lambda \overline{x}_{|_{\ell(\Gamma)}}.0\,]$$

Now, we are going to show that, for any realized collection A in $\mathbf{Col}(\Gamma)$ there are left adjoints to substitution functors of the kind $\mathbf{Col}_{\mathsf{pr}_{[\Gamma,A]}}$ which will be used to interpret the operation of weakening the context Γ to $[\Gamma, A]$. These left adjoints will be used to interpret the strong indexed sum collections of $\mathbf{mTT}^a$.

Lemma 4.33. *Suppose Γ is an object in* **Cont** *and A is an object in* $\mathbf{Col}(\Gamma)$. *Then the functor sending each B in $\mathbf{Col}([\Gamma, A])$ to $\Sigma^{\Gamma}(A, B)$ and each arrow $f := \gamma_{\mathbf{n}} : B \to C$ in $\mathbf{Col}([\Gamma, A])$ to the arrow $\Sigma^{\Gamma}(A, f)$ from $\Sigma^{\Gamma}(A, B)$ to $\Sigma^{\Gamma}(A, C)$ in $\mathbf{Col}([\Gamma])$ represented by $\Lambda\overline{x}_{|\ell(\Gamma)}.\Lambda x.\{\mathbf{p}\}(\{\mathbf{p}_1\}(x), \{\mathbf{n}\}(\overline{x}_{|\ell(\Gamma)}, \{\mathbf{p}_1\}(x), \{\mathbf{p}_2\}(x)))$, in the sense of lemma 4.16, is left adjoint to the functor $\mathbf{Col}_{\mathsf{pr}_{[\Gamma,A]}}$, i. e. there is a bijection (see [20])*

$$Hom_{\mathbf{Col}(\Gamma)}(\Sigma^{\Gamma}(A, B), D) \cong Hom_{\mathbf{Col}([\Gamma,A])}(B, \mathbf{Col}_{\mathsf{pr}_{[\Gamma,A]}}(D))$$

natural in every B in $\mathbf{Col}([\Gamma, A])$ and D in $\mathbf{Col}(\Gamma)$.

We also give the following lemma which will be useful for the interpretation.

Lemma 4.34. *For every Γ in* **Cont** *and for every A in $\mathbf{Col}(\Gamma)$ and B in $\mathbf{Col}([\Gamma, A])$, the object $\Sigma^{\Gamma}(A, B)$ satisfies the following properties. If $\mathsf{p}_1^{\Sigma} := \gamma_{\mathbf{p}_1^{\Gamma}} : \Sigma^{\Gamma}(A, B) \to A$ (see lemma 4.22) in $\mathbf{Col}(\Gamma)$, for every $f : \mathbf{1} \to A$ in $\mathbf{Col}(\Gamma)$ and $g : \mathbf{1} \to \mathbf{Col}_{\widetilde{f}}(B)$ in $\mathbf{Col}(\Gamma)$, there is a unique arrow $\langle f, g\rangle_{\Sigma} : \mathbf{1} \to \Sigma^{\Gamma}(A, B)$ in $\mathbf{Col}(\Gamma)$ for which the following diagrams commute (the first in $\mathbf{Col}(\Gamma)$, the second in* **Cont***)*

$$\begin{array}{ccc} \mathbf{1} & \xrightarrow{\langle f,g\rangle_{\Sigma}} & \Sigma^{\Gamma}(A,B) \\ {\scriptstyle f}\downarrow & \swarrow {\scriptstyle \mathsf{p}_1^{\Sigma}} & \\ A & & \end{array} \qquad \begin{array}{ccc} [\Gamma, \mathbf{1}] & \xrightarrow{\langle f,g\rangle_{\Sigma}} & [\Gamma, \Sigma^{\Gamma}(A,B)] \\ \downarrow{\scriptstyle g} & & \downarrow{\scriptstyle \simeq} \\ [\Gamma, \mathbf{Col}_{\widetilde{f}}(B)] & \xrightarrow{q(\widetilde{f}, [\Gamma,A,B])} & [\Gamma, A, B] \end{array}$$

where $\simeq$ is the isomorphism from $[\Gamma, \Sigma^{\Gamma}(A, B)]$ to $[\Gamma, A, B]$ defined in lemma 4.12.

Conversely, for every $h : \mathbf{1} \to \Sigma^{\Gamma}(A, B)$ in $\mathbf{Col}(\Gamma)$, there is a unique arrow

$$\mathsf{p}_2^{\Sigma}(h) : \mathbf{1} \to \mathbf{Col}_{\widetilde{\mathsf{p}_1^{\Sigma}\circ h}}(B)$$

in $\mathbf{Col}(\Gamma)$ for which the following diagram commutes in **Cont**

$$\begin{array}{ccc} [\Gamma, \mathbf{1}] & \xrightarrow{h} & [\Gamma, \Sigma^{\Gamma}(A,B)] \\ \downarrow{\scriptstyle \mathsf{p}_2^{\Sigma}(h)} & & \downarrow{\scriptstyle \simeq} \\ [\Gamma, \mathbf{Col}_{\widetilde{\mathsf{p}_1^{\Sigma}\circ h}}(B)] & \xrightarrow{q(\widetilde{\mathsf{p}_1^{\Sigma}\circ h}, [\Gamma,A,B])} & [\Gamma, A, B] \end{array}$$

where $\widetilde{f}$ and $\widetilde{\mathsf{p}_1^{\Sigma} \circ h}$ are as in definition 4.32.

Now, we are going to show that, for any realized collection A in $\mathbf{Col}(\Gamma)$ there are right adjoints to substitution functors of the kind $\mathbf{Col}_{\mathsf{pr}_{[\Gamma,A]}}$. These right adjoints will be used to interpret the dependent product sets of $\mathbf{mTT}^a$.

Definition 4.35. *Let Γ be an object of* **Cont** *with $\ell(\Gamma) = n$, $A(\overline{x}_{|n})$ an object of $\mathbf{Col}(\Gamma)$ and $B(\overline{x}_{|n+1})$ an object of $\mathbf{Col}([\Gamma, A(\overline{x}_{|n})])$. We define $\Pi^\Gamma(A(\overline{x}_{|n}), B(\overline{x}_{|n+1}))$ as follows:*

1. $x \,\varepsilon\, \Pi^\Gamma(A(\overline{x}_{|n}), B(\overline{x}_{|n+1})) \equiv^{def}$

$$\overline{x}_{|n} \,\varepsilon\, \Gamma \wedge \forall t \forall s\, (t \sim_{A(\overline{x}_{|n})} s \rightarrow \{x\}(t) \sim_{B(\overline{x}_{|n}, t)} \{x\}(s));$$

2. $x \sim_{\Pi^\Gamma(A(\overline{x}_{|n}), B(\overline{x}_{|n+1}))} y \equiv^{def} x \,\varepsilon\, \Pi^\Gamma(A(\overline{x}_{|n}), B(\overline{x}_{|n+1})) \wedge$

$$y \,\varepsilon\, \Pi^\Gamma(A(\overline{x}_{|n}), B(\overline{x}_{|n+1})) \wedge \forall t\, (t \,\varepsilon\, A(\overline{x}_{|n}) \rightarrow \{x\}(t) \sim_{B(\overline{x}_{|n}, t)} \{y\}(t)).$$

Lemma 4.36. *Suppose Γ is an object in* **Cont** *and A is an object in $\mathbf{Col}(\Gamma)$. Then the functor sending each object B in $\mathbf{Col}([\Gamma, A])$ to $\Pi^\Gamma(A, B)$ and each arrow $f := \gamma_{\mathbf{n}} : B \rightarrow C$ in $\mathbf{Col}([\Gamma, A])$ to the arrow $\Pi^\Gamma(A, f)$ from $\Pi^\Gamma(A, B)$ to $\Pi^\Gamma(A, C)$ in $\mathbf{Col}([\Gamma])$ represented by $\Lambda \overline{x}_{|\ell(\Gamma)}.\Lambda x.\Lambda y.\{\mathbf{n}\}(\overline{x}_{|\ell(\Gamma)}, y, \{x\}(y))$, in the sense of lemma 4.16, is right adjoint to the functor $\mathbf{Col}_{\mathsf{pr}_{[\Gamma,A]}}$, i. e. there is a bijection (see [20])*

$$Hom_{\mathbf{Col}(\Gamma)}(D, \Pi^\Gamma(A, B)) \cong Hom_{\mathbf{Col}([\Gamma,A])}(\mathbf{Col}_{\mathsf{pr}_{[\Gamma,A]}}(D), B)$$

natural in every B in $\mathbf{Col}([\Gamma, A])$ and D in $\mathbf{Col}(\Gamma)$.

Corollary 4.37. *For every Γ in* **Cont**, *for every A and C in $\mathbf{Col}(\Gamma)$ and for every B in $\mathbf{Col}([\Gamma, A])$, the object $\Pi^\Gamma(A, B)$ satisfies the following universal property: there is an arrow ev^Γ_Π from $\mathbf{Col}_{\mathsf{pr}_{[\Gamma,A]}}(\Pi^\Gamma(A, B))$ to B in $\mathbf{Col}([\Gamma, A])$ such that for every $f : \mathbf{Col}_{\mathsf{pr}_{[\Gamma,A]}}(C) \rightarrow B$ in $\mathbf{Col}([\Gamma, A])$, there exists a unique arrow $\mathsf{Cur}_\Pi(f) : C \rightarrow \Pi^\Gamma(A, B)$ in $\mathbf{Col}(\Gamma)$ for which the following diagram commutes in $\mathbf{Col}([\Gamma, A])$:*

$$\begin{array}{ccc}
\mathbf{Col}_{\mathsf{pr}_{[\Gamma,A]}}(C) & \xrightarrow{\quad f \quad} & B \\
{\scriptstyle \mathbf{Col}_{\mathsf{pr}_{[\Gamma,A]}}(\mathsf{Cur}_\Pi(f))} \Big\downarrow & \nearrow {\scriptstyle \mathsf{ev}^\Gamma_\Pi} & \\
\mathbf{Col}_{\mathsf{pr}_{[\Gamma,A]}}(\Pi^\Gamma(A, B)) & &
\end{array}$$

Observe that the substitution functor $\mathbf{Col}_f$ along any morphism f of **Cont** preserves left and right adjoints described above as follows:

Lemma 4.38. *Suppose $f : \Gamma' \to \Gamma$ in* **Cont**, *A is an object of* $\mathbf{Col}(\Gamma)$, $f' := q(f, [\Gamma, A])$ *and B is an object of* $\mathbf{Col}([\Gamma, A])$. *Then*

1. $\mathbf{Col}_f(\Sigma^{\Gamma}(A, B)) = \Sigma^{\Gamma'}(\mathbf{Col}_f(A), \mathbf{Col}_{f'}(B))$
2. $\mathbf{Col}_f(\Pi^{\Gamma}(A, B)) = \Pi^{\Gamma'}(\mathbf{Col}_f(A), \mathbf{Col}_{f'}(B))$

Note here that left adjoints and right adjoints to substitution functors of the kind $\mathbf{Col}_{\mathsf{pr}_{[\Gamma,A]}}$ provide respectively binary products and exponentials as follows:

Lemma 4.39. *If Γ is an object of* **Cont** *and A, B are objects of* $\mathbf{Col}(\Gamma)$, *then*

$$\Sigma^{\Gamma}(A, \mathbf{Col}_{\mathsf{pr}_{[\Gamma,A]}}(B))) = A \times^{\Gamma} B \qquad \Pi^{\Gamma}(A, \mathbf{Col}_{\mathsf{pr}_{[\Gamma,A]}}(B))) = A \Rightarrow^{\Gamma} B$$

The following lemma will be useful in the interpretation of $\mathbf{mTT}^a$-eliminators for disjoint sums, natural numbers and lists.

Lemma 4.40. *If Γ is an object of* **Cont**, *$A_1, ..., A_n, B$ are objects of* $\mathbf{Col}(\Gamma)$ *and*

1. $\widetilde{A_1} := A_1$
2. $\widetilde{A_{i+1}} := \mathbf{Col}_{\mathsf{pr}^{(i)}_{[\Gamma,\widetilde{A_1},...,\widetilde{A_i}]}}(A_{i+1})$ *for* $i = 1, .., n-1$
3. $\widetilde{B} := \mathbf{Col}_{\mathsf{pr}^{(n)}_{[\Gamma,\widetilde{A_1},...,\widetilde{A_n}]}}(B)$

and $f := \gamma_{\mathbf{n}} : \mathbf{1} \to \widetilde{B}$ *in* $\mathbf{Col}([\Gamma, \widetilde{A_1}, ..., \widetilde{A_n}])$, *then*

$$f^{\Gamma}_{\natural} := \gamma_{\mathbf{n}'} : ((A_1 \times^{\Gamma} A_2) \times .... \times^{\Gamma} A_n) \to B$$

where $\mathbf{n}'$ is defined as

$$\Lambda \overline{x}_{|\ell(\Gamma)}.\Lambda x.\{\mathbf{n}\}(\overline{x}_{|\ell(\Gamma)}, \{\mathbf{p}^n_1\}(x), ..., \{\mathbf{p}^n_n\}(x), 0)$$

is a well defined morphism in $\mathbf{Col}(\Gamma)$.

The left and right adjoints to the substitution functors of the kind $\mathbf{Col}_{\mathsf{pr}_{[\Gamma,A]}}$ are enough to provide left and right adjoints to substitution functors along any arrow in **Cont** (see for example [26] and loc.cit. for a proof):

Corollary 4.41. *For any arrow f in* **Cont**, *the substitution functor* $\mathbf{Col}_f$ *enjoys left and right adjoints satisfying Beck-Chevalley conditions.*

Moreover, the category **Cont** *is locally cartesian closed (see for instance [36] for a definition).*

4.4 Families of realized propositions as an indexed category

Here we are going to define an indexed category of realized propositions equipped with the structure of a first-order Lawvere hyperdoctrine

$$\mathbf{Prop} : \mathbf{Cont}^{op} \to \mathbf{Cat}$$

on the category of realized contexts. This will be used to interpret generic propositions of $\mathbf{mTT}^a$.

We start by giving a lemma characterizing a proof-irrelevant dependent realized collection, namely a realized collection with at most one element:

Lemma 4.42. *Let Γ be an object of* **Cont** *and let P be an object of* $\mathbf{Col}(\Gamma)$. *Then the following conditions are equivalent:*

1. *for every object A in* $\mathbf{Col}(\Gamma)$, *if $f, g : A \to P$ are arrows in* $\mathbf{Col}(\Gamma)$, *then $f = g$;*
2. $\pi_1^{P,P} = \pi_2^{P,P} : P \times^\Gamma P \to P$;
3. *P is* proof-irrelevant, *i.e.* $x \, \varepsilon \, P \wedge y \, \varepsilon \, P \vdash_{\widehat{ID_1}} x \sim_P y$.

Now we define the notion of *a family of realized propositions* as a proof-irrelevant dependent realized collection:

Definition 4.43. *Let Γ be an object of* **Cont**. *A* family of realized propositions *on Γ (or a* realized proposition depending on Γ*) is a proof-irrelevant object of* $\mathbf{Col}(\Gamma)$ *as in lemma 4.42.*

Definition 4.44. *Let* $\mathbf{Prop}(\Gamma)$ *be the full subcategory of* $\mathbf{Col}(\Gamma)$ *whose objects are families of realized propositions on Γ.*

Observe that $\mathbf{Prop}(\Gamma)$ is a preorder, as a consequence of point 1. in lemma 4.42. Hence we put:

Definition 4.45. *If Γ is an object of* **Cont** *and P and Q are in* $\mathbf{Prop}(\Gamma)$, *we write*

$$P \sqsubseteq^\Gamma Q$$

if there is an arrow in $\mathbf{Prop}(\Gamma)$ *from P to Q. Moreover, if the existing arrow is called f then we may write*

$$f : P \sqsubseteq^\Gamma Q$$

Moreover, observe also that, contrary to what happens in **mTT**, in our indexed category of dependent realized collections it is possible to transform any dependent realized collection into a dependent realized proposition by quotienting it under the trivial relation:

Definition 4.46. *If* Γ *is an object of* **Cont** *and* A *is an object of* $\mathbf{Col}(\Gamma)$*, then the* proof-irrelevant quotient $\mathbf{Pir}(A)$ of A *is the object of* $\mathbf{Prop}(\Gamma)$ *defined by the following conditions:*

1. $x \,\varepsilon\, \mathbf{Pir}(A) \equiv^{def} x \,\varepsilon\, A$

2. $x \sim_{\mathbf{Pir}(A)} y \equiv^{def} x \,\varepsilon\, A \wedge y \,\varepsilon\, A$

Actually, the above operation defines a reflector (see [20] for a definition) from realized collections to propositions:

Lemma 4.47. *For every object* Γ *of* **Cont**

$$\mathbf{Pir} : \mathbf{Col}(\Gamma) \longrightarrow \mathbf{Prop}(\Gamma)$$

defined as $\mathbf{Pir}(A)$ *for any object* A *of* $\mathbf{Col}(\Gamma)$ *and as* $\gamma_{\mathbf{n}} : \mathbf{Pir}(A) \to \mathbf{Pir}(B)$ *for every* $f := \gamma_{\mathbf{n}} : A \to B$ *in* $\mathbf{Col}(\Gamma)$*, is a reflector of the embedding functor of* $\mathbf{Prop}(\Gamma)$ *into* $\mathbf{Col}(\Gamma)$*. This means that there is a bijection*

$$Hom_{\mathbf{Prop}(\Gamma)}(\mathbf{Pir}(A), P) \cong Hom_{\mathbf{Col}(\Gamma)}(A, P)$$

natural in every object A *in* $\mathbf{Col}(\Gamma)$ *and object* P *in* $\mathbf{Prop}(\Gamma)$.

As a consequence, we get that each category of dependent propositions is an Heyting prealgebra:

Corollary 4.48. $\mathbf{Prop}(\Gamma)$ *is an Heyting prealgebra, i. e. it is a preorder with all binary infima and suprema, bottom and top elements and all Heyting implications i. e. it is a cartesian closed preorder category with finite coproducts.*

Proof. In order to show that $\mathbf{Prop}(\Gamma)$ is an Heyting prealgebra it is sufficient to show that it has binary infima and suprema, a bottom element, a top element and Heyting implications. A bottom element is given by $\bot^{\Gamma} := \mathbf{0}^{\Gamma}$, a top element is given by $\top^{\Gamma} := \mathbf{1}^{\Gamma}$, a binary supremum, a binary infimum and a Heyting implication for P and Q in $\mathbf{Prop}(\Gamma)$ are $P \sqcap^{\Gamma} Q := P \times^{\Gamma} Q$, $P \sqcup^{\Gamma} Q := \mathbf{Pir}(P +^{\Gamma} Q)$ and $P \to^{\Gamma} Q := P \Rightarrow^{\Gamma} Q$ respectively. □

Observe that a substitution functor $\mathbf{Prop}_f$ along any arrow f in **Cont** is inherited from that of **Col**:

Lemma 4.49. *Suppose* $f : \Gamma \to \Gamma'$ *in* **Cont** *and suppose* P *is in* $\mathbf{Prop}(\Gamma')$*, then* $\mathbf{Col}_f(P)$ *is in* $\mathbf{Prop}(\Gamma)$*. Moreover,* $\mathbf{Col}_f|_{\mathbf{Prop}(\Gamma')}$ *is a morphism of Heyting prealgebras, i. e. it preserves* $\bot$*,* $\top$*,* $\sqcap$*,* $\sqcup$ *and* $\to$*.*

Hence we are ready to define **Prop** as a *sub-indexed category* of **Col**:

Definition 4.50. *We call* $\mathbf{Prop} : \mathbf{Cont}^{op} \to \mathbf{Cat}$ *the indexed category defined by the assignments* $A \mapsto \mathbf{Prop}(A)$ *and* $f \mapsto \mathbf{Prop}_f := \mathbf{Col}_f|_{\mathbf{Prop}(\mathsf{cod}(f))}$ *where* $\mathsf{cod}(f)$ *denotes the codomain of* f.

Now we describe left and right adjoints to substitution functors which are necessary to interpret existential and universal quantifiers of $\mathbf{mTT}^a$ respectively:

Definition 4.51. *Suppose* Γ *is an object of* **Cont**, A *is an object of* $\mathbf{Col}(\Gamma)$ *and* P *is an object of* $\mathbf{Prop}([\Gamma, A])$, *then*

1. $\exists^{\Gamma}(A, P) := \mathbf{Pir}(\Sigma^{\Gamma}(A, P))$
2. $\forall^{\Gamma}(A, P) := \mathbf{Pir}(\Pi^{\Gamma}(A, P))$

Observe that in $\mathbf{Prop}(\Gamma)$ there are also the propositional equalities of $\mathbf{Col}(\Gamma)$ and these are preserved by substitution functors:

Lemma 4.52. *If* Γ *is an object of* **Cont** *and* $f, g : \mathbf{1}^{\Gamma} \to A$ *in* $\mathbf{Col}(\Gamma)$, *then* $\mathbf{Eq}^{\Gamma}(A, f, g)$ *is an object of* $\mathbf{Prop}(\Gamma)$. *Morerover, if* $\top^{\Gamma} \sqsubseteq^{\Gamma} \mathbf{Eq}^{\Gamma}(A, f, g)$, *then* f *is equal to* g *in* $\mathbf{Col}(\Gamma)$.

Lemma 4.53. *Suppose* $f : \Gamma' \to \Gamma$ *is an arrow of* **Cont**, A *is an object of* $\mathbf{Col}(\Gamma)$ *and* $g, g' : \mathbf{1}^{\Gamma} \to A$ *in* $\mathbf{Col}(\Gamma)$. *Then*

$$\mathbf{Col}_f(\mathbf{Eq}^{\Gamma}(A, g, g')) = \mathbf{Eq}^{\Gamma'}(\mathbf{Col}_f(A), \mathbf{Col}_f(g), \mathbf{Col}_f(g'))$$

Finally, observe that there exist left and right adjoints to substitution functors and that Beck-Chevalley conditions hold for them:

Lemma 4.54. *For every* $f : \Gamma \to \Gamma'$ *in* **Cont**, $\mathbf{Prop}_f : \mathbf{Prop}(\Gamma') \to \mathbf{Prop}(\Gamma)$ *has a left and right adjoint* $\exists_f : \mathbf{Prop}(\Gamma) \to \mathbf{Prop}(\Gamma')$ *and* $\forall_f : \mathbf{Prop}(\Gamma) \to \mathbf{Prop}(\Gamma')$ *respectively, i. e.* $\exists_f$ *and* $\forall_f$ *are preorder morphims for which for every* $P \in \mathbf{Prop}(\Gamma')$ *and* $Q \in \mathbf{Prop}(\Gamma)$:

1. $Q \sqsubseteq^{\Gamma} \mathbf{Prop}_f(P)$ *if and only if* $\exists_f(Q) \sqsubseteq^{\Gamma'} P$,
2. $\mathbf{Prop}_f(P) \sqsubseteq^{\Gamma} Q$ *if and only if* $P \sqsubseteq^{\Gamma'} \forall_f(Q)$.

Moreover these adjoints satisfy the Beck-Chevalley condition: for every pullback square in **Cont**

$$\begin{array}{ccc} \Gamma' & \xrightarrow{f'} & \Delta' \\ {\scriptstyle g'}\downarrow & & \downarrow{\scriptstyle g} \\ \Gamma & \xrightarrow{f} & \Delta \end{array}$$

and for every P in $\mathbf{Prop}(\Delta')$ *the following conditions hold*[7]*:*

1. $\exists_{g'}(\mathbf{Prop}_{f'}(P)) \sqsubseteq^{\Gamma} \mathbf{Prop}_f(\exists_g(P))$ *and* $\mathbf{Prop}_f(\exists_g(P)) \sqsubseteq^{\Gamma} \exists_{g'}(\mathbf{Prop}_{f'}(P))$*;*

2. $\forall_{g'}(\mathbf{Prop}_{f'}(P)) \sqsubseteq^{\Gamma} \mathbf{Prop}_f(\forall_g(P))$ *and* $\mathbf{Prop}_f(\forall_g(P)) \sqsubseteq^{\Gamma} \forall_{g'}(\mathbf{Prop}_{f'}(P))$*;*

Proof. Note that if A is an object of $\mathbf{Col}(\Gamma)$ and P is an object of $\mathbf{Prop}([\Gamma, A])$, then we can define $\exists_{\mathsf{pr}_{[\Gamma,A]}}(P)$ and $\forall_{\mathsf{pr}_{[\Gamma,A]}}(P)$ as the objects $\exists^{\Gamma}(A, P)$ and $\forall^{\Gamma}(A, P)$ defined in 4.51 respectively. □

From lemmas 4.47, 4.49 and 4.54 we conclude

Corollary 4.55. **Prop** *is a hyperdoctrine in the sense of [37] and its posetal reflection is a first-order hyperdoctrine in the sense of [32].*

4.5 Realized sets and small realized propositions

Here we are going to define a notion of realized set and of small realized proposition in order to define a sub-indexed category **Set** of **Col**

$$\mathbf{Set} : \mathbf{Cont}^{op} \to \mathbf{Cat}$$

and a sub-indexed category of **Prop**

$$\mathbf{Prop}_s : \mathbf{Cont}^{op} \to \mathbf{Cat}$$

which will be used to interpret $\mathbf{mTT}^a$-sets and $\mathbf{mTT}^a$-small propositions respectively. In order to interpret the $\mathbf{mTT}^a$-collection of small propositions and that of sets of section 2.3 following the interpretation in [24], both indexed categories will need to enjoy a classifier: the fibres of **Set** will need to be classified by an object US of $\mathbf{Col}([\ ])$ via a natural bijection in Γ

$$\mathbf{Set}(\Gamma) \simeq \mathbf{Cont}(\Gamma, \mathsf{US})$$

and the fibres of $\mathbf{Prop}_s$ will need to be classified by an object USP of $\mathbf{Col}([\ ])$ via a natural bijection in Γ

$$\mathbf{Prop}_s(\Gamma) \simeq \mathbf{Cont}(\Gamma, \mathsf{USP})$$

Actually, we will define both indexed categories by using their classifiers: a realized set depending on Γ will be defined as the dependent realized collection made of elements of a code in US over Γ and, analogously, a small realized proposition will

[7] It is sufficient that one of the two conditions holds as the other follows by adjunction.

be defined as the proof-irrelevant dependent realized collection of elements of a code in USP over Γ. In turn the objects US and USP will be defined as those used in [24] to interpret the collection of sets and the collection of small propositions respectively. Both objects are realized collections according to the terminology used here.

We describe now the construction of US and USP which will make use of fixpoint formulas of $\widehat{ID_1}$ as in [24]. To this purpose, we start by recalling the definition of Kleene realizability for Heyting Arithmetic since it will be used to define the notion of element both of a set and of a proposition.

Definition 4.56. *For every formula φ of HA the formula $x \Vdash_k \varphi$ (x realizes φ) is defined according to the following clauses by external induction on the formation of formulas (x is a variable which is not free in φ).*

1. $x \Vdash_k t = s$ *is* $t = s$
2. $x \Vdash_k (\varphi \wedge \varphi')$ *is* $p_1(x) \Vdash_k \varphi \wedge p_2(x) \Vdash_k \varphi'$
3. $x \Vdash_k (\varphi \vee \varphi')$ *is* $(p_1(x) = 0 \wedge p_2(x) \Vdash_k \varphi) \vee (p_1(x) = 1 \wedge p_2(x) \Vdash_k \varphi')$
4. $x \Vdash_k (\varphi \rightarrow \varphi')$ *is* $\forall t\, (t \Vdash_k \varphi \rightarrow \{x\}(t) \Vdash_k \varphi')$
5. $x \Vdash_k \forall y\, \varphi$ *is* $\forall y\, (\{x\}(y) \Vdash_k \varphi)$
6. $x \Vdash_k \exists y\, \varphi$ *is* $p_2(x) \Vdash_k \varphi[p_1(x)/y]$

Then, we define the following *formulas in $\widehat{ID_1}$ as fixpoints*:

1. $Set(x)$ intended to state that x *is a code for a set of* $\mathbf{mTT}^a$;
2. $x\,\overline{\varepsilon}\,y$ intended to state that x *is an element of the set of* $\mathbf{mTT}^a$ *coded by* y;
3. $x\,\overline{\not\varepsilon}\,y$ intended to state that x *is not an element of the set of* $\mathbf{mTT}^a$ *coded by* y;
4. $x \equiv_z y$ intended to state that x *and* y *are equal elements in the set of* $\mathbf{mTT}^a$ *coded by* z;
5. $x \not\equiv_z y$ intended to state that x *and* y *are not equal elements of the set of* $\mathbf{mTT}^a$ *coded by* z.

We will use such formulas to encode with natural numbers a realizability interpretation of $\mathbf{mTT}^a$-sets in $\widehat{ID_1}$: we use natural numbers to represent both realizers and (codes for) sets and we introduce a membership relation $x\,\overline{\varepsilon}\,z$ between natural numbers (which extends the notion of Kleene realizability) and an equivalence relation $x \equiv_z y$ between numbers (realizers) of a (code of a) set z, which will represent the equality between its elements. These clauses are similar to those presented in Beeson's book ([8]) for the first-order fragment of Martin-Löf type theory with one universe, except that here we need to deal with $\mathbf{mTT}^a$-sets which include an extra notion of proposition, that of small proposition, defined primitively. As in Beeson's book we define also the formal negations $x \not\bar{\varepsilon}\, z$ and $x \not\equiv_z y$ in order to give admissible clauses defining the properties of our new formulas. Then, we will encode the set constructors N_0, N_1, N, Σ, Π, $+$, List, $\bot$, $\wedge$, $\vee$, $\rightarrow$, $\exists$, $\forall$, Eq. In order to mimic the dependency, we define a family of sets on a given set as (a code for) a recursive function defined on the elements of a (code for a) set and producing codes for sets as outputs provided that some coherence requirements are fulfilled. Formally, we introduce the formula $Fam(y, x)$ (*y is a family of sets on the set x*) in order to capture this idea:

$$Fam(y,x) \equiv^{def} Set(x) \wedge \forall t\,(t \not\bar{\varepsilon}\, x \vee Set(\{y\}(t))) \wedge \forall t \forall s\,(t \not\equiv_x s \vee \{y\}(t) =_{ext\#} \{y\}(s))$$

where $x =_{ext\#} y$ is defined as

$$\forall t\,((t\,\bar{\varepsilon}\,x \;\vee\; t \not\bar{\varepsilon}\, y) \;\wedge\; (t\,\bar{\varepsilon}\,y \;\vee\; t \not\bar{\varepsilon}\, x)) \wedge \forall t \forall s\,((t \equiv_x s \;\vee\; t \not\equiv_y s) \;\wedge\; (t \equiv_y s \;\vee\; t \not\equiv_x s))$$

We then declare that *y is a family of small propositions* with the abbreviation $Fam_p(y, x)$ defined formally as

$$Fam(y,x) \wedge \forall t\,(t \not\bar{\varepsilon}\; x \;\vee\; p_1(\{y\}(t)) > 5)$$

where the last condition means that $\{y\}(t)$ is a small proposition (see later explanations).

For every constructor κ the clauses for the definitions of the fixpoint formulas are described by using extra new formulas as follows

1. $Set(\kappa^{\#})$ if $Cond(\kappa)$
2. $x\,\bar{\varepsilon}\,\kappa^{\#}$ if $Cond(\kappa) \wedge P^{\kappa}_{\bar{\varepsilon}}(x)$
3. $x \not\bar{\varepsilon}\, \kappa^{\#}$ if $Cond(\kappa) \wedge \overline{P^{\kappa}_{\bar{\varepsilon}}(x)}$
4. $x \equiv_{\kappa^{\#}} y$ if $Cond(\kappa) \wedge P^{\kappa}_{\bar{\varepsilon}}(x) \wedge P^{\kappa}_{\bar{\varepsilon}}(y) \wedge P^{\kappa}_{\equiv}(x, y)$
5. $x \not\equiv_{\kappa^{\#}} y$ if $Cond(\kappa) \wedge (\overline{P^{\kappa}_{\bar{\varepsilon}}(x)} \vee \overline{P^{\kappa}_{\bar{\varepsilon}}(y)} \vee \overline{P^{\kappa}_{\equiv}(x, y)})$

where for formulas φ in the language of Peano arithmetic enriched with predicate symbols $\bar{\varepsilon}$, $\not\bar{\varepsilon}$, $\equiv$ and $\not\equiv$ and without any occurrence of $\rightarrow$, the formula $\overline{\varphi}$ represents *the negation of ϕ* and is defined according to the following clauses:

1. for primitive formulas ψ of Peano arithmetic $\overline{\psi} := \neg\psi$

2. $\overline{t\,\varepsilon\, u}$ is $t \not\varepsilon\ u$, $\overline{t \not\varepsilon\ u}$ is $t\,\varepsilon\, u$, $\overline{t \equiv_u s}$ is $t \not\equiv_u s$ and $\overline{t \not\equiv_u s}$ is $t \equiv_u s$

3. $\overline{\varphi \wedge \varphi'}$ is $\overline{\varphi} \vee \overline{\varphi'}$ and $\overline{\varphi \vee \varphi'}$ is $\overline{\varphi} \wedge \overline{\varphi'}$

4. $\overline{\exists x\, \varphi}$ is $\forall x\, \overline{\varphi}$ and $\overline{\forall x\, \varphi}$ is $\exists x\, \overline{\varphi}$

Finally, the extra formulas $\kappa^{\#}$, $Cond(\kappa)$, P^{κ}_{ε} and $P^{\kappa}_{\equiv}$ are defined in the following tables. $\kappa^{\#}$ makes explicit the encoding of sets built using the constructor κ. $Cond(\kappa)$ is intended to give the constraints which must be respected to define a set through the constructor κ and finally P^{κ}_{ε} and $P^{\kappa}_{\equiv}$ give the clauses for membership and equality in a set obtained through the constructor κ, respectively.

κ	$\kappa^{\#}$	$Cond(\kappa)$	$P^{\kappa}_{\varepsilon}(x)$
N_0 $\perp$	$p(1,0)$ $p(6,0)$	$\top$	$\perp$
N_1	$p(1,1)$	$\top$	$x = 0$
N	$p(1,2)$	$\top$	$x = x$
List	$p(5,a)$	$Set(a)$	$\forall i\,(i \geq lh(x) \vee (x)_i\,\varepsilon\, a)$
$\wedge$	$p(7,p(a,b))$	$Set(a) \wedge Set(b) \wedge$ $p_1(a) > 5 \wedge p_1(b) > 5$	$p_1(x)\,\varepsilon\, a \wedge p_2(x)\,\varepsilon\, b$
Σ $\exists$	$p(2,p(a,b))$ $p(10,p(a,b))$	$Fam(b,a)$ $Fam_p(b,a)$	$p_1(x)\,\varepsilon\, a \wedge p_2(x)\,\varepsilon\, \{b\}(p_1(x))$

κ	$\kappa^{\#}$	$Cond(\kappa)$	$P^{\kappa}_{\overline{\varepsilon}}(x)$
$\rightarrow$	p(9,p(a,b))	$Set(a) \wedge Set(b) \wedge$ $p_1(a) > 5 \wedge p_1(b) > 5$	$\forall t\,(t \not\overline{\varepsilon}\, a \vee \{x\}(t) \,\overline{\varepsilon}\, b)$
Π $\forall$	$p(3, p(a, b))$ $p(11, p(a, b))$	$Fam(b, a)$ $Fam_p(b, a)$	$\forall t\,(t \not\overline{\varepsilon}\, a \vee \{x\}(t) \,\overline{\varepsilon}\, \{b\}(t)) \wedge$ $\forall t \forall s (t \not\equiv_a s \vee \{x\}(t) \equiv_{\{b\}(t)} \{x\}(s))$
$+$ $\vee$	$p(4, p(a, b))$ $p(8, p(a, b))$	$Set(a) \wedge Set(b)$ $Set(a) \wedge Set(b) \wedge$ $p_1(a) > 5 \wedge p_1(b) > 5$	$(p_1(x) = 0 \wedge p_2(x) \,\overline{\varepsilon}\, a) \vee$ $(p_1(x) = 1 \wedge p_2(x) \,\overline{\varepsilon}\, b)$
Eq	$p(12, p(a, p(b, c)))$	$Set(a) \wedge b \,\overline{\varepsilon}\, a \wedge c \,\overline{\varepsilon}\, a$	$b \equiv_a c$

κ	$P^{\kappa}_{\overline{\equiv}}(x)$
N_0	$\bot$
N_1, N	$x = y$
List	$lh(x) = lh(y) \wedge \forall i\,(i \geq lh(x) \vee (x)_i \equiv_a (y)_i)$
Σ	$p_1(x) \equiv_a p_1(y) \wedge p_2(x) \equiv_{\{b\}(p_1(x))} p_2(y)$
Π	$\forall t\,(t \not\overline{\varepsilon}\, a \vee \{x\}(t) \equiv_{\{b\}(t)} \{y\}(t))$
$+$	$p_1(x) = p_1(y) \wedge$ $((p_1(x) = 0 \wedge p_2(x) \equiv_a p_2(y)) \vee (p_1(x) = 1 \wedge p_2(x) \equiv_b p_2(y)))$
$\bot, \wedge,$ $\vee, \rightarrow,$ $\exists, \forall,$ Eq	$\top$

Note that, as already anticipated, families x of small propositions are characterized as those families of sets satisfying the condition $p_1(x) > 5$.

To be more precise, first we define an admissible formula $\varphi(x, X)$ as follows

$$\varphi(x,X) \equiv^{def} \bigvee_\kappa \exists a\, \exists b\, \exists c\, ($$

$$(x = p(20,\kappa) \wedge \mathbf{Cond}(\kappa)) \vee$$
$$\exists y\, (x = p(21, p(y, \kappa^{\#})) \wedge \mathbf{Cond}(\kappa) \wedge \mathbf{P}^{\kappa}_{\overline{\varepsilon}}(y)) \vee$$
$$\exists y\, (x = p(22, p(y, \kappa^{\#})) \wedge \mathbf{Cond}(\kappa) \wedge \overline{\mathbf{P}^{\kappa}_{\overline{\varepsilon}}(y)}) \vee$$
$$\exists y\, \exists z\, (x = p(23, p(\kappa^{\#}, p(y,z))) \wedge \mathbf{Cond}(\kappa) \wedge \mathbf{P}^{\kappa}_{\overline{\varepsilon}}(y) \wedge \mathbf{P}^{\kappa}_{\overline{\varepsilon}}(z) \wedge \mathbf{P}^{\kappa}_{\overline{\equiv}}(y,z)) \vee$$
$$\exists y\, \exists z\, (x = p(24, p(\kappa^{\#}, p(y,z))) \wedge \mathbf{Cond}(\kappa) \wedge (\overline{\mathbf{P}^{\kappa}_{\overline{\varepsilon}}(y)} \vee \overline{\mathbf{P}^{\kappa}_{\overline{\varepsilon}}(z)} \vee \overline{\mathbf{P}^{\kappa}_{\overline{\equiv}}(y,z)})))$$

where the disjunction $\bigvee_\kappa$ is the finite disjunction indexed by the constructors κ in the previous tables and where the boldface versions of $Cond(\kappa)$, $P^{\kappa}_{\overline{\varepsilon}}(x)$ and $P^{\kappa}_{\overline{\equiv}}(y, z)$ are obtained by substituting $Set(u)$, $t \,\overline{\varepsilon}\, u$, $t \not\overline{\varepsilon}\, u$, $s \equiv_u t$ and $s \not\equiv_u t$ with $p(20, u) \,\varepsilon\, X$, $p(21, p(t, u)) \,\varepsilon\, X$, $p(22, p(t, u)) \,\varepsilon\, X$, $p(23, p(u, p(s, t))) \,\varepsilon\, X$ and $p(24, p(u, p(s, t))) \,\varepsilon\, X$ respectively in the original formulas.

For the sake of example let us show what is the subformula of $\varphi(x, X)$ corresponding to κ equal to N:

$$\exists a\, \exists b\, \exists c\, ((x = p(20, p(1,2)) \wedge \top) \vee$$
$$\exists y\, (x = p(21, p(y, p(1,2))) \wedge \top \wedge y = y) \vee$$
$$\exists y\, (x = p(22, p(y, p(1,2))) \wedge \top \wedge \neg y = y) \vee$$
$$\exists y\, \exists z\, (x = p(23, p(p(1,2), p(y,z))) \wedge \top \wedge y = y \wedge z = z \wedge y = z) \vee$$
$$\exists y\, \exists z\, (x = p(24, p(p(1,2), p(y,z))) \wedge \top \wedge (\neg y = y \vee \neg z = z \vee \neg y = z)))$$

Then we consider the fixpoint formula $P_\varphi(x)$ corresponding to $\varphi(x, X)$ and we define

1. $Set(x) \equiv^{def} P_\varphi(p(20, x))$
2. $x \,\overline{\varepsilon}\, y \equiv^{def} P_\varphi(p(21, p(x, y)))$
3. $x \not\overline{\varepsilon}\, y \equiv^{def} P_\varphi(p(22, p(x, y)))$
4. $x \equiv_z y \equiv^{def} P_\varphi(p(23, p(z, p(x, y))))$
5. $x \not\equiv_z y \equiv^{def} P_\varphi(p(24, p(z, p(x, y))))$

In this framework we can define the following formulas:

- $Coh(x) \equiv^{def} \forall t\,(t\,\bar{\varepsilon}\,x \leftrightarrow \neg(t \not\bar{\varepsilon}\, x)) \wedge \forall t \forall s\,(t \equiv_x s \leftrightarrow \neg(t \not\equiv_x s))$ stating that the formulas $t \not\bar{\varepsilon}\, x$ and $t \not\equiv_x s$ defined by fixpoint behave really like negations of $t\,\bar{\varepsilon}\,x$ and $t \equiv_x s$ respectively;
- $Wd(x) \equiv^{def} \forall t \forall s\,(t \equiv_x s \rightarrow t\,\bar{\varepsilon}\,x \wedge s\,\bar{\varepsilon}\,x)$ stating that the relation $t \equiv_x s$ is well defined on x;
- $Ref(x) \equiv^{def} \forall t\,(t\,\bar{\varepsilon}\,x \leftrightarrow t \equiv_x t)$ stating that the relation $t \equiv_x s$ is reflexive on x;
- $Sym(x) \equiv^{def} \forall t \forall s\,(t \equiv_x s \rightarrow s \equiv_x t)$ stating that the relation $t \equiv_x s$ is symmetric;
- $Tra(x) \equiv^{def} \forall t \forall s \forall u\,(t \equiv_x s \wedge s \equiv_x u \rightarrow t \equiv_x u)$ stating that the relation $t \equiv_x s$ is transitive;
- $EqR(x) \equiv^{def} Ref(x) \wedge Sym(x) \wedge Tra(x)$ stating that the relation $t \equiv_x s$ is an equivalence relation on x;
- $PrIrr(x) \equiv^{def} \forall t \forall s\,(t\,\bar{\varepsilon}\,x \wedge s\,\bar{\varepsilon}\,x \leftrightarrow t \equiv_x s)$ stating that the relation $t \equiv_x s$ is trivial on x (i. e. x is proof-irrelevant);
- $x =_{ext} y \equiv^{def} \forall t \forall s\,(t \equiv_x s \leftrightarrow t \equiv_y s)$ stating that two sets are defined *extensionally equal* if they share the same equivalent (equal) elements.

Notice that the following hold:

- $\widehat{ID_1} \vdash Ref(x) \wedge Ref(y) \wedge x =_{ext} y \rightarrow \forall t\,(t\,\bar{\varepsilon}\,x \leftrightarrow t\,\bar{\varepsilon}\,y)$ namely, two reflexive sets are *extensionally equal* if and only if they *share the same elements*;
- $\widehat{ID_1} \vdash PrIrr(x) \rightarrow EqR(x)$
- $\widehat{ID_1} \vdash EqR(x) \rightarrow Wd(x)$

4.5.1 The classifier of realized sets and that of small realized propositions

Definition 4.57. *We define the collection* US *as the universe of codes for sets with extensional equality:*

- $|\mathsf{US}| := \{x \mid Set(x) \wedge Coh(x) \wedge EqR(x)\}$

- $x \sim_{\mathsf{US}} y$ *is* $x \,\varepsilon\, \mathsf{US} \wedge y \,\varepsilon\, \mathsf{US} \wedge x =_{ext} y$

Definition 4.58. *We define the collection* USP *as the universe of codes for small propositions:*

- $|\mathsf{USP}| := \{x \,|\, Set(x) \wedge p_1(x) > 5 \wedge Coh(x) \wedge PrIrr(x)\}$
- $x \sim_{\mathsf{USP}} y$ *is* $x \,\varepsilon\, \mathsf{USP} \wedge y \,\varepsilon\, \mathsf{USP} \wedge x =_{ext} y$

Definition 4.59. *For every object* Γ *in* **Cont** *we define the following families of collections in* $\mathbf{Col}(\Gamma)$*:*

$$\mathsf{US}^{\Gamma} := \mathbf{Col}_{!_{\Gamma,[\,]}}(\mathsf{US}) \qquad\qquad \mathsf{USP}^{\Gamma} := \mathbf{Col}_{!_{\Gamma,[\,]}}(\mathsf{USP})$$

Definition 4.60. *For any object* Γ *of* **Cont** *we define* τ^{Γ} *as the collection depending on* $\overline{x}_{\ell(\Gamma)+1}$ *determined by the following conditions:*

- $x \,\varepsilon\, \tau^{\Gamma}$ *is* $x \,\overline{\varepsilon}\, x_{\ell(\Gamma)+1} \wedge \overline{x}_{|\ell(\Gamma)} \,\varepsilon\, \Gamma \wedge x_{\ell(\Gamma)+1} \,\varepsilon\, \mathsf{US}$
- $x \sim_{\tau^{\Gamma}} y$ *is* $x \equiv_{x_{\ell(\Gamma)+1}} y \wedge \overline{x}_{|\ell(\Gamma)} \,\varepsilon\, \Gamma \wedge x_{\ell(\Gamma)+1} \,\varepsilon\, \mathsf{US}$

Lemma 4.61. *Suppose* Γ *is an object of* **Cont**. *Then* τ^{Γ} *is an object of* $\mathbf{Col}([\Gamma, \mathsf{US}^{\Gamma}])$ *and the following are well defined arrows in* $\mathbf{Col}(\Gamma)$*:*

- $\widehat{\mathsf{N}_0}^{\Gamma} := \gamma_{\Lambda \overline{x}_{|\ell(\Gamma)}.\Lambda x.\{\mathbf{p}\}(1,0)}$ *and* $\widehat{\mathsf{N}_1}^{\Gamma} := \gamma_{\Lambda \overline{x}_{|\ell(\Gamma)}.\Lambda x.\{\mathbf{p}\}(1,1)}$ *and* $\widehat{\mathsf{N}}^{\Gamma} := \gamma_{\Lambda \overline{x}_{|\ell(\Gamma)}.\Lambda x.\{\mathbf{p}\}(1,2)}$ *from* $\mathbf{1}^{\Gamma}$ *to* US^{Γ}
- $\widehat{\Sigma}^{\Gamma} := \gamma_{\Lambda \overline{x}_{|\ell(\Gamma)}.\Lambda x.\{\mathbf{p}\}(2,x)}$ *and* $\widehat{\Pi}^{\Gamma} := \gamma_{\Lambda \overline{x}_{|\ell(\Gamma)}.\Lambda x.\{\mathbf{p}\}(3,x)}$ *from* $\Sigma^{\Gamma}(\mathsf{US}^{\Gamma}, \tau^{\Gamma} \Rightarrow^{[\Gamma,\mathsf{US}^{\Gamma}]} \mathsf{US}^{[\Gamma,\mathsf{US}^{\Gamma}]})$ *to* US^{Γ}
- $\widehat{+}^{\Gamma} := \gamma_{\Lambda \overline{x}_{|\ell(\Gamma)}.\Lambda x.\{\mathbf{p}\}(4,x)}$ *from* $\mathsf{US}^{\Gamma} \times^{\Gamma} \mathsf{US}^{\Gamma}$ *to* US^{Γ}
- $\widehat{\mathsf{List}}^{\Gamma} := \gamma_{\Lambda \overline{x}_{|\ell(\Gamma)}.\Lambda x.\{\mathbf{p}\}(5,x)}$ *from* US^{Γ} *to* US^{Γ}
- $\widehat{\bot}^{\Gamma} := \gamma_{\Lambda \overline{x}_{|\ell(\Gamma)}.\Lambda x.\{\mathbf{p}\}(6,0)} : \mathbf{1}^{\Gamma} \to \mathsf{USP}^{\Gamma}$
- $\widehat{\wedge}^{\Gamma} := \gamma_{\Lambda \overline{x}_{|\ell(\Gamma)}.\Lambda x.\{\mathbf{p}\}(7,x)}$ *and* $\widehat{\vee}^{\Gamma} := \gamma_{\Lambda \overline{x}_{|\ell(\Gamma)}.\Lambda x.\{\mathbf{p}\}(8,x)}$ *and* $\widehat{\Rightarrow}^{\Gamma} := \gamma_{\Lambda \overline{x}_{|\ell(\Gamma)}.\Lambda x.\{\mathbf{p}\}(9,x)}$ *from* $\mathsf{USP}^{\Gamma} \times^{\Gamma} \mathsf{USP}^{\Gamma}$ *to* USP^{Γ}

- $\widehat{\exists}^{\Gamma} := \gamma_{\Lambda \overline{x}_{|\ell(\Gamma)}.\Lambda x.\{\mathbf{p}\}(10,x)}$ *and* $\widehat{\forall}^{\Gamma} := \gamma_{\Lambda \overline{x}_{|\ell(\Gamma)}.\Lambda x.\{\mathbf{p}\}(11,x)}$ *from* $\Sigma^{\Gamma}(\mathsf{US}^{\Gamma}, \tau^{\Gamma} \Rightarrow^{[\Gamma,\mathsf{US}^{\Gamma}]} \mathsf{USP}^{[\Gamma,\mathsf{US}^{\Gamma}]})$ *to* USP^{Γ}
- $\widehat{\mathsf{Eq}}^{\Gamma} := \gamma_{\Lambda \overline{x}_{|\ell(\Gamma)}.\Lambda x.\{\mathbf{p}\}(12,x)}$ *from* $\Sigma^{\Gamma}(\mathsf{US}^{\Gamma}, \tau^{\Gamma} \times^{[\Gamma,\mathsf{US}^{\Gamma}]} \tau^{\Gamma})$ *to* USP^{Γ}
- $\sigma^{\Gamma} := \gamma_{\Lambda \overline{x}_{|\ell(\Gamma)}.\Lambda x.x}$ *from* USP^{Γ} *to* US^{Γ}

4.6 Dependent realized sets and small realized propositions and their indexed categories

Here we finally give the definitions of realized sets and of small realized propositions by using their classifiers.

Note that, for any context Γ, any generalized element of US over Γ (i. e. an arrow from Γ to US) in **Cont**, or equivalently any global element of US^{Γ} in $\mathbf{Col}(\Gamma)$ gives rise to a realized collection in $\mathbf{Col}(\Gamma)$:

Lemma 4.62. *Let Γ be an object of* **Cont** *with $\ell(\Gamma) = n$.*

Suppose $f = \gamma_{\mathbf{n}_f} : \mathbf{1}^{\Gamma} \to \mathsf{US}^{\Gamma}$ (recall the notation in lemma 4.16) in $\mathbf{Col}(\Gamma)$. Then the collection $\tau_s^{\Gamma}(f)$ of $\widehat{ID_1}$ depending on $\overline{x}_{|n}$ defined by

1. $x \,\varepsilon\, \tau_s^{\Gamma}(f) \equiv^{def} \overline{x}_{|n} \,\varepsilon\, \Gamma \wedge x \,\overline{\varepsilon}\, \{\mathbf{n}_f\}(\overline{x}_{|n}, 0)$
2. $x \sim_{\tau_s^{\Gamma}(f)} y \equiv^{def} \overline{x}_{|n} \,\varepsilon\, \Gamma \wedge x \equiv_{\{\mathbf{n}_f\}(\overline{x}_{|n},0)} y$

is a well defined object of $\mathbf{Col}(\Gamma)$. Moreover, for arrows $f, g : \mathbf{1}^{\Gamma} \to \mathsf{US}^{\Gamma}$ in $\mathbf{Col}(\Gamma)$, if $\tau_s^{\Gamma}(f)$ is equal to $\tau_s^{\Gamma}(g)$, then f and g are equal arrows in $\mathbf{Col}(\Gamma)$.

Note that any global element of USP^{Γ} in $\mathbf{Col}(\Gamma)$, or equivalently any generalized element of USP over Γ in **Cont**, gives rise to a realized proposition in $\mathbf{Prop}(\Gamma)$:

Lemma 4.63. *Let Γ be an object of* **Cont** *with $\ell(\Gamma) = n$.*

Suppose $f = \gamma_{\mathbf{n}_f} : \mathbf{1}^{\Gamma} \to \mathsf{USP}^{\Gamma}$, then the collection $\tau_{sp}^{\Gamma}(f)$ of $\widehat{ID_1}$ depending on $\overline{x}_{|n}$ defined by

1. $x \,\varepsilon\, \tau_{sp}^{\Gamma}(f) \equiv^{def} \overline{x}_{|n} \,\varepsilon\, \Gamma \wedge x \,\overline{\varepsilon}\, \{\mathbf{n}_f\}(\overline{x}_{|n}, 0)$
2. $x \sim_{\tau_{sp}^{\Gamma}(f)} y \equiv^{def} \overline{x}_{|n} \,\varepsilon\, \Gamma \wedge x \equiv_{\{\mathbf{n}_f\}(\overline{x}_{|n},0)} y$

is a well defined object of $\mathbf{Prop}(\Gamma)$. Moreover, for arrows $f, g : \mathbf{1}^{\Gamma} \to \mathsf{USP}^{\Gamma}$ in $\mathbf{Col}(\Gamma)$, if $\tau_{sp}^{\Gamma}(f)$ is equal to $\tau_{sp}^{\Gamma}(g)$, then f and g are equal arrows in $\mathbf{Col}(\Gamma)$.

Finally, we define a realized set depending on a context Γ as the collection of elements of a global element of US^Γ in $\mathbf{Col}(\Gamma)$ or, equivalently, of a generalized element of US over Γ in $\mathbf{Cont}$:

Definition 4.64. *If Γ is an object of* $\mathbf{Cont}$*, a* realized set depending on Γ *(or a* family of realized sets on Γ*) is a realized collection of the form $\tau_s^\Gamma(f)$ for an arrow $f : \mathbf{1}^\Gamma \to \mathsf{US}^\Gamma$ in $\mathbf{Col}(\Gamma)$.*

Analogously, we define a small realized proposition depending on a context Γ as the collection of elements of a global element of USP^Γ in $\mathbf{Col}(\Gamma)$ or, equivalently, of a generalized element of USP over Γ in $\mathbf{Cont}$:

Definition 4.65. *If Γ is an object of* $\mathbf{Cont}$*, a* small realized proposition depending on Γ *(or a* family of realized small propositions on Γ*) is a realized collection of the form $\tau_{sp}^\Gamma(f)$ for an arrow $f : \mathbf{1}^\Gamma \to \mathsf{USP}^\Gamma$ in $\mathbf{Col}(\Gamma)$.*

Now we are ready to define the indexed category of realized sets and that of small realized propositions. We start by defining their fibres as follows:

Definition 4.66. *If Γ is an object of $\mathbf{Cont}$, we define $\mathbf{Set}(\Gamma)$ as the full subcategory of $\mathbf{Col}(\Gamma)$ whose objects are realized sets depending on Γ.*

Moreover, if A is an object of $\mathbf{Set}(\Gamma)$, we write $\mathbf{en}_s^\Gamma(A)$ for the arrow satisfying $A = \tau_s^\Gamma(\mathbf{en}_s^\Gamma(A))$.

Definition 4.67. *If Γ is an object of $\mathbf{Cont}$, we define $\mathbf{Prop}_s(\Gamma)$ as the full subcategory of $\mathbf{Col}(\Gamma)$ whose objects are small realized propositions depending on Γ.*

The substitution functors for $\mathbf{Set}$ and $\mathbf{Prop}_s$ will be both inherited from those of $\mathbf{Col}$:

Lemma 4.68. *If $f : \Gamma' \to \Gamma$ in $\mathbf{Cont}$ and A is an object of $\mathbf{Set}(\Gamma)$ (resp. of $\mathbf{Prop}_s(\Gamma)$), then $\mathbf{Col}_f(A)$ is an object of $\mathbf{Set}(\Gamma')$ (resp. of $\mathbf{Prop}_s(\Gamma')$).*

Definition 4.69. *The pair of assignments*

$$\Gamma \mapsto \mathbf{Set}(\Gamma) \qquad f \mapsto \mathbf{Set}_f := \mathbf{Col}_f|_{\mathbf{Set}(\mathsf{cod}(f))}$$

where $\mathsf{cod}(f)$ is the codomain of f, defines an indexed category,

$$\mathbf{Set} : \mathbf{Cont}^{op} \to \mathbf{Cat}$$

Definition 4.70. *The pair of assignments*

$$\Gamma \mapsto \mathbf{Prop}_s(\Gamma) \qquad f \mapsto \mathbf{Prop}_{s,f} := \mathbf{Col}_f|_{\mathbf{Prop}_s(\mathsf{cod}(f))}$$

where $\mathsf{cod}(f)$ is the codomain of f, defines an indexed category

$$\mathbf{Prop}_s : \mathbf{Cont}^{op} \to \mathbf{Cat}$$

The indexed category of small realized propositions is a sub-indexed category of that of realized sets:

Lemma 4.71. *If* Γ *is an object of* **Cont***, every object in* $\mathbf{Prop}_s(\Gamma)$ *is also in* $\mathbf{Set}(\Gamma)$.

The following lemma is instrumental to show that the fibres of **Set** and of $\mathbf{Prop}_s$ are closed under finite limits, finite coproducts, function spaces and under left and right adjoints to substitution functors along morphisms of the kind $\mathsf{pr}_{[\Gamma,A]}$ for any A in $\mathbf{Set}(\Gamma)$ (they are not closed under left and right adjoints to substitution functors along any morphism of **Cont** for predicativity reasons!). Moreover, from this lemma it also follows that each fibre of **Set** is closed under list objects and contains the natural numbers object of the corresponding fibre of **Col**.

Lemma 4.72. *Let* Γ *be an object of* **Cont**. *Then*

1. $\bot^\Gamma$ *is in* $\mathbf{Prop}_s(\Gamma)$;
2. $\mathbf{0}^\Gamma$, $\mathbf{1}^\Gamma$ *and* $\mathbf{N}^\Gamma$ *are in* $\mathbf{Set}(\Gamma)$;
3. *if* A *and* B *are in* $\mathbf{Set}(\Gamma)$, *then* $A \times^\Gamma B$, $A +^\Gamma B$ *and* $A \Rightarrow^\Gamma B$ *are in* $\mathbf{Set}(\Gamma)$;
4. *if* A *is in* $\mathbf{Set}(\Gamma)$, *then* $\mathbf{List}^\Gamma(A)$ *is in* $\mathbf{Set}(\Gamma)$;
5. *if* A *is in* $\mathbf{Set}(\Gamma)$ *and* B *is in* $\mathbf{Set}([\Gamma, A])$, *then* $\Pi^\Gamma(A, B)$ *and* $\Sigma^\Gamma(A, B)$ *are in* $\mathbf{Set}(\Gamma)$;
6. *if* P *and* Q *are in* $\mathbf{Prop}_s(\Gamma)$, *then* $P \sqcup^\Gamma Q$, $P \sqcap^\Gamma Q$ *and* $P \to^\Gamma Q$ *are in* $\mathbf{Prop}_s(\Gamma)$;
7. *if* A *is in* $\mathbf{Set}(\Gamma)$ *and* P *is in* $\mathbf{Prop}_s([\Gamma, A])$, *then* $\forall^\Gamma(A, P)$ *and* $\exists^\Gamma(A, P)$ *are in* $\mathbf{Prop}_s(\Gamma)$;
8. *if* A *is in* $\mathbf{Set}(\Gamma)$, *for arrows* $f, g : \mathbf{1}^\Gamma \to A$ *in* $\mathbf{Col}(\Gamma)$, *then* $\mathbf{Eq}^\Gamma(A, f, g)$ *is in* $\mathbf{Prop}_s(\Gamma)$.

The following lemma will be useful to validate the equality rules of the collection of small propositions in $\mathbf{mTT}^a$.

Lemma 4.73. *Let* Γ *be an object of* **Cont**, *let* $f : \mathbf{1} \to \mathsf{US}^\Gamma$, $p, p' : \mathbf{1} \to \mathsf{USP}^\Gamma$ *and* $g, g' : \mathbf{1} \to \tau_s^\Gamma(f)$ *be arrows in* $\mathbf{Col}(\Gamma)$ *and let* $h : \mathbf{1} \to \mathsf{USP}^{[\Gamma, \tau_s^\Gamma(f)]}$ *be an arrow in* $\mathbf{Col}([\Gamma, \tau_s^\Gamma(f)])$. *Then in* $\mathbf{Prop}_s(\Gamma)$ *(recall the notation in lemma 4.40):*

1. $\tau_{sp}^\Gamma(\widehat{\bot}^\Gamma)$ *coincides with* $\bot^\Gamma$;

2. $\tau_{sp}^{\Gamma}(\widehat{\wedge}^{\Gamma} \circ \langle p, p' \rangle)$ *coincides with* $\tau_{sp}^{\Gamma}(p) \sqcap^{\Gamma} \tau_{sp}^{\Gamma}(p')$;

3. $\tau_{sp}^{\Gamma}(\widehat{\vee}^{\Gamma} \circ \langle p, p' \rangle)$ *coincides with* $\tau_{sp}^{\Gamma}(p) \sqcup^{\Gamma} \tau_{sp}^{\Gamma}(p')$;

4. $\tau_{sp}^{\Gamma}(\widehat{\Rightarrow}^{\Gamma} \circ \langle p, p' \rangle)$ *coincides with* $\tau_{sp}^{\Gamma}(p) \rightarrow^{\Gamma} \tau_{sp}^{\Gamma}(p')$;

5. $\tau_{sp}^{\Gamma}(\widehat{\exists}^{\Gamma} \circ \langle f, \mathsf{Cur}(h_{\natural}^{\Gamma} \circ \pi_2^{\mathbf{1}, \tau_s^{\Gamma}(f)}) \rangle_{\Sigma})$ *coincides with* $\exists^{\Gamma}(\tau_s^{\Gamma}(f), \tau_{sp}^{[\Gamma, \tau_s^{\Gamma}(f)]}(h))$;

6. $\tau_{sp}^{\Gamma}(\widehat{\forall}^{\Gamma} \circ \langle f, \mathsf{Cur}(h_{\natural}^{\Gamma} \circ \pi_2^{\mathbf{1}, \tau_s^{\Gamma}(f)}) \rangle_{\Sigma})$ *coincides with* $\forall^{\Gamma}(\tau_s^{\Gamma}(f), \tau_{sp}^{[\Gamma, \tau_s^{\Gamma}(f)]}(h))$;

7. $\tau_{sp}^{\Gamma}(\widehat{\mathsf{Eq}}^{\Gamma} \circ \langle f, \langle g, g' \rangle \rangle_{\Sigma})$ *coincides with* $\mathsf{Eq}(\tau_s^{\Gamma}(f), g, g')$;

4.7 Structure of the effective pretripos

What shown so far, together with well known results in categorical logic (see [32], [37], [26]), allows to prove the following:

Theorem 4.74. *The functor*

$$\mathbf{Col} : \mathbf{Cont}^{op} \rightarrow \mathbf{Cat}$$

is an indexed category whose fibres $\mathbf{Col}(\Gamma)$ *for any* Γ *in* **Cont** *are finitely complete cartesian closed categories with finite coproducts and list objects. Moreover, for any morphism* f *in* **Cont** *the substitution functor* $\mathbf{Col}_f$ *preserves the mentioned fibre structure and it has both left and right adjoints satisfying Beck-Chevalley conditions. Finally, the fibre on the terminal object* $\mathbf{Col}([\,])$ *is equivalent to* **Cont** *itself making it a locally cartesian closed category.*

The functor

$$\mathbf{Prop} : \mathbf{Cont}^{op} \rightarrow \mathbf{Cat}$$

is an indexed full subcategory of **Col** *which is also a hyperdoctrine according to the notion defined in [37], and its posetal reflection it is a first order hyperdoctrine in the sense of [32].*

The functor

$$\mathbf{Set} : \mathbf{Cont}^{op} \rightarrow \mathbf{Cat}$$

is an indexed full subcategory of **Col** *whose fibres are also finitely complete cartesian closed categories with finite coproducts and list objects. Moreover, for any morphism* f *in* **Cont**, *the substitution functor* $\mathbf{Set}_f$ *preserves the mentioned fibre structure and, for any* $f = \mathsf{pr}_{[\Gamma, A]}$ *with* A *in* $\mathbf{Set}(\Gamma)$, *it has both left and right adjoints satisfying Beck-Chevalley conditions.*

The functor

$$\mathbf{Prop}_s : \mathbf{Cont}^{op} \to \mathbf{Cat}$$

is an indexed full subcategory both of **Prop** *and of* **Set** *whose fibres are Heyting pre-algebras. Moreover, for any morphism f in* **Cont**, *the substitution functor* $(\mathbf{Prop}_s)_f$ *preserves the mentioned fibre structure and for any $f = \mathsf{pr}_{[\Gamma,A]}$ with A in* $\mathbf{Set}(\Gamma)$ *it has both left and right adjoints satisfying Beck-Chevalley conditions.*

Furthermore, for every object Γ in **Cont** *the object* US *allows to represent the functor* **Set** *in the sense that there is a bijection*

$$\mathbf{Cont}(\Gamma, \mathsf{US}) \simeq \mathbf{Set}(\Gamma)$$

natural in Γ and the object USP *allows to represent* $\mathbf{Prop}_s$ *in the sense that there is a bijection*

$$\mathbf{Cont}(\Gamma, \mathsf{USP}) \simeq \mathbf{Prop}_s(\Gamma)$$

natural in Γ.

Finally all the embeddings in the below diagram preserve the relevant mentioned structures of each indexed category:

$$\begin{array}{ccc} \mathbf{Set} & \hookrightarrow & \mathbf{Col} \\ \uparrow & & \uparrow \\ \mathbf{Prop}_s & \hookrightarrow & \mathbf{Prop} \end{array}$$

Definition 4.75. *The 5-tuple* $(\mathbf{Cont}, \mathbf{Col}, \mathbf{Set}, \mathbf{Prop}, \mathbf{Prop}_s)$ *is called the* effective pretripos for **mTT**.

We will see later that the principle of formal Church thesis will be validated in the effective pretripos for **mTT**.

5 The interpretation of the Minimalist Foundation

Here we give a partial interpretation $\mathcal{I}$ of precontexts and of types and terms in precontext of the fully annotated syntax of $\mathbf{mTT}^a$ in our effective pretripos for **mTT** following Streicher's technique in [38]. We call the resulting model $\mathcal{R}$.

Definition 5.1. *The validity of judgements in the model ($\mathcal{R} \vDash J$) is defined as follows:*

$\mathcal{R} \models$	*if*
$\Gamma\, context$	$\mathcal{I}(\Gamma)$ *is well defined and is an object of* **Cont**
$B\, col\, [\Gamma]$	$\mathcal{R} \models \Gamma\, context$ *and* $\mathcal{I}(B\, [\Gamma])$ *is a well defined object of* $\mathbf{Col}(\mathcal{I}(\Gamma))$
$A\, set\, [\Gamma]$	$\mathcal{R} \models A\, col\, [\Gamma]$ *and* $\mathcal{I}(A\, [\Gamma])$ *is an object of* $\mathbf{Set}(\mathcal{I}(\Gamma))$
$\phi\, prop\, [\Gamma]$	$\mathcal{R} \models \phi\, col\, [\Gamma]$ *and* $\mathcal{I}(\phi\, [\Gamma])$ *is an object of* $\mathbf{Prop}(\mathcal{I}(\Gamma))$
$\phi\, prop_s\, [\Gamma]$	$\mathcal{R} \models \phi\, col\, [\Gamma]$ *and* $\mathcal{I}(\phi\, [\Gamma])$ *is an object of* $\mathbf{Prop}_s(\mathcal{I}(\Gamma))$
$A = B\ type\, [\Gamma]$ $type \in$ $\{col, set, prop, prop_s\}$	$\mathcal{R} \models A\, type$ *and* $\mathcal{R} \models B\, type$ *and* $\mathcal{I}(A[\Gamma])$ *and* $\mathcal{I}(B[\Gamma])$ *are equal objects of* $\mathbf{Col}(\mathcal{I}(\Gamma))$
$a \in A\, [\Gamma]$	$\mathcal{R} \models A\, col\, [\Gamma]$ *and* $\mathcal{I}(a[\Gamma])$ *is well defined* *and* $\mathcal{I}(a[\Gamma]) : \mathbf{1}^{\mathcal{I}(\Gamma)} \to \mathcal{I}(A[\Gamma])$ *is in* $\mathbf{Col}(\mathcal{I}(\Gamma))$
$a = b \in A\, [\Gamma]$	$\mathcal{R} \models a \in A$ *and* $\mathcal{R} \models b \in A$ *and* $\mathcal{I}(a[\Gamma])$ *and* $\mathcal{I}(b[\Gamma])$ *are equal arrows of* $\mathbf{Col}(\mathcal{I}(\Gamma))$

Definition 5.2. *If* $\mathbf{mTT}^a \vdash \phi\, prop\, [\Gamma]$ *we will say that* $\mathcal{R}$ *validates* ϕ *in context* Γ*, also written* $\mathcal{R} \models \phi\, [\Gamma]$*, if we have that* $\mathcal{R} \models \phi\, prop\, [\Gamma]$ *and* $\top^{\mathcal{I}(\Gamma)} \sqsubseteq^{\mathcal{I}(\Gamma)} \mathcal{I}(\phi[\Gamma])$ *in* $\mathbf{Prop}(\mathcal{I}(\Gamma))$.

In the next subsections we will omit superscripts and subscripts in the categorical notation when they will be clear from the context.

5.1 Precontexts

We interpret precontexts as objects of **Cont** as follows:

$\mathcal{I}([\,]) := [\,] \in Ob(\mathbf{Cont})$;

$\mathcal{I}([\Gamma, x \in A]) := [\mathcal{I}(\Gamma), \mathcal{I}(A[\Gamma])]$ provided that $\mathcal{I}(\Gamma)$ is a well defined object of **Cont** and $\mathcal{I}(A[\Gamma])$ is a well defined object of $\mathbf{Col}(\Gamma)$.

5.2 Variables

If $\Gamma := [x_1 \in A_1, ..., x_n \in A_n]$, then variables in context are defined as arrows in $\mathbf{Col}(\Gamma)$ as follows

$$1 \xrightarrow{\mathcal{I}(x_1 \in A_1[\Gamma]) := \gamma_{\pi_1^{n+1}}} \mathbf{Col}_{\mathsf{pr}^{(n)}_{\mathcal{I}(\Gamma)}}(\mathcal{I}(A_1[\,]))$$

$$1 \xrightarrow{\mathcal{I}(x_{i+1} \in A_{i+1}[\Gamma]) := \gamma_{\pi_{i+1}^{n+1}}} \mathbf{Col}_{\mathsf{pr}^{(n-i)}_{\mathcal{I}(\Gamma)}}(\mathcal{I}(A_{i+1}[x_1 \in A_1, ..., x_i \in A_i]))$$

if $1 \leq i \leq n-1$.

provided that $\mathcal{I}(\Gamma)$ is a well defined object of **Cont**.

5.3 Basic sets

We interpret the emptyset, the singleton and the natural numbers type as follows:

$$\mathcal{I}(\mathsf{N}_0[\Gamma]) := \mathbf{0}^{\mathcal{I}(\Gamma)} \qquad \mathcal{I}(\mathsf{N}_1[\Gamma]) := \mathbf{1}^{\mathcal{I}(\Gamma)} \qquad \mathcal{I}(\mathsf{N}[\Gamma]) := \mathbf{N}^{\mathcal{I}(\Gamma)}$$

provided that $\mathcal{I}(\Gamma)$ is a well defined object of **Cont**.

The interpretation of the emptyset eliminator $\mathsf{emp}_0^A(a)[\Gamma]$ is defined as the composed arrow in the following commuting diagram in $\mathbf{Col}(\mathcal{I}(\Gamma))$

$$\begin{array}{ccc} \mathbf{1} & \xrightarrow{\mathcal{I}(a[\Gamma])} & \mathbf{0} \\ & \searrow^{\mathcal{I}(\mathsf{emp}_0^A(a)[\Gamma])} & \downarrow {!_{\mathbf{0},\mathcal{I}(A[\Gamma])}} \\ & & \mathcal{I}(A[\Gamma]) \end{array}$$

provided that $\mathcal{I}(\Gamma)$ is a well defined object of **Cont**, $\mathcal{I}(A[\Gamma])$ is a well defined object of $\mathbf{Col}(\mathcal{I}(\Gamma))$ and $\mathcal{I}(a[\Gamma])$ is a well defined arrow from $\mathbf{1}$ to $\mathbf{0}$ in $\mathbf{Col}(\mathcal{I}(\Gamma))$.

The interpretation of the singleton constant $\star[\Gamma]$ is $\mathcal{I}(\star[\Gamma]) := id_{\mathbf{1}} : \mathbf{1} \to \mathbf{1}$ in $\mathbf{Col}(\mathcal{I}(\Gamma))$, provided that $\mathcal{I}(\Gamma)$ is a well defined object of **Cont**.

The interpretation of the singleton eliminator $\mathsf{El}_{\mathsf{N}_1}^A(b, a)[\Gamma]$ is defined as the composed arrow in the following commuting diagram in $\mathbf{Col}(\mathcal{I}(\Gamma))$

$$\begin{array}{ccc} \mathbf{1} & \xrightarrow{\mathcal{I}(b[\Gamma])} & \mathbf{1} \\ & \searrow^{\mathcal{I}(\mathsf{El}_{\mathsf{N}_1}^A(b,a)[\Gamma])} & \downarrow {\mathcal{I}(a[\Gamma])} \\ & & \mathcal{I}(A[\Gamma]) \end{array}$$

provided that $\mathcal{I}(\Gamma)$ is a well defined object of **Cont**, $\mathcal{I}(A[\Gamma])$ is a well defined object of $\mathbf{Col}(\mathcal{I}(\Gamma))$, $\mathcal{I}(b[\Gamma])$ is a well defined arrow from $\mathbf{1}$ to $\mathbf{1}$ in $\mathbf{Col}(\mathcal{I}(\Gamma))$ and $\mathcal{I}(a[\Gamma])$ is a well defined arrow from $\mathbf{1}$ to $\mathcal{I}(A[\Gamma])$ in $\mathbf{Col}(\mathcal{I}(\Gamma))$.

The interpretation of the constant $0[\Gamma]$ is defined as $\mathcal{I}(0[\Gamma]) := \mathsf{z} : \mathbf{1} \to \mathbf{N}$ in $\mathbf{Col}(\mathcal{I}(\Gamma))$, provided that $\mathcal{I}(\Gamma)$ is a well defined object of **Cont**.

The interpretation of the successor constructor $\mathsf{succ}(a)[\Gamma]$ is defined as

$$\mathcal{I}(\mathsf{succ}(a)[\Gamma]) := \mathsf{s} \circ \mathcal{I}(a[\Gamma]) : \mathbf{1} \to \mathbf{N}$$

in $\mathbf{Col}(\mathcal{I}(\Gamma))$ according with the notation in remark 4.30, provided that $\mathcal{I}(\Gamma)$ is a well defined object of **Cont** and $\mathcal{I}(a[\Gamma])$ is a well defined arrow from $\mathbf{1}$ to $\mathbf{N}$ in $\mathbf{Col}(\mathcal{I}(\Gamma))$.

The interpretation of the natural numbers eliminator $\mathsf{El}_{\mathsf{N}}^{A}(a, b, (x)\, c)[\Gamma]$ is defined as

$$\mathcal{I}(\mathsf{El}_{\mathsf{N}}^{A}(a, b, (x)\, c)[\Gamma]) := rec(\,\mathcal{I}(b[\Gamma]), \mathcal{I}(c[\Gamma, x \in A])_{\natural}^{\mathcal{I}(\Gamma)}\,) \circ \mathcal{I}(a[\Gamma]) : \mathbf{1} \to \mathcal{I}(A[\Gamma])$$

in $\mathbf{Col}(\mathcal{I}(\Gamma))$ according with the notation in remark 4.30 and lemma 4.40, provided that $\mathcal{I}(\Gamma)$ is a well defined object of **Cont**, $\mathcal{I}(A[\Gamma])$ is a well defined object of $\mathbf{Col}(\mathcal{I}(\Gamma))$, $\mathcal{I}(a[\Gamma])$ is a well defined arrow from $\mathbf{1}$ to $\mathbf{N}$ in $\mathbf{Col}(\mathcal{I}([\Gamma]))$, $\mathcal{I}(b[\Gamma])$ is a well defined arrow from $\mathbf{1}$ to $\mathcal{I}(A[\Gamma])$ in $\mathbf{Col}(\mathcal{I}([\Gamma]))$ and $\mathcal{I}(c[\Gamma, x \in A])$ is a well defined arrow from $\mathbf{1}$ to $\mathbf{Col}_{\mathsf{pr}}(\mathcal{I}(A[\Gamma]))$ in $\mathbf{Col}(\mathcal{I}([\Gamma, x \in A]))$.

5.4 Dependent sums

We interpret the dependent sum as follows:

$$\mathcal{I}((\Sigma x \in A)B[\Gamma]) := \Sigma^{\mathcal{I}(\Gamma)}(\mathcal{I}(A[\Gamma]), \mathcal{I}(B[\Gamma, x \in A]))$$

provided that $\mathcal{I}(\Gamma)$ is a well defined object of **Cont**, $\mathcal{I}(A[\Gamma])$ is a well defined object of $\mathbf{Col}(\mathcal{I}(\Gamma))$ and $\mathcal{I}(B[\Gamma, x \in A])$ is a well defined object of $\mathbf{Col}(\mathcal{I}([\Gamma, x \in A]))$. The interpretation of the pairing of the dependent sum $\langle a, b\rangle^{A,(x)B}[\Gamma]$ is defined as

$$\mathcal{I}(\langle a, b\rangle^{A,(x)B}[\Gamma]) := \langle \mathcal{I}(a[\Gamma]), \mathcal{I}(b[\Gamma])\rangle_{\Sigma} : \mathbf{1} \to \Sigma^{\mathcal{I}(\Gamma)}(\mathcal{I}(A[\Gamma]), \mathcal{I}(B[\Gamma, x \in A]))$$

with reference to lemma 4.34 provided that $\mathcal{I}(\Gamma)$ is a well defined object of **Cont**, $\mathcal{I}(A[\Gamma])$ is a well defined object of $\mathbf{Col}(\mathcal{I}(\Gamma))$, $\mathcal{I}(B[\Gamma, x \in A])$ is a well defined object of $\mathbf{Col}(\mathcal{I}([\Gamma, x \in A]))$, $\mathcal{I}(a[\Gamma])$ is a well defined arrow from $\mathbf{1}$ to $\mathcal{I}(A[\Gamma])$ in $\mathbf{Col}(\mathcal{I}([\Gamma]))$ and $\mathcal{I}(b[\Gamma])$ is a well defined arrow from $\mathbf{1}$ to $\mathbf{Col}_{\widetilde{\mathcal{I}(a[\Gamma])}}(\,\mathcal{I}(B[\Gamma, x \in A])\,)$ in $\mathbf{Col}(\mathcal{I}([\Gamma]))$.

The interpretations of the projections of the dependent sum $\pi_1^{A,(x)B}(c)[\Gamma]$ and $\pi_2^{A,(x)B}(c)[\Gamma]$ are defined as follows

$$\mathcal{I}(\pi_1^{A,(x)B}(c)[\Gamma]) := \mathsf{p}_1^{\Sigma} \circ \mathcal{I}(c[\Gamma]) : \mathbf{1} \to \mathcal{I}(A[\Gamma])$$

$$\mathcal{I}(\pi_2^{A,(x)B}(c)[\Gamma]) := \mathsf{p}_2^{\Sigma}(\mathcal{I}(c[\Gamma])) : \mathbf{1} \to \mathbf{Col}_{\widetilde{\mathsf{p}_1^{\Sigma} \circ \mathcal{I}(c[\Gamma])}}(\mathcal{I}(B[\Gamma]))$$

with reference to lemma 4.34 provided that $\mathcal{I}(\Gamma)$ is a well defined object of **Cont**, $\mathcal{I}(A[\Gamma])$ is a well defined object of $\mathbf{Col}(\mathcal{I}(\Gamma))$, $\mathcal{I}(B[\Gamma, x \in A])$ is a well defined object of $\mathbf{Col}(\mathcal{I}([\Gamma, x \in A]))$ and $\mathcal{I}(c[\Gamma])$ is a well defined arrow in $\mathbf{Col}(\mathcal{I}(\Gamma))$ from $\mathbf{1}$ to $\Sigma^{\mathcal{I}(\Gamma)}(\mathcal{I}(A[\Gamma]), \mathcal{I}(B[\Gamma, x \in A]))$.

5.5 Dependent products

We interpret the dependent product as follows:

$$\mathcal{I}((\Pi x \in A)B[\Gamma]) := \Pi^{\mathcal{I}(\Gamma)}(\mathcal{I}(A[\Gamma]), \mathcal{I}(B[\Gamma, x \in A]))$$

provided that $\mathcal{I}(\Gamma)$ is a well defined object of **Cont**, $\mathcal{I}(A[\Gamma])$ is a well defined object of $\mathbf{Col}(\mathcal{I}(\Gamma))$ and $\mathcal{I}(B[\Gamma, x \in A])$ is a well defined object of $\mathbf{Col}(\mathcal{I}([\Gamma, x \in A]))$.

The interpretation of the lambda-abstraction $(\lambda x)^{A,B}b[\Gamma]$ is defined as

$$\mathcal{I}((\lambda x)^{A,B}b[\Gamma]) := \mathsf{Cur}_{\Pi}(\mathcal{I}(b[\Gamma, x \in A])) : \mathbf{1} \to \Pi^{\mathcal{I}(\Gamma)}(\mathcal{I}(A), \mathcal{I}(B[\Gamma, x \in A]))$$

in $\mathbf{Col}(\mathcal{I}(\Gamma))$ where $\mathsf{Cur}_{\Pi}(\mathcal{I}(b[\Gamma, x \in A]))$ is the arrow (see corollary 4.37) making the following diagram commute in $\mathbf{Col}([\mathcal{I}(\Gamma), \mathcal{I}(A[\Gamma])])$

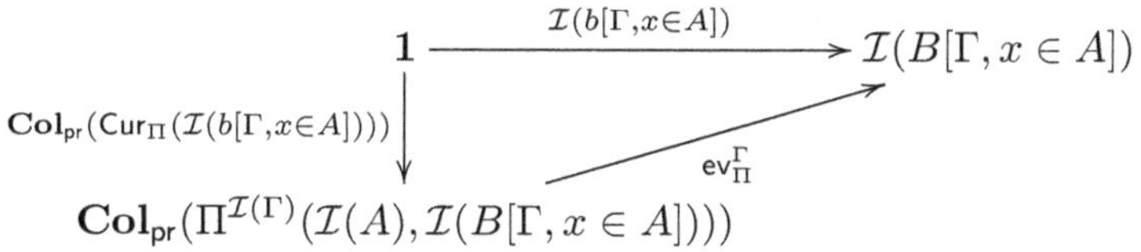

provided that $\mathcal{I}(\Gamma)$ is a well defined object of **Cont**, $\mathcal{I}(A[\Gamma])$ is a well defined object of $\mathbf{Col}(\mathcal{I}(\Gamma))$, $\mathcal{I}(B[\Gamma, x \in A])$ is a well defined object of $\mathbf{Col}(\mathcal{I}([\Gamma, x \in A]))$ and $\mathcal{I}(b[\Gamma, x \in A])$ is a well defined arrow from $\mathbf{1}$ to $\mathcal{I}(B[\Gamma, x \in A])$ in $\mathbf{Col}(\mathcal{I}([\Gamma, x \in A]))$.

The interpretation of the application $\mathsf{Ap}^{A,(x)B}(c, a)[\Gamma]$ is defined as the unique arrow directed towards the pullback of $\widetilde{\mathcal{I}(a[\Gamma])}$ along $\mathsf{pr}_{[\mathcal{I}(\Gamma), \mathcal{I}(A[\Gamma]), \mathcal{I}(B[\Gamma, x\in A])]}$ (which is the middle rectangle in the following diagram defined as in lemma 4.20) making the following diagram in **Cont** commute (with the notation of lemma 4.40)

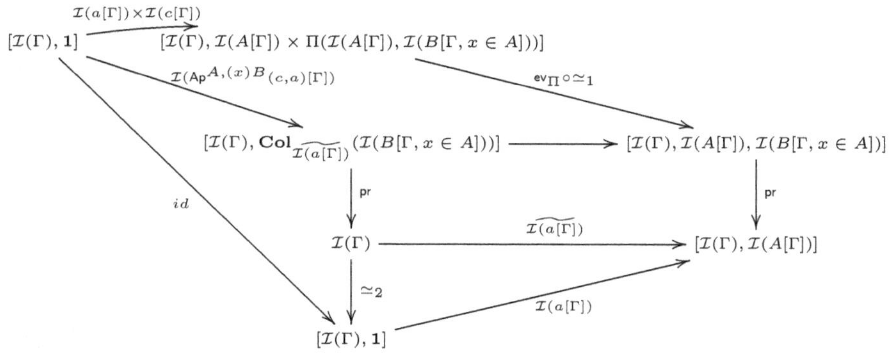

where $\simeq_1$ is the isomorphism from $[\mathcal{I}(\Gamma), \mathcal{I}(A[\Gamma]) \times \Pi(\mathcal{I}(A[\Gamma]), \mathcal{I}(B[\Gamma, x \in A]))]$ to $[\mathcal{I}(\Gamma), \mathcal{I}(A[\Gamma]), \mathbf{Col}_{\mathbf{pr}_{[\Gamma, \mathcal{I}(A[\Gamma])]}}\Pi(\mathcal{I}(A[\Gamma]), \mathcal{I}(B[\Gamma, x \in A]))]$ in **Cont** defined as in lemma 4.12 thanks to lemma 4.39 and $\simeq_2$ is the inverse of the isomorphism $\mathbf{pr}_{[\Gamma, \mathbf{1}]}$.

This arrow exists thanks to corollary 4.37, provided that $\mathcal{I}(\Gamma)$ is a well defined object of **Cont**, $\mathcal{I}(A[\Gamma])$ is a well defined object of $\mathbf{Col}(\mathcal{I}(\Gamma))$, $\mathcal{I}(B[\Gamma, x \in A])$ is a well defined object of $\mathbf{Col}(\mathcal{I}([\Gamma, x \in A]))$, $\mathcal{I}(a[\Gamma])$ is a well defined arrow from $\mathbf{1}$ to $\mathcal{I}(A[\Gamma])$ in $\mathbf{Col}(\mathcal{I}([\Gamma]))$ and, finally, $\mathcal{I}(c[\Gamma])$ is a well defined arrow from $\mathbf{1}$ to $\Pi(\mathcal{I}(A[\Gamma]), \mathcal{I}(B[\Gamma, x \in A]))$ in $\mathbf{Col}(\mathcal{I}([\Gamma]))$.

5.6 Disjoint sums

We interpret the disjoint sum as follows:

$$\mathcal{I}(A + B[\Gamma]) := \mathcal{I}(A[\Gamma]) +^{\mathcal{I}(\Gamma)} \mathcal{I}(B[\Gamma])$$

provided that $\mathcal{I}(\Gamma)$ is a well defined object of **Cont** and $\mathcal{I}(A[\Gamma])$ and $\mathcal{I}(B[\Gamma])$ are well defined objects of $\mathbf{Col}(\mathcal{I}(\Gamma))$.

The interpretation of the first injection of the disjoint sum $\mathsf{inl}^{A,B}(a)[\Gamma]$ is defined as the composed arrow making the following diagram commute in $\mathbf{Col}(\mathcal{I}(\Gamma))$

$$\begin{array}{ccc}
\mathbf{1} & \xrightarrow{\mathcal{I}(a[\Gamma])} & \mathcal{I}(A[\Gamma]) \\
& {\scriptstyle \mathcal{I}(\mathsf{inl}^{A,B}(a)[\Gamma])} \searrow & \downarrow {\scriptstyle \mathsf{j}_1} \\
& & \mathcal{I}(A[\Gamma]) + \mathcal{I}(B[\Gamma])
\end{array}$$

provided that $\mathcal{I}(\Gamma)$ is a well defined object of **Cont**, $\mathcal{I}(A[\Gamma])$ and $\mathcal{I}(B[\Gamma])$ are well defined objects of $\mathbf{Col}(\mathcal{I}(\Gamma))$ and $\mathcal{I}(a[\Gamma])$ is a well defined arrow from $\mathbf{1}$ to $\mathcal{I}(A[\Gamma])$ in $\mathbf{Col}(\mathcal{I}(\Gamma))$.

The interpretation of the second injection of the disjoint sum $\mathsf{inr}^{A,B}(b)[\Gamma]$ is defined as the composed arrow making the following diagram commute in $\mathbf{Col}(\mathcal{I}(\Gamma))$

$$\begin{array}{ccc}
\mathbf{1} & \xrightarrow{\mathcal{I}(b[\Gamma])} & \mathcal{I}(B[\Gamma]) \\
& {\scriptstyle \mathcal{I}(\mathsf{inr}^{A,B}(b)[\Gamma])} \searrow & \downarrow {\scriptstyle \mathsf{j}_2} \\
& & \mathcal{I}(A[\Gamma]) + \mathcal{I}(B[\Gamma])
\end{array}$$

provided that $\mathcal{I}(\Gamma)$ is a well defined object of **Cont**, $\mathcal{I}(A[\Gamma])$ and $\mathcal{I}(B[\Gamma])$ are well defined objects of $\mathbf{Col}(\mathcal{I}(\Gamma))$ and $\mathcal{I}(b[\Gamma])$ is a well defined arrow from $\mathbf{1}$ to $\mathcal{I}(B[\Gamma])$ in $\mathbf{Col}(\mathcal{I}(\Gamma))$.

The interpretation of the eliminator of the disjoint sum $\mathsf{El}_{+}^{A,B,C}(c, (x)\, d, (y)\, e)[\Gamma]$ is defined as $f \circ \mathcal{I}(c[\Gamma])$ in the following commuting diagram in $\mathbf{Col}(\mathcal{I}(\Gamma))$ (with the

notation of lemma 4.40)

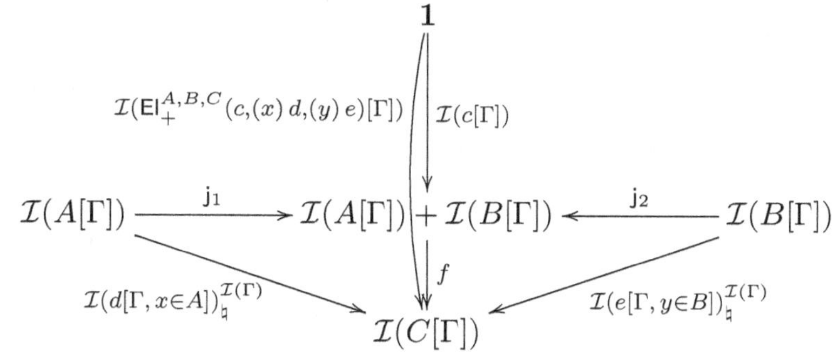

where the existence and uniqueness of f is guaranteed by lemma 4.26, provided that $\mathcal{I}(\Gamma)$ is a well defined object of **Cont**, $\mathcal{I}(A[\Gamma])$, $\mathcal{I}(B[\Gamma])$ and $\mathcal{I}(C[\Gamma])$ are well defined objects of $\mathbf{Col}(\mathcal{I}(\Gamma))$, $\mathcal{I}(c[\Gamma])$ is a well defined arrow from $\mathbf{1}$ to $\mathcal{I}(A[\Gamma]) + \mathcal{I}(B[\Gamma])$ in $\mathbf{Col}(\mathcal{I}(\Gamma))$, $\mathcal{I}(d[\Gamma, x \in A])$ is a well defined arrow from $\mathbf{1}$ to $\mathbf{Col}_{\mathsf{pr}}(\mathcal{I}(C[\Gamma]))$ in $\mathbf{Col}(\mathcal{I}([\Gamma, x \in A]))$ and, finally, $\mathcal{I}(e[\Gamma, y \in B])$ is a well defined arrow from $\mathbf{1}$ to $\mathbf{Col}_{\mathsf{pr}}(\mathcal{I}(C[\Gamma]))$ in $\mathbf{Col}(\mathcal{I}([\Gamma, y \in B]))$.

5.7 Lists

We interpret the type of lists on a type as follows:

$$\mathcal{I}(\mathsf{List}(A)[\Gamma]) := \mathbf{List}^{\mathcal{I}(\Gamma)}(\mathcal{I}(A[\Gamma]))$$

provided that $\mathcal{I}(\Gamma)$ is a well defined object of **Cont** and $\mathcal{I}(A[\Gamma])$ is a well defined object of $\mathbf{Col}(\mathcal{I}(\Gamma))$.

The interpretation of the empty list $\epsilon^A[\Gamma]$ is defined as

$$\mathcal{I}(\epsilon^A[\Gamma]) := \epsilon : \mathbf{1} \to \mathbf{List}(\mathcal{I}(A[\Gamma]))$$

in $\mathbf{Col}(\mathcal{I}(\Gamma))$ provided that $\mathcal{I}(\Gamma)$ is a well defined object of **Cont** and $\mathcal{I}(A[\Gamma])$ is a well defined object of $\mathbf{Col}(\mathcal{I}(\Gamma))$.

The interpretation of the list constructor $\mathsf{cons}^A(b, a)[\Gamma]$ is defined as the composed arrow making the following diagram commute in $\mathbf{Col}(\mathcal{I}(\Gamma))$

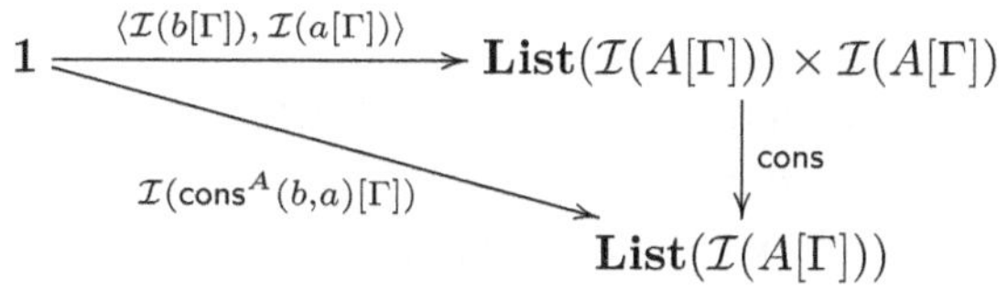

provided that $\mathcal{I}(\Gamma)$ is a well defined object of **Cont** and $\mathcal{I}(A[\Gamma])$ is a well defined object of $\mathbf{Col}(\mathcal{I}(\Gamma))$ and $\mathcal{I}(b[\Gamma])$ is a well defined arrow from $\mathbf{1}$ to $\mathbf{List}(\mathcal{I}(A[\Gamma]))$ in $\mathbf{Col}(\mathcal{I}(\Gamma))$ while $\mathcal{I}(a[\Gamma])$ is a well defined arrow from $\mathbf{1}$ to $\mathcal{I}(A[\Gamma])$.

The interpretation of the list eliminator $\mathsf{El}_{\mathsf{List}}^{A,B}(a, b, (x, y)\, c)[\Gamma]$ is defined as the composed arrow $f \circ \mathcal{I}(a[\Gamma])$ making the following diagram commute in $\mathbf{Col}(\mathcal{I}(\Gamma))$ (with notation as in lemma 4.40)

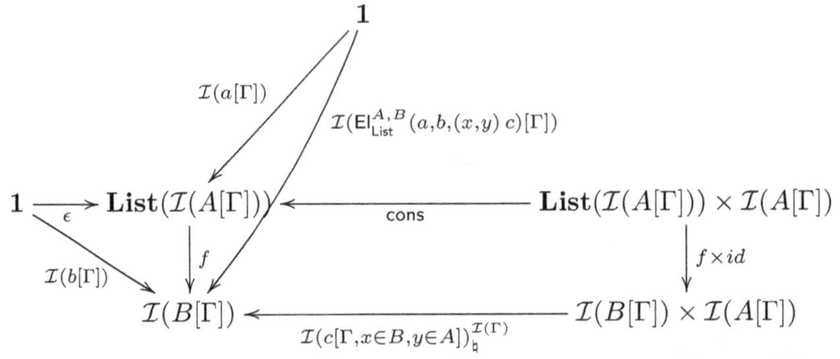

where the existence and uniqueness of f is guaranteed by lemma 4.28, provided that $\mathcal{I}(\Gamma)$ is a well defined object of **Cont**, $\mathcal{I}(A[\Gamma])$ and $\mathcal{I}(B[\Gamma])$ are well defined objects of $\mathbf{Col}(\mathcal{I}(\Gamma))$, $\mathcal{I}(a[\Gamma])$ is a well defined arrow from $\mathbf{1}$ to $\mathbf{List}(\mathcal{I}(A[\Gamma]))$ in $\mathbf{Col}(\mathcal{I}(\Gamma))$ and $\mathcal{I}(b[\Gamma])$ is a well defined arrow from $\mathbf{1}$ to $\mathcal{I}(B[\Gamma])$ in $\mathbf{Col}(\mathcal{I}(\Gamma))$ and $\mathcal{I}(c[\Gamma, x \in B, y \in A])$ is a well defined arrow from $\mathbf{1}$ to $\mathbf{Col}_{\mathsf{pr}^{(2)}}(\mathcal{I}(B[\Gamma]))$ in $\mathbf{Col}([\mathcal{I}(\Gamma), \mathcal{I}(B[\Gamma]), \mathbf{Col}_{\mathsf{pr}}(\mathcal{I}(A[\Gamma]))])$.

5.8 Collection of small propositions

The collection of small propositions is interpreted as follows:

$$\mathcal{I}(\mathsf{Prop}_s[\Gamma]) := \mathsf{USP}^{\mathcal{I}(\Gamma)}$$

provided that $\mathcal{I}(\Gamma)$ is a well defined object of **Cont**.

Recalling lemma 4.61, we define the interpretation of terms as follows.

The interpretation of the falsum code $\widehat{\bot}[\Gamma]$ is defined as

$$\mathcal{I}(\widehat{\bot}[\Gamma]) := \widehat{\bot} : \mathbf{1} \to \mathsf{USP}^{\mathcal{I}(\Gamma)}$$

provided that $\mathcal{I}(\Gamma)$ is a well defined object of **Cont**.

The interpretations of the conjunction code $a \mathbin{\widehat{\wedge}} b\,[\Gamma]$, the disjunction code $a \mathbin{\widehat{\vee}} b\,[\Gamma]$ and the implication code $a \mathbin{\widehat{\rightarrow}} b\,[\Gamma]$ are defined as the composed arrows making the following diagrams commute in $\mathbf{Col}(\mathcal{I}(\Gamma))$

$$
\begin{array}{ccc}
\mathsf{USP}^{\mathcal{I}(\Gamma)} \times \mathsf{USP}^{\mathcal{I}(\Gamma)} & \xrightarrow{\ \widehat{\wedge}\ } & \mathsf{USP}^{\mathcal{I}(\Gamma)} \\
{\scriptstyle\langle\mathcal{I}(a[\Gamma]),\mathcal{I}(b[\Gamma])\rangle}\uparrow & \nearrow{\scriptstyle\mathcal{I}(a\widehat{\wedge}b[\Gamma])} & \\
\mathbf{1} & &
\end{array}
\qquad
\begin{array}{ccc}
\mathsf{USP}^{\mathcal{I}(\Gamma)} \times \mathsf{USP}^{\mathcal{I}(\Gamma)} & \xrightarrow{\ \widehat{\vee}\ } & \mathsf{USP}^{\mathcal{I}(\Gamma)} \\
{\scriptstyle\langle\mathcal{I}(a[\Gamma]),\mathcal{I}(b[\Gamma])\rangle}\uparrow & \nearrow{\scriptstyle\mathcal{I}(a\widehat{\vee}b[\Gamma])} & \\
\mathbf{1} & &
\end{array}
$$

$$
\begin{array}{ccc}
\mathsf{USP}^{\mathcal{I}(\Gamma)} \times \mathsf{USP}^{\mathcal{I}(\Gamma)} & \xrightarrow{\ \widehat{\rightarrow}\ } & \mathsf{USP}^{\mathcal{I}(\Gamma)} \\
{\scriptstyle\langle\mathcal{I}(a[\Gamma]),\mathcal{I}(b[\Gamma])\rangle}\uparrow & \nearrow{\scriptstyle\mathcal{I}(a\widehat{\rightarrow}b[\Gamma])} & \\
\mathbf{1} & &
\end{array}
$$

provided that $\mathcal{I}(\Gamma)$ is a well defined object of $\mathbf{Cont}$ and $\mathcal{I}(a[\Gamma])$ and $\mathcal{I}(b[\Gamma])$ are well defined arrows from $\mathbf{1}$ to $\mathsf{USP}^{\mathcal{I}(\Gamma)}$ in $\mathbf{Col}(\mathcal{I}(\Gamma))$.

The interpretations of the existential quantification code $(\widehat{\exists x\in A})b[\Gamma]$ and the universal quantification code $(\widehat{\forall x\in A})b[\Gamma]$ are defined as the composed arrows making the following diagrams commute in $\mathbf{Col}(\mathcal{I}(\Gamma))$

$$
\begin{array}{ccc}
\mathbf{1} & \xrightarrow{\ \mathbf{fam}_p(\mathcal{I}(b[\Gamma,\,x\in A]))\ } & \Sigma(\mathsf{US}^{\mathcal{I}(\Gamma)},\tau^{\mathcal{I}(\Gamma)} \Rightarrow \mathsf{USP}^{[\mathcal{I}(\Gamma),\mathsf{US}^{\mathcal{I}(\Gamma)}]}) \\
 & {\scriptstyle\mathcal{I}((\widehat{\exists x\in A})b[\Gamma])}\searrow & \downarrow{\scriptstyle\widehat{\exists}} \\
 & & \mathsf{USP}
\end{array}
$$

$$
\begin{array}{ccc}
\mathbf{1} & \xrightarrow{\ \mathbf{fam}_p(\mathcal{I}(b[\Gamma,\,x\in A]))\ } & \Sigma(\mathsf{US}^{\mathcal{I}(\Gamma)},\tau^{\mathcal{I}(\Gamma)} \Rightarrow \mathsf{USP}^{[\mathcal{I}(\Gamma),\mathsf{US}^{\mathcal{I}(\Gamma)}]}) \\
 & {\scriptstyle\mathcal{I}((\widehat{\forall x\in A})b[\Gamma])}\searrow & \downarrow{\scriptstyle\widehat{\forall}} \\
 & & \mathsf{USP}
\end{array}
$$

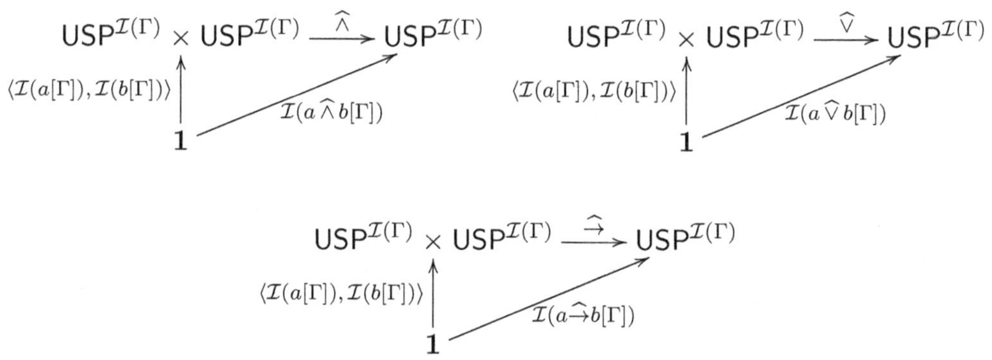

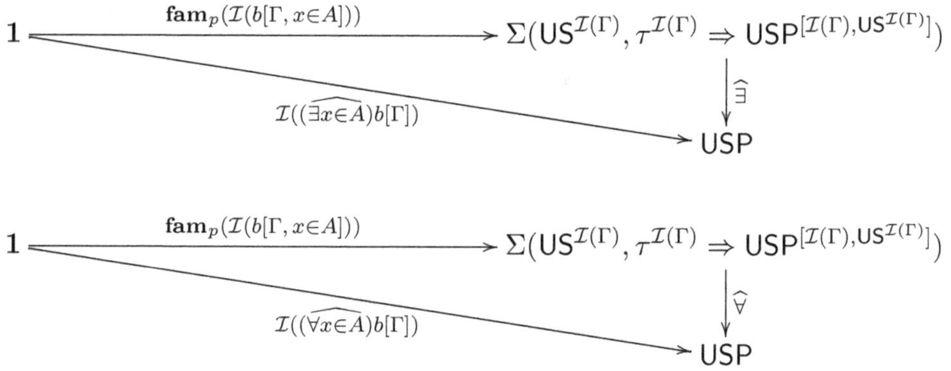

where $\mathbf{fam}_p(\mathcal{I}(b[\Gamma, x \in A]))$ is defined using the notation in definition 4.66 and lemma 4.40 as

$$
\langle \mathbf{en}_s^{\mathcal{I}(\Gamma)}(\mathcal{I}(A[\Gamma])), \mathsf{Cur}(\mathcal{I}(b[\Gamma, x \in A])_{\flat}^{\mathcal{I}(\Gamma)} \circ \pi_2^{1,\mathcal{I}(A[\Gamma])})\rangle_\Sigma
$$

provided that $\mathcal{I}(\Gamma)$ is a well defined object in $\mathbf{Cont}$, $\mathcal{I}(A[\Gamma])$ is a well defined object of $\mathbf{Set}(\mathcal{I}(\Gamma))$ and $\mathcal{I}(b[\Gamma, x \in A])$ is a well defined arrow from $\mathbf{1}$ to $\mathsf{USP}^{\mathcal{I}([\Gamma,x\in A])}$ in $\mathbf{Col}(\mathcal{I}([\Gamma, x \in A]))$.

The interpretations of the propositional equality code $\widehat{\mathsf{Eq}}(A, a, b)[\Gamma]$ is defined as the composed arrow making the following diagram commute in $\mathbf{Col}(\mathcal{I}(\Gamma))$

$$
\begin{array}{ccc}
\mathbf{1} & \xrightarrow{\langle\, \mathbf{en}_s^{\mathcal{I}(\Gamma)}(\mathcal{I}(A[\Gamma]))\ ,\ \langle\mathcal{I}(a[\Gamma]), \mathcal{I}(b[\Gamma])\rangle\,\rangle_\Sigma} & \Sigma(\mathsf{US}^{\mathcal{I}(\Gamma)}, \tau^{\mathcal{I}(\Gamma)} \times \tau^{\mathcal{I}(\Gamma)}) \\
 & {\scriptstyle \mathcal{I}(\widehat{\mathsf{Eq}}(A,a,b)[\Gamma])} \searrow & \downarrow{\scriptstyle \widehat{\mathsf{Eq}}} \\
 & & \mathsf{USP}^{\mathcal{I}(\Gamma)}
\end{array}
$$

provided that $\mathcal{I}(\Gamma)$ is a well defined object of $\mathbf{Cont}$, $\mathcal{I}(A[\Gamma])$ is well defined object of $\mathbf{Set}(\mathcal{I}(\Gamma))$ and $\mathcal{I}(a[\Gamma])$ and $\mathcal{I}(b[\Gamma])$ are well defined arrows from $\mathbf{1}$ to $\mathcal{I}(A[\Gamma])$ in $\mathbf{Col}(\mathcal{I}(\Gamma))$.

5.9 Collection of sets

We interpret the collection of sets as follows

$$
\mathcal{I}(\mathsf{Set}[\Gamma]) := \mathsf{US}^{\mathcal{I}(\Gamma)}
$$

provided that $\mathcal{I}(\Gamma)$ is a well defined object of $\mathbf{Cont}$.

Recalling lemma 4.61, we define the interpretation of terms as follows.

The interpretation of the empty set code $\widehat{\mathsf{N}}_0$, the singleton code $\widehat{\mathsf{N}}_1$ and the natural numbers set code $\widehat{\mathsf{N}}$ are defined as follows: $\mathcal{I}(\widehat{\mathsf{N}}_0[\Gamma]) := \widehat{\mathsf{N}}_0 : \mathbf{1} \to \mathsf{US}^{\mathcal{I}(\Gamma)}$, $\mathcal{I}(\widehat{\mathsf{N}}_1[\Gamma]) := \widehat{\mathsf{N}}_1 : \mathbf{1} \to \mathsf{US}^{\mathcal{I}(\Gamma)}$ and $\mathcal{I}(\widehat{\mathsf{N}}[\Gamma]) := \widehat{\mathsf{N}} : \mathbf{1} \to \mathsf{US}^{\mathcal{I}(\Gamma)}$ in $\mathbf{Col}(\mathcal{I}(\Gamma))$, all provided that $\mathcal{I}(\Gamma)$ is a well defined object of $\mathbf{Cont}$.

The interpretation of the disjoint sum code $a \mathbin{\widehat{+}} b[\Gamma]$ is defined as the composed arrow making the following diagram commute in $\mathbf{Col}(\mathcal{I}(\Gamma))$

$$
\begin{array}{ccc}
\mathsf{US}^{\mathcal{I}(\Gamma)} \times \mathsf{US}^{\mathcal{I}(\Gamma)} & \xrightarrow{\widehat{+}} & \mathsf{US}^{\mathcal{I}(\Gamma)} \\
{\scriptstyle \langle\mathcal{I}(a[\Gamma]), \mathcal{I}(b[\Gamma])\rangle}\uparrow & \nearrow{\scriptstyle \mathcal{I}(a \mathbin{\widehat{+}} b[\Gamma])} & \\
\mathbf{1} & &
\end{array}
$$

provided that $\mathcal{I}(\Gamma)$ is a well defined object of $\mathbf{Cont}$ and $\mathcal{I}(a[\Gamma])$ and $\mathcal{I}(b[\Gamma])$ are well defined arrows from $\mathbf{1}$ to $\mathsf{US}^{\mathcal{I}(\Gamma)}$ in $\mathbf{Col}(\mathcal{I}(\Gamma))$.

The interpretation of the list set code $\widehat{\mathsf{List}}(a)[\Gamma]$ is defined as the composed arrow making the following diagram commute in $\mathbf{Col}(\mathcal{I}(\Gamma))$

$$
\begin{array}{ccc}
\mathsf{US}^{\mathcal{I}(\Gamma)} & \xrightarrow{\widehat{\mathsf{List}}} & \mathsf{US}^{\mathcal{I}(\Gamma)} \\
{\scriptstyle \mathcal{I}(a[\Gamma])}\uparrow & \nearrow{\scriptstyle \mathcal{I}(\widehat{\mathsf{List}}(a)[\Gamma])} & \\
\mathbf{1} & &
\end{array}
$$

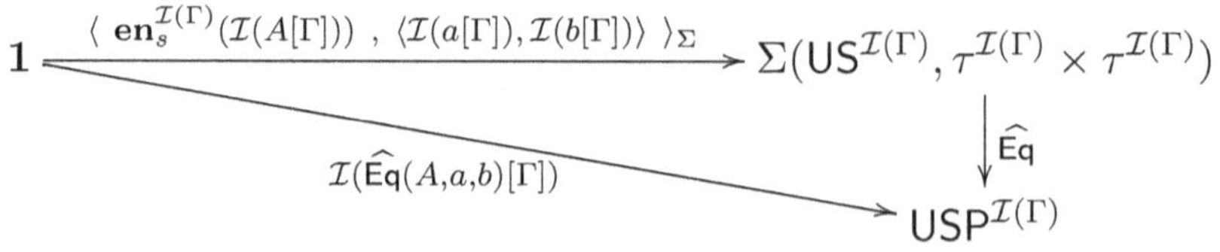

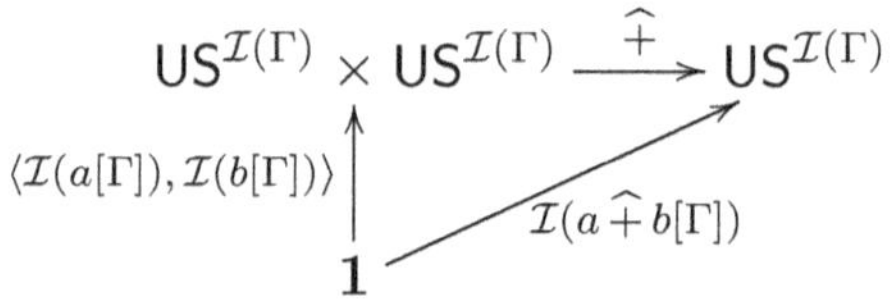

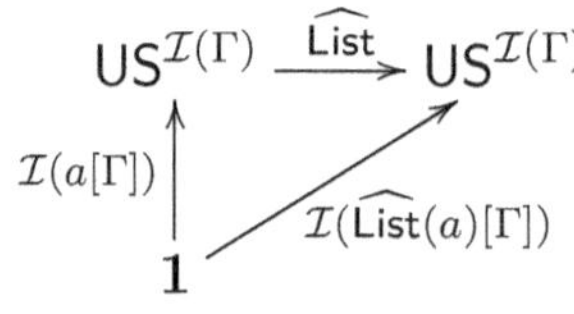

provided that $\mathcal{I}(\Gamma)$ is a well defined object of **Cont** and $\mathcal{I}(a[\Gamma])$ is a well defined arrow from $\mathbf{1}$ to $\mathsf{US}^{\mathcal{I}(\Gamma)}$ in $\mathbf{Col}(\mathcal{I}(\Gamma))$.

The interpretation of the dependent sum code $(\widehat{\Sigma x \in A})b[\Gamma]$ and the dependent product code $(\widehat{\Pi x \in A})b[\Gamma]$ are defined as the composed arrows making the following diagrams commute in $\mathbf{Col}(\mathcal{I}(\Gamma))$

$$\begin{array}{ccc}
\mathbf{1} & \xrightarrow{\mathbf{fam}(\mathcal{I}(b[\Gamma,x\in A]))} & \Sigma(\mathsf{US}^{\mathcal{I}(\Gamma)}, \tau^{\mathcal{I}(\Gamma)} \Rightarrow \mathsf{US}^{[\mathcal{I}(\Gamma),\mathsf{US}^{\mathcal{I}(\Gamma)}]}) \\
& \searrow^{\mathcal{I}((\widehat{\Sigma x\in A})b[\Gamma])} & \downarrow \widehat{\Sigma} \\
& & \mathsf{US}^{\mathcal{I}(\Gamma)}
\end{array}$$

$$\begin{array}{ccc}
\mathbf{1} & \xrightarrow{\mathbf{fam}(\mathcal{I}(b[\Gamma,x\in A]))} & \Sigma(\mathsf{US}^{\mathcal{I}(\Gamma)}, \tau^{\mathcal{I}(\Gamma)} \Rightarrow \mathsf{US}^{[\mathcal{I}(\Gamma),\mathsf{US}^{\mathcal{I}(\Gamma)}]}) \\
& \searrow^{\mathcal{I}((\widehat{\Pi x\in A})b[\Gamma])} & \downarrow \widehat{\Pi} \\
& & \mathsf{US}^{\mathcal{I}(\Gamma)}
\end{array}$$

where $\mathbf{fam}(\mathcal{I}(b[\Gamma, x \in A]))$ is defined using the notation in definition 4.66 and lemma 4.40 as

$$\langle \mathbf{en}_s^{\mathcal{I}(\Gamma)}(\mathcal{I}(A[\Gamma])), \mathsf{Cur}(\mathcal{I}(b[\Gamma, x \in A])_{\natural}^{\mathcal{I}(\Gamma)} \circ \pi_2^{\mathbf{1},\mathcal{I}(A[\Gamma])})\rangle_\Sigma$$

provided that $\mathcal{I}(\Gamma)$ is a well defined object of **Cont**, $\mathcal{I}(A[\Gamma])$ is well defined object of $\mathbf{Set}(\mathcal{I}(\Gamma))$ and $\mathcal{I}(b[\Gamma, x \in A])$ is a well defined arrow from $\mathbf{1}$ to $\mathsf{US}^{\mathcal{I}([\Gamma,x\in A])}$ in $\mathbf{Col}(\mathcal{I}([\Gamma, x \in A]))$.

The interpretation of the small proposition code $\sigma(a)[\Gamma]$ is defined as the composed arrow making the following diagram commute in $\mathbf{Col}(\mathcal{I}(\Gamma))$

$$\begin{array}{ccc}
\mathbf{1} & \xrightarrow{\mathcal{I}(a[\Gamma])} & \mathsf{USP}^{\mathcal{I}(\Gamma)} \\
& \searrow^{\mathcal{I}(\sigma(a)[\Gamma])} & \downarrow \sigma \\
& & \mathsf{US}^{\mathcal{I}(\Gamma)}
\end{array}$$

provided that $\mathcal{I}(\Gamma)$ is a well defined object of **Cont** and $\mathcal{I}(a[\Gamma])$ is a well defined arrow from $\mathbf{1}$ to $\mathsf{USP}^{\mathcal{I}(\Gamma)}$ in $\mathbf{Col}(\mathcal{I}(\Gamma))$.

5.10 Collection of propositional functions

We interpret the collection of propositional functions as follows:

$$\mathcal{I}(A \Rightarrow \mathsf{Prop}_s[\Gamma]) := \mathcal{I}(A[\Gamma]) \Rightarrow^{\mathcal{I}(\Gamma)} \mathsf{USP}^{\mathcal{I}(\Gamma)}$$

provided that $\mathcal{I}(\Gamma)$ is a well defined object of **Cont** and $\mathcal{I}(A[\Gamma])$ is a well defined object of $\mathbf{Col}(\mathcal{I}(\Gamma))$.

The interpretation of the propositional function lambda-abstraction $(\lambda x)^{A}_{\Rightarrow} b[\Gamma]$ is defined as the unique arrow (see lemma 4.27) making the following diagram commute in $\mathbf{Col}(\mathcal{I}(\Gamma))$ (with notation in lemma 4.40)

$$\begin{array}{ccc}
\mathcal{I}(A[\Gamma]) & \xrightarrow{\mathcal{I}(b[\Gamma, x\in A])^{\mathcal{I}(\Gamma)}_{\natural}} & \mathsf{USP}^{\mathcal{I}(\Gamma)} \\
\downarrow{\scriptstyle \langle !, id\rangle} & & \uparrow{\scriptstyle \mathsf{ev}} \\
\mathbf{1} \times \mathcal{I}(A[\Gamma]) & \xrightarrow{\mathcal{I}((\lambda x)^{A}_{\Rightarrow} b[\Gamma]) \times id} & (\mathcal{I}(A[\Gamma]) \Rightarrow \mathsf{USP}^{\mathcal{I}(\Gamma)}) \times \mathcal{I}(A[\Gamma])
\end{array}$$

provided that $\mathcal{I}(\Gamma)$ is a well defined object of **Cont**, $\mathcal{I}(A[\Gamma])$ is a well defined object of $\mathbf{Col}(\mathcal{I}(\Gamma))$ and $\mathcal{I}(b[\Gamma, x \in A])$ is a well defined arrow from $\mathbf{1}$ to $\mathsf{USP}^{\mathcal{I}([\Gamma, x\in A])}$ in $\mathbf{Col}(\mathcal{I}([\Gamma, x \in A]))$.

The interpretation of the propositional function application $\mathsf{Ap}^{A}_{\Rightarrow}(c, a)[\Gamma]$ is defined as the composed arrow making the following diagram commute in $\mathbf{Col}(\mathcal{I}(\Gamma))$

$$\begin{array}{ccc}
\mathbf{1} & \xrightarrow{\langle \mathcal{I}(c[\Gamma]), \mathcal{I}(a[\Gamma])\rangle} & (\mathcal{I}(A[\Gamma]) \Rightarrow \mathsf{USP}^{\mathcal{I}(\Gamma)}) \times \mathcal{I}(A[\Gamma]) \\
 & \searrow{\scriptstyle \mathcal{I}(\mathsf{Ap}^{A}_{\Rightarrow}(c,a)[\Gamma])} & \downarrow{\scriptstyle \mathsf{ev}} \\
 & & \mathsf{USP}^{\mathcal{I}(\Gamma)}
\end{array}$$

provided that $\mathcal{I}(\Gamma)$ is a well defined object of **Cont**, $\mathcal{I}(A[\Gamma])$ is a well defined object of $\mathbf{Col}(\mathcal{I}(\Gamma))$ and $\mathcal{I}(c[\Gamma])$ is a well defined arrow from $\mathbf{1}$ to $\mathcal{I}(A[\Gamma]) \Rightarrow \mathsf{USP}^{\mathcal{I}(\Gamma)}$ in $\mathbf{Col}(\mathcal{I}(\Gamma))$ and $\mathcal{I}(a[\Gamma])$ is a well defined arrow from $\mathbf{1}$ to $\mathcal{I}(A[\Gamma])$ in $\mathbf{Col}(\mathcal{I}(\Gamma))$.

5.11 Falsum

We interpret falsum as follows:

$$\mathcal{I}(\bot[\Gamma]) := \bot^{\mathcal{I}(\Gamma)}$$

provided that $\mathcal{I}(\Gamma)$ is a well defined object of **Cont**.

The interpretation of the falsum eliminator $\mathsf{r}_0^{A}(a)[\Gamma]$ is given by

$$\mathcal{I}(\mathsf{r}_0^{A}(a)[\Gamma]) : \top \sqsubseteq \mathcal{I}(A[\Gamma])$$

in $\mathbf{Prop}(\mathcal{I}(\Gamma))$ provided that $\mathcal{I}(\Gamma)$ is a well defined object in **Cont**, $\mathcal{I}(A[\Gamma])$ is a well defined object of $\mathbf{Prop}(\mathcal{I}(\Gamma))$ and $\mathcal{I}(a[\Gamma]) : \top \sqsubseteq \bot$ is well defined in $\mathbf{Prop}(\mathcal{I}(\Gamma))$.

5.12 Conjunction

We interpret the conjunction as follows:

$$\mathcal{I}(A \wedge B[\Gamma]) := \mathcal{I}(A[\Gamma]) \sqcap_{\mathcal{I}(\Gamma)} \mathcal{I}(B[\Gamma])$$

provided that $\mathcal{I}(\Gamma)$ is a well defined object of **Cont** and $\mathcal{I}(A[\Gamma])$ and $\mathcal{I}(B[\Gamma])$ are well defined objects of $\mathbf{Prop}(\mathcal{I}(\Gamma))$.

The interpretation of the conjunction pairing $\langle a, b\rangle_{\wedge}^{A,B}[\Gamma]$ is defined as

$$\mathcal{I}(\langle a, b\rangle_{\wedge}^{A,B}[\Gamma]) : \top \sqsubseteq \mathcal{I}(A[\Gamma]) \sqcap \mathcal{I}(B[\Gamma])$$

in $\mathbf{Prop}(\mathcal{I}(\Gamma))$ provided that $\mathcal{I}(\Gamma)$ is a well defined object of **Cont** and $\mathcal{I}(A[\Gamma])$ and $\mathcal{I}(B[\Gamma])$ are well defined objects of $\mathbf{Prop}(\mathcal{I}(\Gamma))$ and $\mathcal{I}(a[\Gamma]) : \top \sqsubseteq \mathcal{I}(A[\Gamma])$ and $\mathcal{I}(b[\Gamma]) : \top \sqsubseteq \mathcal{I}(B[\Gamma])$ are well defined in $\mathbf{Prop}(\mathcal{I}(\Gamma))$.

The interpretations of the conjunction projections $\pi_{\wedge,1}^{A,B}(c)[\Gamma]$ and $\pi_{\wedge,2}^{A,B}(c)[\Gamma]$ are defined as

$$\mathcal{I}(\pi_{\wedge,1}^{A,B}(c)[\Gamma]) : \top \sqsubseteq \mathcal{I}(A[\Gamma]) \qquad \mathcal{I}(\pi_{\wedge,2}^{A,B}(c)[\Gamma]) : \top \sqsubseteq \mathcal{I}(B[\Gamma])$$

in $\mathbf{Prop}(\mathcal{I}(\Gamma))$ provided that $\mathcal{I}(\Gamma)$ is a well defined object of **Cont**, $\mathcal{I}(A[\Gamma])$ and $\mathcal{I}(B[\Gamma])$ are well defined objects of $\mathbf{Prop}(\mathcal{I}(\Gamma))$ and $\mathcal{I}(c[\Gamma]) : \top \sqsubseteq \mathcal{I}(A[\Gamma]) \sqcap \mathcal{I}(B[\Gamma])$ is well defined in $\mathbf{Prop}(\mathcal{I}(\Gamma))$.

5.13 Disjunction

We interpret the disjunction as follows:

$$\mathcal{I}(A \vee B[\Gamma]) := \mathcal{I}(A[\Gamma]) \sqcup_{\mathcal{I}(\Gamma)} \mathcal{I}(B[\Gamma])$$

provided that $\mathcal{I}(\Gamma)$ is a well defined object of **Cont** and $\mathcal{I}(A[\Gamma])$ and $\mathcal{I}(B[\Gamma])$ are well defined objects of $\mathbf{Prop}(\mathcal{I}(\Gamma))$.

The interpretations of disjunction injections $\mathsf{inl}_{\vee}^{A,B}(a)[\Gamma]$ and $\mathsf{inr}_{\vee}^{A,B}(b)[\Gamma]$ are defined as

$$\mathcal{I}(\mathsf{inl}_{\vee}^{A,B}(a)[\Gamma]) : \top \sqsubseteq \mathcal{I}(A[\Gamma]) \sqcup \mathcal{I}(B[\Gamma]) \qquad \mathcal{I}(\mathsf{inr}_{\vee}^{A,B}(b)[\Gamma]) : \top \sqsubseteq \mathcal{I}(A[\Gamma]) \sqcup \mathcal{I}(B[\Gamma])$$

in $\mathbf{Prop}(\mathcal{I}(\Gamma))$ provided that $\mathcal{I}(\Gamma)$ is a well defined object of **Cont** and $\mathcal{I}(A[\Gamma])$ and $\mathcal{I}(B[\Gamma])$ are both well defined objects of $\mathbf{Prop}(\mathcal{I}(\Gamma))$ and finally, when interpreting the first injection $\mathcal{I}(a[\Gamma]) : \top \sqsubseteq \mathcal{I}(A[\Gamma])$ is also assumed to be well defined in $\mathbf{Prop}(\mathcal{I}(\Gamma))$, and when interpreting the second injection $\mathcal{I}(b[\Gamma]) : \top \sqsubseteq \mathcal{I}(B[\Gamma])$ is also assumed to be well defined in $\mathbf{Prop}(\mathcal{I}(\Gamma))$.

The interpretation of the disjunction eliminator $\mathsf{El}_{\vee}^{A,B,C}(c,(x)\,d,(y)\,e)[\Gamma]$ is defined as

$$\mathcal{I}(\mathsf{El}_{\vee}^{A,B,C}(c,(x)\,d,(y)\,e)[\Gamma]) : \top \sqsubseteq \mathcal{I}(C[\Gamma])$$

in $\mathbf{Prop}(\mathcal{I}(\Gamma))$ provided that $\mathcal{I}(\Gamma)$ is a well defined object of **Cont**, $\mathcal{I}(A[\Gamma])$ and $\mathcal{I}(B[\Gamma])$ are well defined objects of $\mathbf{Prop}(\mathcal{I}(\Gamma))$ and $\mathcal{I}(c[\Gamma]) : \top \sqsubseteq \mathcal{I}(A[\Gamma]) \sqcup \mathcal{I}(B[\Gamma])$ is well defined in $\mathbf{Prop}(\mathcal{I}(\Gamma))$ and $\mathcal{I}(d[\Gamma, x \in A]) : \top \sqsubseteq \mathbf{Prop}_{\mathsf{pr}}(\mathcal{I}(C[\Gamma]))$ is well defined in $\mathbf{Prop}(\mathcal{I}([\Gamma, x \in A]))$ and $\mathcal{I}(e[\Gamma, y \in B]) : \top \sqsubseteq \mathbf{Prop}_{\mathsf{pr}}(\mathcal{I}(C[\Gamma]))$ is well defined in $\mathbf{Prop}(\mathcal{I}([\Gamma, y \in B]))$.

5.14 Implication

We interpret implication as follows:

$$\mathcal{I}(A \to B[\Gamma]) := \mathcal{I}(A[\Gamma]) \to_{\mathcal{I}(\Gamma)} \mathcal{I}(B[\Gamma])$$

provided that $\mathcal{I}(\Gamma)$ is a well defined object of **Cont** and $\mathcal{I}(A[\Gamma])$ and $\mathcal{I}(B[\Gamma])$ are well defined objects of $\mathbf{Prop}(\mathcal{I}(\Gamma))$.

The interpretation of the implication lambda-abstraction $(\lambda x)_{\to}^{A,B}(b)[\Gamma]$ is defined as

$$\mathcal{I}((\lambda x)_{\to}^{A,B}(b)[\Gamma]) : \top \sqsubseteq \mathcal{I}(A[\Gamma]) \to \mathcal{I}(B[\Gamma])$$

in $\mathbf{Prop}(\mathcal{I}(\Gamma))$ provided that $\mathcal{I}(\Gamma)$ is a well defined object of **Cont**, $\mathcal{I}(A[\Gamma])$ and $\mathcal{I}(B[\Gamma])$ are well defined objects of $\mathbf{Prop}(\mathcal{I}(\Gamma))$ and

$$\mathcal{I}(b[\Gamma, x \in A]) : \top \sqsubseteq \mathbf{Prop}_{\mathsf{pr}}(\mathcal{I}(B[\Gamma]))$$

is well defined in $\mathbf{Prop}(\mathcal{I}([\Gamma, x \in A]))$.

The interpretation of the implication application $\mathsf{Ap}_{\to}^{A,B}(c,a)[\Gamma]$ is defined as

$$\mathcal{I}(\mathsf{Ap}_{\to}^{A,B}(c,a)[\Gamma]) : \top \sqsubseteq \mathcal{I}(B[\Gamma])$$

in $\mathbf{Prop}(\mathcal{I}(\Gamma))$ provided that $\mathcal{I}(\Gamma)$ is a well defined object of **Cont**, $\mathcal{I}(A[\Gamma])$ and $\mathcal{I}(B[\Gamma])$ are well defined objects of $\mathbf{Prop}(\mathcal{I}(\Gamma))$ and $\mathcal{I}(c[\Gamma]) : \top \sqsubseteq \mathcal{I}(A[\Gamma]) \to \mathcal{I}(B[\Gamma])$ and $\mathcal{I}(a[\Gamma]) : \top \sqsubseteq \mathcal{I}(A[\Gamma])$ are well defined in $\mathbf{Prop}(\mathcal{I}([\Gamma]))$.

5.15 Existential quantifier

We interpret the existential quantifier as follows:

$$\mathcal{I}((\exists x \in A)B[\Gamma]) := \exists^{\mathcal{I}(\Gamma)}(\mathcal{I}(A[\Gamma]), \mathcal{I}(B[\Gamma, x \in A]))$$

provided that $\mathcal{I}(\Gamma)$ is a well defined object of **Cont**, $\mathcal{I}(A[\Gamma])$ is a well defined object of $\mathbf{Col}(\mathcal{I}(\Gamma))$ and $\mathcal{I}(B[\Gamma, x \in A])$ is a well defined object of $\mathbf{Prop}(\mathcal{I}([\Gamma, x \in A]))$.

The interpretation of the existential quantifier pairing $\langle a, b\rangle_{\exists}^{A,(x)B}[\Gamma]$ is defined as

$$\mathcal{I}(\langle a, b\rangle_{\exists}^{A,(x)B}[\Gamma]) : \top^{\mathcal{I}(\Gamma)} \sqsubseteq_{\mathcal{I}(\Gamma)} \exists^{\mathcal{I}(\Gamma)}(\mathcal{I}(A[\Gamma]), \mathcal{I}(B[\Gamma, x \in A]))$$

in $\mathbf{Prop}(\mathcal{I}(\Gamma))$ provided that $\mathcal{I}(\Gamma)$ is a well defined object of **Cont**, $\mathcal{I}(A[\Gamma])$ is a well defined objects of $\mathbf{Col}(\Gamma)$ and $\mathcal{I}(B[\Gamma, x \in A])$ is a well defined object of $\mathbf{Prop}(\mathcal{I}([\Gamma, x \in A]))$ and furthermore, $\mathcal{I}(a[\Gamma])$ is a well defined arrow from $\mathbf{1}$ to $\mathcal{I}(A[\Gamma])$ in $\mathbf{Col}(\mathcal{I}([\Gamma]))$ and $\mathcal{I}(b[\Gamma]) : \top \sqsubseteq \mathbf{Prop}_{\widetilde{\mathcal{I}(a[\Gamma])}}(\mathcal{I}(B[\Gamma, x \in A]))$ is well defined in $\mathbf{Prop}(\mathcal{I}([\Gamma]))$ (see 4.32 for notation).

The interpretation of the existential quantifier eliminator $\mathsf{El}_{\exists}^{A,(x)B,C}(a, (x, y)\, b)[\Gamma]$ is defined as

$$\mathcal{I}(\mathsf{El}_{\exists}^{A,(x)B,C}(a, (x, y)\, b)[\Gamma]) : \top^{\mathcal{I}(\Gamma)} \sqsubseteq_{\mathcal{I}(\Gamma)} \mathcal{I}(C[\Gamma])$$

in $\mathbf{Prop}(\mathcal{I}(\Gamma))$ provided that $\mathcal{I}(\Gamma)$ is a well defined object of **Cont**, $\mathcal{I}(A[\Gamma])$ is a well defined object of $\mathbf{Col}(\Gamma)$, $\mathcal{I}(C[\Gamma])$ is a well defined object of $\mathbf{Prop}(\Gamma)$ and $\mathcal{I}(B[\Gamma, x \in A])$ is a well defined object of $\mathbf{Prop}(\mathcal{I}([\Gamma, x \in A]))$ and furthermore

$$\mathcal{I}(a[\Gamma]) : \top \sqsubseteq \exists(\mathcal{I}(A[\Gamma]), \mathcal{I}(B[\Gamma, x \in A]))$$

is well defined in $\mathbf{Prop}(\mathcal{I}([\Gamma]))$ and

$$\mathcal{I}(b[\Gamma, x \in A, y \in B]) : \top \sqsubseteq \mathbf{Prop}_{\mathsf{pr}^{(2)}}(\mathcal{I}(C[\Gamma]))$$

is well defined in $\mathbf{Prop}(\mathcal{I}([\Gamma, x \in A, y \in B]))$.

5.16 Universal quantifier

We interpret the universal quantifier as follows:

$$\mathcal{I}((\forall x \in A)B[\Gamma]) := \forall^{\mathcal{I}(\Gamma)}(\mathcal{I}(A[\Gamma]), \mathcal{I}(B[\Gamma, x \in A]))$$

provided that $\mathcal{I}(\Gamma)$ is a well defined object of **Cont**, $\mathcal{I}(A[\Gamma])$ is a well defined object of $\mathbf{Col}(\mathcal{I}(\Gamma))$ and $\mathcal{I}(B[\Gamma, x \in A])$ is a well defined object of $\mathbf{Prop}(\mathcal{I}([\Gamma, x \in A]))$.

The interpretation of the universal quantifier lambda-abstraction $(\lambda x)_{\forall}^{A,B} b[\Gamma]$ is defined as

$$\mathcal{I}((\lambda x)_{\forall}^{A,B} b[\Gamma]) : \top^{\mathcal{I}(\Gamma)} \sqsubseteq_{\mathcal{I}(\Gamma)} \forall^{\mathcal{I}(\Gamma)}(\mathcal{I}(A[\Gamma]), \mathcal{I}(B[\Gamma, x \in A]))$$

in $\mathbf{Prop}(\mathcal{I}(\Gamma))$ provided that $\mathcal{I}(\Gamma)$ is a well defined object of **Cont**, $\mathcal{I}(A[\Gamma])$ is a well defined object of $\mathbf{Col}(\mathcal{I}(\Gamma))$, $\mathcal{I}(B[\Gamma, x \in A])$ is a well defined object of

$\mathbf{Prop}(\mathcal{I}([\Gamma, x \in A]))$ and $\mathcal{I}(b[\Gamma, x \in A]) : \top \sqsubseteq \mathcal{I}(B[\Gamma, x \in A])$ is well defined in $\mathbf{Prop}(\mathcal{I}([\Gamma, x \in A]))$.

The interpretation of the universal quantifier application $\mathsf{Ap}_{\forall}^{A,(x)B}(c, a)[\Gamma]$ is defined in $\mathbf{Prop}(\mathcal{I}(\Gamma))$ with the notation in 4.32 as

$$\mathcal{I}(\mathsf{Ap}_{\forall}^{A,(x)B}(c, a)[\Gamma]) : \top^{\mathcal{I}(\Gamma)} \sqsubseteq_{\mathcal{I}(\Gamma)} \mathbf{Prop}_{\widetilde{\mathcal{I}(a[\Gamma])}}(\mathcal{I}(B[\Gamma, x \in A]))$$

provided that $\mathcal{I}(\Gamma)$ is a well defined object of **Cont**, $\mathcal{I}(A[\Gamma])$ is a well defined object of $\mathbf{Col}(\mathcal{I}(\Gamma))$, $\mathcal{I}(B[\Gamma, x \in A])$ is a well defined object of $\mathbf{Prop}(\mathcal{I}([\Gamma, x \in A]))$ and, furthermore,

$$\mathcal{I}(c[\Gamma]) : \top \sqsubseteq \forall(\mathcal{I}(A[\Gamma]), \mathcal{I}(B[\Gamma, x \in A]))$$

is well defined in $\mathbf{Prop}(\mathcal{I}(\Gamma))$ and $\mathcal{I}(a[\Gamma]) : \mathbf{1} \to \mathcal{I}(A[\Gamma])$ is well defined in $\mathbf{Col}(\mathcal{I}([\Gamma]))$.

5.17 Equality proposition

We interpret the propositional equality as follows:

$$\mathcal{I}(\mathsf{Eq}(A, a, b)[\Gamma]) := \mathbf{Eq}^{\mathcal{I}(\Gamma)}(\mathcal{I}(A[\Gamma]), \mathcal{I}(a[\Gamma]), \mathcal{I}(b[\Gamma]))$$

provided that $\mathcal{I}(\Gamma)$ is a well defined object of **Cont**, $\mathcal{I}(A[\Gamma])$ is a well defined object of $\mathbf{Col}(\mathcal{I}(\Gamma))$ and both $\mathcal{I}(a[\Gamma])$ and $\mathcal{I}(b[\Gamma])$ are well defined arrows from $\mathbf{1}$ to $\mathcal{I}(A[\Gamma])$ in $\mathbf{Col}(\mathcal{I}(\Gamma))$.

The interpretation of the propositional equality term $\mathsf{eq}_A(a)[\Gamma]$ is defined as

$$\mathcal{I}(\mathsf{eq}_A(a)[\Gamma]) : \top^{\mathcal{I}(\Gamma)} \sqsubseteq_{\mathcal{I}(\Gamma)} \mathbf{Eq}^{\mathcal{I}(\Gamma)}(\mathcal{I}(A[\Gamma]), \mathcal{I}(a[\Gamma]), \mathcal{I}(a[\Gamma]))$$

in $\mathbf{Prop}(\mathcal{I}(\Gamma))$ provided that $\mathcal{I}(\Gamma)$ is a well defined object of **Cont**, $\mathcal{I}(A[\Gamma])$ is a well defined object of $\mathbf{Col}(\mathcal{I}(\Gamma))$ and $\mathcal{I}(a[\Gamma])$ is a well defined arrow from $\mathbf{1}$ to $\mathcal{I}(A[\Gamma])$ in $\mathbf{Col}(\mathcal{I}(\Gamma))$.

5.18 Decoding

$$\mathcal{I}(\tau(a)[\Gamma]) := \tau_{sp}^{\mathcal{I}(\Gamma)}(\mathcal{I}(a[\Gamma]))$$

provided that $\mathcal{I}(\Gamma)$ is a well defined object of **Cont** and $\mathcal{I}(a[\Gamma])$ is a well defined arrow from $\mathbf{1}$ to $\mathsf{USP}^{\mathcal{I}(\Gamma)}$ in $\mathbf{Col}(\mathcal{I}(\Gamma))$.

6 Validity theorem

We start with defining a list of arrows useful to interpret telescopic substitutions of dependent type theory in realized contexts Γ of $\widehat{ID_1}$.

Definition 6.1. *Suppose* Γ *and* Γ' *are objects of* **Cont**. *We define simultaneously by induction on the length of the realized context* Γ

1. *a list of arrows* $\overline{a} = [a_1, ..., a_{\ell(\Gamma)}]$ *in* $\mathbf{Col}(\Gamma')$ *with domain* $\mathbf{1}^{\Gamma'}$ *called* instance of substitution *for* Γ *in context* Γ'
2. *an arrow* $\mathbf{sub}(\overline{a},\, \Gamma',\, \Gamma) : \Gamma' \to \Gamma$ *for every instance of substitution* $\overline{a}$

as follows:

1. *the empty list* $[\,]$ *is an instance of substitution for* $[\,]$ *in context* Γ' *and* $\mathbf{sub}([\,],\, \Gamma',\, [\,]) :=!_{\Gamma',[\,]} : \Gamma' \to [\,]$
2. *if* $[\Gamma, B]$ *is an object of* **Cont**, *then* $[\overline{a}, b]$ *is an instance of substitution for* $[\Gamma, B]$ *in context* Γ' *if and only if* $\overline{a}$ *is an instance of substitution for* Γ *in context* Γ' *and* b *is an arrow from* $\mathbf{1}$ *to* $\mathbf{Col}_{\mathbf{sub}(\overline{a},\Gamma',\Gamma)}(B)$ *in* $\mathbf{Col}(\Gamma')$.

   In this case $\mathbf{sub}([\overline{a}, b],\, \Gamma',\, [\Gamma, B])$ *is defined as* $q(\mathbf{sub}(\overline{a},\, \Gamma',\, \Gamma),\, [\Gamma, B]) \circ \widetilde{b}$ *with the notation in 4.32:*

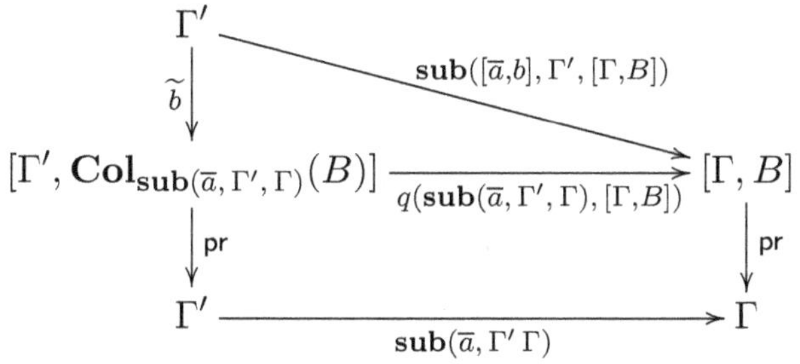

Remark 6.2. *Notice that there is a bijection between lists of arrows which are instances of substitution for* Γ *in context* Γ' *and arrows in* **Cont** *from* Γ' *to* Γ.

The following lemma, which can be proved by induction on the definition of the syntax in precontext, shows that weakening is interpreted as one could expect.

Lemma 6.3 (weakening). *Suppose* $[\Gamma, \Gamma', \Gamma'']$ *is a precontext such that both* $[\Gamma, \Gamma']$ *and* $[\Gamma, \Gamma'']$ *are precontexts. Suppose that the length of* Γ, Γ' *and* Γ'' *are* n, n', n'' *respectively and that* $\mathcal{I}([\Gamma, \Gamma'])$ *and* $\mathcal{I}([\Gamma, \Gamma''])$ *are well defined. Then* $\mathcal{I}([\Gamma, \Gamma', \Gamma''])$ *is well defined and if* $[\Gamma, \Gamma', \Gamma'']$ *is* $[y_1 \in A_1, ..., y_{n+n'+n''} \in A_{n+n'+n''}]$, *then*

1. *the list* $\overline{\mathbf{weak}}$ *defined by*

   $[\mathcal{I}(y_1[\Gamma,\Gamma',\Gamma'']),...,\mathcal{I}(y_n[\Gamma,\Gamma',\Gamma'']),\mathcal{I}(y_{n+n'+1}[\Gamma,\Gamma',\Gamma'']),...,\mathcal{I}(y_{n+n'+n''}[\Gamma,\Gamma',\Gamma''])]$

   is an instance of substitution for $\mathcal{I}([\Gamma,\Gamma''])$ *in context* $\mathcal{I}([\Gamma,\Gamma',\Gamma''])$

2. *if* $\mathcal{I}(A[\Gamma,\Gamma''])$ *is well defined, then* $\mathcal{I}(A[\Gamma,\Gamma',\Gamma''])$ *is well defined and it coincides with*

$$\mathbf{Col}_{\mathbf{sub}(\overline{\mathbf{weak}},\mathcal{I}([\Gamma,\Gamma',\Gamma'']),\mathcal{I}([\Gamma,\Gamma'']))}(\mathcal{I}(A[\Gamma,\Gamma'']))$$

3. *if* $\mathcal{I}(a[\Gamma,\Gamma''])$ *is well defined, then* $\mathcal{I}(a[\Gamma,\Gamma',\Gamma''])$ *is well defined and it coincides with*

$$\mathbf{Col}_{\mathbf{sub}(\overline{\mathbf{weak}},\mathcal{I}([\Gamma,\Gamma',\Gamma'']),\mathcal{I}([\Gamma,\Gamma'']))}(\mathcal{I}(a[\Gamma,\Gamma'']))$$

The next lemma can be proved by induction on the definition of the syntax in precontext and it shows that substitution commutes with the interpretation $\mathcal{I}$, i. e. that one can first perform a substitution in $\mathbf{mTT}^a$ and then interpret the resulting type or term or, equivalently, first interpret terms and types of $\mathbf{mTT}^a$ and then perform the substitution of the interpreted terms.

Lemma 6.4 (Substitution Lemma). *Let* $\Gamma = [x_1 \in A_1, ..., x_n \in A_n]$ *be a precontext with* $n > 0$ *and let* Γ' *be a precontext. Let* $\mathcal{I}(\Gamma)$ *and* $\mathcal{I}(\Gamma')$ *be well defined and suppose that* $\mathcal{I}(a_1[\Gamma'])$,...., $\mathcal{I}(a_n[\Gamma'])$ *are well defined and constitute an instance of subtitution for* $\mathcal{I}(\Gamma)$ *in context* $\mathcal{I}(\Gamma')$. *Then*

1. *if* $\mathcal{I}(B[\Gamma])$ *is well defined in* $\mathbf{Col}(\mathcal{I}(\Gamma))$, *then* $\mathcal{I}(B[a_1/x_1,...,a_n/x_n][\Gamma'])$ *is well defined and it coincides with*

$$\mathbf{Col}_{\mathbf{sub}(\mathcal{I}(\overline{a}[\Gamma']),\mathcal{I}(\Gamma'),\mathcal{I}(\Gamma))}(\mathcal{I}(B[\Gamma]))$$

2. *if* $\mathcal{I}(b[\Gamma])$ *is well defined in* $\mathbf{Col}(\mathcal{I}(\Gamma))$, *then* $\mathcal{I}(b[a_1/x_1,...,a_n/x_n][\Gamma'])$ *is well defined and it coincides with*

$$\mathbf{Col}_{\mathbf{sub}(\mathcal{I}(\overline{a}[\Gamma']),\mathcal{I}(\Gamma'),\mathcal{I}(\Gamma))}(\mathcal{I}(b[\Gamma]))$$

where we denote by $\mathcal{I}(\overline{a}[\Gamma'])$ *the list of the interpretations* $\mathcal{I}(a_i[\Gamma'])$.

What shown so far helps to prove our main theorem:

Theorem 6.5. *The effective pretripos* $(\mathbf{Cont},\mathbf{Col},\mathbf{Set},\mathbf{Prop},\mathbf{Prop}_s)$ *validates all judgements of* $\mathbf{mTT}^a$ *in the sense that:*
for every judgement J *of* $\mathbf{mTT}^a$, *if* $\mathbf{mTT}^a \vdash J$, *then* $\mathcal{R} \vDash J$.

Thanks to proposition 2.1 we also deduce

Corollary 6.6. *The effective pretripos* $(\mathbf{Cont},\mathbf{Col},\mathbf{Set},\mathbf{Prop},\mathbf{Prop}_s)$ *validates all judgements of* $\mathbf{mTT}$ *in the sense that:*
for every judgement J *of* $\mathbf{mTT}$, *if* $\mathbf{mTT} \vdash J$, *then* $\mathcal{R} \vDash J$.

6.1 The validity of CT

Proposition 6.7. *The effective pretripos validates* **CT**, *i.e.* $\mathcal{R} \models \mathbf{CT}$, *where* **CT** *is the formula*

$$(\forall x \in \mathsf{N})\,(\exists y \in \mathsf{N})\,R(x,y) \to (\exists e \in \mathsf{N})\,(\forall x \in \mathsf{N})\,(\exists z \in \mathsf{N})\,(T(e,x,z) \wedge R(x,U(z)))$$

where T *and* U *are respectively the Kleene predicate and the primitive recursive function representing Kleene application in* $\mathbf{mTT}^a$ *respectively.*

Proof. The validity in $\mathcal{R}$ of **CT** can be obtained as a consequence of the validity in $\mathcal{R}$ of the following principles:
Formal Church thesis for type-theoretic functions $\mathbf{CT}_\lambda$ defined as:

$$(\forall f \in (\Pi x \in \mathsf{N})\mathsf{N})\,(\exists e \in \mathsf{N})\,(\forall x \in \mathsf{N})\,(\exists y \in \mathsf{N})$$
$$(T(e,x,y) \wedge \mathsf{Eq}(\mathsf{N}, U(y), \mathsf{Ap}(f,x)))$$

and *the axiom of countable choice* $\mathbf{AC}_{\mathsf{N},\mathsf{N}}$ defined for $\mathbf{mTT}^a \vdash R\,prop\,[x \in \mathsf{N}, y \in \mathsf{N}]$ as

$$(\forall x \in \mathsf{N})\,(\exists y \in \mathsf{N})\,R(x,y) \to (\exists f \in (\Pi x \in \mathsf{N})\,\mathsf{N})\,(\forall x \in \mathsf{N})\,R(x, \mathsf{Ap}(f,x))$$

One can easily show that in $\mathbf{Prop}([\,])$:

$$\mathbf{1} \sqsubseteq \mathcal{I}(\mathbf{CT}_\lambda[\,]).$$

In fact we know by general results on Kleene realizability that there exists a numeral $\mathbf{r}$ for which

$$HA \vdash \exists u\, T(f,x,u) \to (\{\mathbf{r}\}(f,x) \Vdash_k \exists u\, T(f,x,u))$$

Using this remark and proof irrelevance we can show that the interpretation of $\mathbf{CT}_\lambda[\,]$ has a global element determined by the numeral

$$\Lambda z.\Lambda f.\,\{\mathbf{p}\}\,(f,\,\Lambda x.\,\{\mathbf{p}\}\,(\{\mathbf{p}_1\}\,(\{\mathbf{r}\}(f,x)),\,\{\mathbf{p}\}(\{\mathbf{p}_2\}\,(\{\mathbf{r}\}\,(f,x)),\,0)))$$

where the first variable z belongs to $\mathbf{1}$. Moreover $\mathcal{R} \models \mathbf{AC}_{\mathsf{N},\mathsf{N}}$ as equality in N is interpreted as numerical equality. □

It is worth noting that theorem 6.6 shows that $\widehat{ID_1}$ is an upper bound of the proof-theoretic strength of **mTT**. Actually, this is a direct proof of it because in [23] it was observed that **mTT** can be interpreted in first-order Martin-Löf's type theory with one universe, for short MLtt_1, whose proof-theoretic strength is known to be equal to that of $\widehat{ID_1}$(see [8]). Even more our interpretation of **mTT** in $\widehat{ID_1}$ is a modification of that in [8] used to establish that $\widehat{ID_1}$ is an upper bound of MLtt_1. The main difference between our proof and that for MLtt_1 is that ours validates **CT** while that in [8] falsifies **CT**.

It is left to future work to establish whether the proof-theoretic strength of **mTT** and hence of **MF** coincides with that of $\widehat{ID_1}$ as it happens to MLtt_1.

7 Conclusions

We have built here an effective predicative categorical structure, called *effective pretripos* for the intensional level **mTT** of **MF** extended with the formal Church thesis **CT**, in Feferman's predicative classical theory $\widehat{ID_1}$.

This is intended to be a basic categorical structure of realizers for **mTT** useful to build a predicative variant of Hyland's Effective Topos. A predicative effective topos will be obtained by completing our effective pretripos with quotients by means of the elementary quotient completion introduced and studied in [26, 25, 27]. Indeed, such an elementary quotient completion axiomatizes the quotient model used in [23] to interpret the extensional level of **MF** into **mTT** and generalizes the notion of the exact completion on a lex category. Therefore, it appears to be a starting point to generalize the tripos-to-topos construction in [18] predicatively and to validate the extensional level **emTT** of **MF** extended with **CT** when applied to our effective pretripos.

Another goal of our future work will be to make a precise comparison between our categorical structures of realizers for **MF** and the categorical approach to predicative effective models for Aczel's CZF in [41].

References

[1] P. Aczel. The type theoretic interpretation of constructive set theory. In *Logic Colloquium '77 (Proc. Conf., Wrocław, 1977)*, volume 96 of *Stud. Logic Foundations Math.*, Amsterdam-New York, 1978. North-Holland.

[2] P. Aczel. The type theoretic interpretation of constructive set theory: choice principles. In Dirk van Dalen Anne Troelstra, editor, *The L.E.J. Brouwer Centenary Symposium (Noordwijkerhout, 1981)*, volume 110 of *Stud. Logic Foundations Math.*, Amsterdam-New York, 1982. North-Holland.

[3] P. Aczel. The type theoretic interpretation of constructive set theory: inductive definitions. In *Logic, methodology and philosophy of science, VII (Salzburg, 1983)*, volume 114 of *Stud. Logic Foundations Math.*, Amsterdam-New York, 1986. North-Holland.

[4] P. Aczel and M. Rathjen. Notes on constructive set theory. Mittag-Leffler Technical Report No.40, 2001.

[5] A. Asperti, W. Ricciotti, C. Sacerdoti Coen, and E. Tassi. The Matita interactive theorem prover. In *Proceedings of the 23rd International Conference on Automated Deduction (CADE-2011), Wroclaw, Poland*, volume 6803 of *LNCS*, 2011.

[6] Andrea Asperti, Wilmer Ricciotti, and Claudio Sacerdoti Coen. Matita tutorial. *Journal of Formalized Reasoning*, 7(2):91–199, 2014.

[7] G. Barthes, V. Capretta, and O. Pons. Setoids in type theory. *J. Funct. Programming*, 13(2):261–293, 2003. Special issue on "Logical frameworks and metalanguages".

[8] M. Beeson. *Foundations of Constructive Mathematics.* Springer-Verlag, Berlin, 1985.

[9] Y. Bertot and P. Castéran. *Interactive Theorem Proving and Program Development.* Texts in Theoretical Computer Science. Springer Verlag, 2004. ISBN-3-540-20854-2.

[10] A. Bove, P. Dybjer, and U. Norell. A brief overview of Agda - a functional language with dependent types. In S. Berghofer, T. Nipkow, C. Urban, and M. Wenzel, editors, *Theorem Proving in Higher Order Logics, 22nd International Conference, TPHOLs 2009*, volume 5674 of *Lecture Notes in Computer Science*, pages 73–78. Springer, August 2009.

[11] W. Bucholz, S. Feferman, W. Pohlers, and W. Sieg. *Iterated inductive definitions and subsystems of analysis.* Number 897 in Lecture Notes in Mathematics. Springer Verlag, 1981.

[12] Coq development team. *The Coq Proof Assistant Reference Manual: release 8.4pl6.* INRIA, Orsay, France, April 2015.

[13] T. Coquand. Metamathematical investigation of a calculus of constructions. In P. Odifreddi, editor, *Logic in Computer Science*, pages 91–122. Academic Press, 1990.

[14] T. Coquand and C. Paulin-Mohring. Inductively defined types. In P. Martin-Löf and G. Mints, editors, *Proceedings of the International Conference on Computer Logic (Colog '88)*, volume 417 of *Lecture Notes in Computer Science*, pages 50–66, Berlin, Germany, 1990. Springer.

[15] S. Feferman. Iterated inductive fixed-point theories: application to Hancock's conjecture. In *Patras Logic Symposion*, pages 171–196. North Holland, 1982.

[16] E. Griffor and M. Rathjen. The strenght of some martin löf type theories. *Archive for Mathematical Logic*, 33(5):347–385, 1994.

[17] J. M. E. Hyland. The effective topos. In A. S. Troesltra and D. van Dalen, editors, *The L. E. J. Brouwer centenary symposium*, Studies in logic and the foundations of mathematics, pages 165–216. North-Holland, 1982.

[18] J. M. E. Hyland, P. T. Johnstone, and A. M. Pitts. Tripos theory. *Bull. Austral. Math. Soc.*, 88:205–232, 1980.

[19] A. Joyal and I. Moerdijk. *Algebraic set theory*, volume 220 of *Lecture Note Series.* Cambridge University Press, 1995.

[20] S. Mac Lane. *Categories for the working mathematician.*, volume 5 of *Graduate text in Mathematics.* Springer, 1971.

[21] S. Mac Lane and I. Moerdijk. *Sheaves in Geometry and Logic. A first introduction to*

Topos theory. Springer Verlag, 1992.

[22] M. E. Maietti. Modular correspondence between dependent type theories and categories including pretopoi and topoi. *Mathematical Structures in Computer Science*, 15(6):1089–1149, 2005.

[23] M. E. Maietti. A minimalist two-level foundation for constructive mathematics. *Annals of Pure and Applied Logic*, 160(3):319–354, 2009.

[24] M. E. Maietti and S. Maschio. An extensional Kleene realizability model for the Minimalist Foundation. In *20th International Conference on Types for Proofs and Programs, TYPES 2014, May 12-15, 2014, Paris, France*, pages 162–186, 2014.

[25] M. E. Maietti and G. Rosolini. Elementary quotient completion. *Theory and Applications of Categories*, 27(17):445–463, 2013.

[26] M. E. Maietti and G. Rosolini. Quotient completion for the foundation of constructive mathematics. *Logica Universalis*, 7(3):371–402, 2013.

[27] M. E. Maietti and G. Rosolini. Unifying exact completions. *Applied Categorical Structures*, 23(1):43–52, 2015.

[28] M. E. Maietti and G. Sambin. Toward a minimalist foundation for constructive mathematics. In L. Crosilla and P. Schuster, editor, *From Sets and Types to Topology and Analysis: Practicable Foundations for Constructive Mathematics*, number 48 in Oxford Logic Guides, pages 91–114. Oxford University Press, 2005.

[29] M. E. Maietti and G. Sambin. Why topology in the Minimalist Foundation must be pointfree. *Logic and Logical Philosophy*, 22(2):167–199, 2013.

[30] P. Martin-Löf. *Intuitionistic Type Theory. Notes by G. Sambin of a series of lectures given in Padua, June 1980.* Bibliopolis, Naples, 1984.

[31] B. Nordström, K. Petersson, and J. Smith. *Programming in Martin Löf's Type Theory.* Clarendon Press, Oxford, 1990.

[32] A. M. Pitts. Categorical logic. In Oxford University Press, editor, *Handbook of Logic in Computer Science*, volume 5, pages 39–128, 2000.

[33] A. M. Pitts. Tripos theory in retrospect. *Mathematical Structures in Computer Science*, 12:265–279, 2002.

[34] G. Sambin. Two applications of dynamic constructivism: Brouwer's continuity principle and choice sequences in formal topology. In M. van Atten, P. Boldini, M. Bourdeau, and G. Heinzmann, editors, *One Hundred years of Intuitionism (1907-2007). The Cerisy Conference*, pages 301–315. Birkhäuser, 2008.

[35] G. Sambin and S. Valentini. Building up a toolbox for Martin-Löf's type theory: subset theory. In G. Sambin and J. Smith, editors, *Twenty-five years of constructive type theory, Proceedings of a Congress held in Venice, October 1995*, pages 221–244. Oxford U. P., 1998.

[36] R. Seely. Locally cartesian closed categories and type theory. *Math. Proc. Cambr. Phyl. Soc.*, 95:33–48, 1984.

[37] R. A. G. Seely. Hyperdoctrines, natural deduction and the Beck condition. *Zeitschr. f. Math. Logik. und Grundlagen d. Math.*, 29:505–542, 1983.

[38] T. Streicher. *Semantics of type theory.* Birkhäuser, 1991.

[39] T. Streicher. Independence of the induction principle and the axiom of choice in the pure calculus of constructions. *Theoretical Computer Science*, 103(2):395–408, 1992.

[40] A. S. Troelstra and D. van Dalen. Constructivism in mathematics, an introduction, vol. I. In *Studies in logic and the foundations of mathematics.* North-Holland, 1988.

[41] B. van den Berg and I. Moerdijk. Aspects of predicative algebraic set theory II: Realizability. *Theoretical Computer Science*, Theoretical Computer 412:1916–1940, 2011.

Received 28 July 2015

A Translation Theorem For Restricted R-Formulas

Joan Rand Moschovakis
Occidental College (Emerita) and $\mu\pi\lambda\alpha$
joan.rand@gmail.com

Abstract

The three-sorted formal system **RLS** described in [5] is like **RLS($\prec$)** in [4] but without the $\prec$. **IRLS** is a strictly intuitionistic subsystem of **RLS**. This note gives a natural, syntactically defined translation φ mapping each restricted formula E with only number and lawlike sequence variables free, to a formula $\varphi(E)$ containing only number and lawlike sequence variables, such that **IRLS** proves $E \leftrightarrow \varphi(E)$. If E contains no choice sequence variables then $\varphi(E)$ is E.

1 The systems RLS, IRLS, R, IR and C

1.1 A three-sorted language $\mathcal{L}$

The language, extending the two-sorted language of [2] and [1], contains three sorts of variables with or without subscripts, also used as metavariables:

$i, j, k, l, m, n, w, x, y, z$ over natural numbers,

a, b, c, d, e, g, h over lawlike sequences,

$\alpha, \beta, \gamma, \ldots$ over arbitrary choice sequences;

finitely many constants f_0 $(= 0)$, f_1 $(= ')$ (successor), f_2 $(= +)$, f_3 $(= \cdot)$, f_4 $(= exp)$, f_5, $\ldots$, f_p for primitive recursive functions and functionals; the binary predicate constant $=$ (between terms); Church's λ denoting function abstraction; parentheses (,) denoting function application; and the logical symbols $\&$, $\vee, \rightarrow, \neg$ and quantifiers $\forall, \exists$ over each sort of variable.

I thank Sean Walsh and Kai Wehmeyer of UC Irvine, and the organizers of the 2014 Chiemsee Summer School, for giving me new opportunities to talk about this subject, resulting in this theorem. I am also very grateful to an anonymous referee whose careful reading led to many improvements.

Terms (of type 0) and *functors* (of type 1) are defined inductively. Number variables and 0 are terms. Sequence variables of both sorts, and unary function constants, are functors. If f_i is a k_i, m_i-ary function constant, $u_1, \ldots, u_{k_i}$ are functors and $t_1, \ldots, t_{m_i}$ are terms, then $\mathrm{f}_i(u_1, \ldots, u_{k_i}, t_1, \ldots, t_{m_i})$ is a term. If u is a functor and t is a term then $(u)(t)$ (also written $u(t)$) is a term. If t is a term and x is a number variable then $\lambda x(t)$ (also written $\lambda x.t$) is a functor.

Prime formulas are of the form $s = t$ where s, t are terms. If u, v are functors then $u = v$ abbreviates $\forall x\,(u(x) = v(x))$. Composite formulas are formed as usual, with parentheses determining scopes.

Terms and functors with no occurrences of arbitrary choice sequence variables are *R-terms* and *R-functors* respectively. Formulas with no *free* occurrences of arbitrary choice sequence variables are *R-formulas.*

1.2 The logical axioms and rules

The logical basis is intuitionistic three-sorted predicate logic, extending the rules and axiom schemas in [2] to formulas, terms and functors of $\mathcal{L}$ as defined above, with new rules and axiom schemas for lawlike sequence variables and R-functors:

9^R. $C \to A(b)$ / $C \to \forall b A(b)$ if b is not free in C.

10^R. $\forall b A(b) \to A(u)$ if u is an R-functor free for b in $A(b)$.

11^R. $A(u) \to \exists b A(b)$ if u is an R-functor free for b in $A(b)$.

12^R. $A(b) \to C$ / $\exists b A(b) \to C$ if b is not free in C.

1.3 Axioms for 3-sorted intuitionistic number theory

Equality axioms assert that $=$ is an equivalence relation and $x = y \to \alpha(x) = \alpha(y)$, so $\forall x(a(x) = a(x))$ is provable (since lawlike sequence variables are functors), so $\forall a \exists \beta \forall x(a(x) = \beta(x))$ follows by the instance $\forall x(a(x) = \gamma(x)) \to \exists \beta \forall x(a(x) = \beta(x))$ of axiom schema 11^F from [2]. Just as Brouwer's infinitely proceeding sequences include all the sharp arrows, every lawlike sequence is (equal to) a choice sequence.

By a similar argument, if u is an R-functor in which the variable b does not occur then $\exists b \forall x(b(x) = u(x))$ is provable, so every R-functor denotes a lawlike sequence.

For terms $r(x), t$ with t free for x in $r(x)$, the λ-reduction axiom schema is

$$(\lambda x.r(x))(t) = r(t),$$

where $r(t)$ is the result of substituting t for all free occurrences of x in $r(x)$.

The mathematical axioms include the assertions that 0 ($= \mathrm{f}_0$) is not a successor and the successor function ($= \mathrm{f}_1$) is one-to-one, the defining equations for the primitive recursive function and functional constants $\mathrm{f}_2, \ldots, \mathrm{f}_p$ ([2], [1]) and the mathematical induction schema extended to $\mathcal{L}$. For the countable axiom of choice

$$\mathrm{AC}_{01}. \qquad \forall x \exists \alpha A(x, \alpha) \rightarrow \exists \alpha \forall x A(x, \lambda y.\alpha(2^x \cdot 3^y))$$

the x must be distinct from y, and free for α in $A(x, \alpha)$.

Finite sequences are coded primitive recursively as in [2], so $\langle x_0, \ldots, x_k \rangle = \Pi_0^k p_i^k$ where p_i is the *ith* prime with $p_0 = 2$, and $(y)_i$ is the exponent of p_i in the prime factorization of y. Let $Seq(y)$ abbreviate $\forall i < lh(y)((y)_i > 0)$ where $lh(y)$ is the number of nonzero exponents in the prime factorization of y. The empty sequence is coded by $\langle\,\rangle = 1$, and if $k \geq 0$ then $\langle x_0 + 1, \ldots, x_k + 1 \rangle$ codes the finite sequence $(x_0, \ldots, x_k)$. If $Seq(y)$ and $Seq(z)$ then $y * z$ codes the concatenation of the sequences coded by y and z.

The finite initial segment of length n of a choice sequence α is coded by $\overline{\alpha}(n)$, where $\overline{\alpha}(0) = 1$ and $\overline{\alpha}(n+1) = \langle \alpha(0) + 1, \ldots, \alpha(n) + 1 \rangle$. Other useful abbreviations are $\alpha \in w$ for $\overline{\alpha}(lh(w)) = w$, $w \sqsubseteq y$ for $Seq(y)$ & $\forall i < lh(w)((w)_i = (y)_i)$, and $w \sqsubset y$ for $w \sqsubseteq y$ & $lh(y) > lh(w)$. If $Seq(w)$ then $w * \alpha = \beta$ where $\beta \in w$ and $\beta(lh(w) + n) = \alpha(n)$; if $\neg Seq(w)$ then $w * \alpha = \alpha$. Note that $w * \alpha$ is a functor and $w * a$ is an R-functor.

1.4 Bar induction

Kleene formulated Brouwer's "bar theorem" as an axiom schema, in four versions which are all equivalent using AC_{01} (or even AC_{00}!), and included it in his basic system **B**. The version we assume (now for the three-sorted language) is[1]

$$\text{BI!} \quad \forall \alpha \exists ! x R(\overline{\alpha}(x)) \ \&\ \forall w(Seq(w)\ \&\ R(w) \rightarrow A(w)) \\ \&\ \forall w(Seq(w)\ \&\ \forall n A(w * \langle n \rangle) \rightarrow A(w)) \rightarrow A(1).$$

This schema (for the two-sorted language without lawlike sequence variables) completed Kleene's *basic system* **B**, which is neutral in the sense that it is correct both intuitionistically and classically.

1.5 *R*-lawless sequences, restricted quantification and lawlike comprehension

Intuitively, a lawless sequence should not be predictable by any lawlike process, but this negative condition is not enough to satisfy Kreisel's axioms. Instead, call

[1]In general, $\exists ! x A(x)$ abbreviates $\exists x A(x)$ & $\forall x \forall y (A(x)$ & $A(y) \rightarrow x = y)$.

a choice sequence β a *predictor* if β maps finite sequence codes to finite sequence codes, and call a choice sequence α *R-lawless* if every lawlike predictor correctly predicts α somewhere. Formally, $RLS(\alpha)$ abbreviates

$$\forall b[Pred(b) \rightarrow \exists x \ \alpha \in \overline{\alpha}(x) * b(\overline{\alpha}(x))],$$

where $Pred(b)$ abbreviates $\forall w(Seq(w) \rightarrow Seq(b(w)))$.

Since each prediction affects only finitely many values, this positive condition leaves room for (indeed, insures) plenty of chaotic behavior *if there are only countably many lawlike predictors.* The usual diagonal argument guarantees that there is no lawlike enumeration of the lawlike sequences, but a classical model with countably many lawlike sequences is described in [4].

Troelstra's extension principle, which claims that every continuous partial function defined on all lawless sequences has a continuous total extension, fails for R-lawless sequences, since $\forall\alpha[RLS(\alpha) \rightarrow \exists n\alpha(n) = 1]$ but the function assigning to each R-lawless α the least n such that $\alpha(n) = 1$ cannot be extended continuously to all choice sequences. And while Kreisel and Troelstra considered any two distinct lawless sequences to be independent, a stronger condition for independence is needed here.

Two R-lawless sequences α, β will be called *independent* if their fair merge $[\alpha, \beta]$ is lawless, and similarly for $\alpha_0, \ldots, \alpha_k$ where $[\alpha_0, \ldots, \alpha_k]((k+1)n + i) = \alpha_i(n)$ for $0 \leq i \leq k$ and all n. This natural notion of independence for lawless sequences was proposed by M. Fourman at the Brouwer Centenary Conference in 1981.

The class of *restricted* formulas is defined inductively: Each formula E with no arbitrary choice sequence quantifiers is *restricted.* If A is *restricted* and contains free no arbitrary choice sequence variables other than α, then $\forall\alpha[RLS(\alpha) \rightarrow A]$ and $\exists\alpha[RLS(\alpha) \ \& \ A]$ are *restricted.* If $k > 0$ and A is *restricted* with no arbitrary choice sequence variables other than $\alpha_0, \ldots, \alpha_k$ occurring free, then for $i = 0, \ldots, k$ the formulas $\forall\alpha_i[RLS([\alpha_0, \ldots, \alpha_k]) \rightarrow A]$ and $\exists\alpha_i[RLS([\alpha_0, \ldots, \alpha_k]) \ \& \ A]$ are *restricted.* No other formulas are *restricted.*

There is a lawlike function-comprehension schema

$$\mathrm{AC}^R_{00}! \qquad \forall x \exists! y A(x,y) \rightarrow \exists b \forall x A(x, b(x))$$

where $A(x, y)$ is any restricted R-formula and b is free for y in $A(x, y)$. By this axiom, the lawlike sequences are closed under "recursive in."[2]

[2]While a restricted formula can have free occurrences of arbitrary choice sequence variables, a restricted R-formula cannot. If $A(x, y)$ is a restricted R-formula, the informal abbreviation $\mu y A(x, y)$ may be allowed under either of the assumptions $\exists! y A(x, y)$ or $\exists y(A(x, y) \ \& \ \forall z < y \neg A(x, z))$.

For restricted R-formulas $A(x,a)$ the lawlike comprehension schema entails

$$\mathrm{AC}^R_{01}! \qquad \forall x \exists ! a A(x,a) \to \exists b \forall x A(x, \lambda y. b(2^x \cdot 3^y)),$$

with the obvious conditions on the variables.

1.6 Axioms for R-lawless sequences

These are Kreisel's and Troelstra's axioms from [3] and [7], adapted to Kleene's convention for coding continuous functions, with inequality of lawless sequences replaced by independence. There are two *density axioms*:

$$\text{RLS1.} \qquad \forall w(Seq(w) \to \exists \alpha [RLS(\alpha) \;\&\; \alpha \in w]),$$

$$\text{RLS2.} \qquad \forall w(Seq(w) \to \forall \alpha [RLS(\alpha) \to \exists \beta [RLS([\alpha, \beta]) \;\&\; \beta \in w]]).$$

Kreisel's principle of *open data* is stated as follows, on condition that $A(\alpha)$ is restricted and has no other arbitrary choice sequence variables free, and β is free for α in $A(\alpha)$:

$$\text{RLS3.} \quad \forall \alpha [RLS(\alpha) \to (A(\alpha) \to \exists w(Seq(w) \;\&\; \alpha \in w \;\&\; \forall \beta [RLS(\beta) \to (\beta \in w \to A(\beta))]))].$$

Effective continuous choice for lawless sequences is the schema

$$\text{RLS4.} \quad \forall \alpha [RLS(\alpha) \to \exists b A(\alpha, b)] \to \exists e \exists b \forall \alpha [RLS(\alpha) \to \exists ! y e(\overline{\alpha}(y)) > 0 \;\&\; \forall y (e(\overline{\alpha}(y)) > 0 \to A(\alpha, \lambda x. b(\langle e(\overline{\alpha}(y)) \dot{-} 1, x \rangle))))]$$

where $A(\alpha, b)$ is restricted with no arbitrary choice sequence variables but α free, and e, y, α are free for b in $A(\alpha, b)$.[3]

1.7 The restricted law of excluded middle

For $A(\alpha)$ restricted, with no choice sequence variables free except possibly α, **RLS** also has the axiom schema

$$\text{RLEM.} \qquad \forall \alpha [RLS(\alpha) \to A(\alpha) \vee \neg A(\alpha)].$$

By an easy argument, RLS3 and the restricted LEM entail the following principle of *closed data* with the same restrictions on $A(\alpha)$ as for RLS3:

$$\text{RLS5.} \quad \forall \alpha [RLS(\alpha) \to (\forall w(\alpha \in w \to \exists \beta [RLS(\beta) \;\&\; \beta \in w \;\&\; A(\beta)]) \to A(\alpha))].$$

[3] In general, $e(\alpha) \simeq n$ abbreviates $\exists x(e(\overline{\alpha}(x)) = n+1 \;\&\; \forall y < x e(\overline{\alpha}(y)) = 0)$.

In a strictly intuitionistic system without RLEM, RLS5 may or may not be taken as an additional axiom schema. With RLS1, RLEM entails the law of excluded middle for all formulas with only lawlike and number variables. Observe that RLS1, RLS2, and all instances of RLS3, RLS4, RLEM and RLS5 are restricted R-formulas.

1.8 Five axiomatic systems

In addition to Kleene's basic formal system **B** for neutral analysis we consider five other formal systems. All but one are consistent with full intuitionistic analysis **FIM** as formalized in [2].[4]

IRLS extends **B** to the three-sorted language and adds axioms RLS1,2 and axiom schemas $\mathrm{AC}^R_{00}!$ and RLS3,4. **IRLS** expresses a strictly intuitionistic theory of lawlike and relatively lawless sequences in the context of full intuitionistic analysis.

RLS is the three-sorted semi-intuitionistic system **IRLS** + RLEM.[5]

Lawlike classical analysis **R** is the two-sorted subsystem of **RLS** obtained by restricting the language to number and lawlike sequence variables, omitting RLS1-4 and BI!, replacing RLEM by $A \vee \neg A$ for formulas of the two-sorted language, replacing AC_{01} and $\mathrm{AC}^R_{00}!$ by AC^R_{01} (like $\mathrm{AC}^R_{01}!$ but without the !) for formulas of the two-sorted language, and restating the equality axioms and primitive recursive definitions of function constants using lawlike instead of arbitrary choice sequence variables.

Constructive analysis **IR** is the two-sorted intuitionistic subsystem of **R** obtained by omitting $A \vee \neg A$. Note that **IR** has no version of bar induction.

Classical analysis **C** is the two-sorted system obtained from Kleene's **B** by strengthening the logic to classical logic. A lawlike version $\mathrm{BI}!^R$ of BI!, with lawlike sequence variables replacing arbitrary choice sequence variables, is provable in **R**. Thus **C** and **R** are notational variants, as are **B** and **IR** + $\mathrm{BI}!^R$.

1.9 Closure properties of RLS: Lemma

The three-sorted subsystem **IRS** *of* **IRLS** *obtained by omitting* RLS1-4, *but retaining* $\mathrm{AC}^R_{00}!$, *proves*

(i) $\forall\alpha[RLS(\alpha) \leftrightarrow \forall w(Seq(w) \rightarrow RLS(w * \alpha))]$.

[4]The relative consistency of a common extension of **RLS** and **FIM** is established in [4] under the assumption that a definably well-ordered subset of ω^ω is countable.

[5]The translation theorem will show that **RLS** can also be axiomatized by **IRLS** plus the law of excluded middle for strictly lawlike formulas, so **RLS** is indeed semi-intuitionistic.

(ii) $\forall b(\forall x\forall y(b(x) = b(y) \to x = y)\ \&\ \forall y(\exists x\, b(x) = y\ \vee\ \neg\exists x\, b(x) = y) \to \forall\alpha[RLS(\alpha) \to RLS(\alpha \circ b)])$.

(iii) $\forall b(Pred(b) \to \forall n\forall\alpha[RLS(\alpha) \to \exists m(m \geq n\ \&\ \alpha \in \overline{\alpha}(m) * b(\overline{\alpha}(m)))])$.

Proofs. This is a formal version of Lemma 2 of [4]. For (i $\to$) assume $Seq(w)$ and $Pred(b)$. Then $\forall x\exists! y((Seq(x) \to y = b(w * x))\ \&\ (\neg Seq(x) \to y = 0))$, so by $\mathrm{AC}^{R}_{00}!$ there is a c such that $Pred(c)$ and $\forall x(Seq(x) \to c(x) = b(w * x))$, so if $RLS(\alpha)$ then $\exists z\, \alpha \in \overline{\alpha}(z) * c(\overline{\alpha}(z))$ and hence $\exists z(w * \alpha \in \overline{w * \alpha}(lh(w) + z) * b(\overline{w * \alpha}(lh(w) + z)))$. For (i $\leftarrow$) take $w = 1 = \langle\,\rangle$, so $w * \alpha = \alpha$. The proofs of (ii), (iii) similarly formalize the proofs of (ii), (iii) of Lemma 2 of [4]. □

1.10 Axioms RLS1-3 reconsidered

Lemma 1.9 guarantees that the following schemas RLS1′, RLS2′ and RLS3′ are equivalent over **IRS** to RLS1, RLS2 and RLS3 respectively.

RLS1′. $\exists\alpha RLS(\alpha)$

RLS2′. $\forall\alpha[RLS(\alpha) \leftrightarrow \exists\beta RLS([\alpha, \beta])]$

RLS3′. $\forall\alpha[RLS(\alpha) \to (A(\alpha) \leftrightarrow \exists w(Seq(w)\,\&\,\alpha \in w\,\&\,\forall\beta[RLS(\beta) \to A(w * \beta)]))]$

under the same conditions on $A(\alpha)$ as for RLS3. The next section suggests a way to simplify RLS4 as well.

2 The translation theorem

2.1 Theorem

Every restricted formula E of the three-sorted language with no arbitrary choice sequence variables free is equivalent in **IRLS** *to a formula $\varphi(E)$ of the two-sorted language with only number and lawlike sequence variables. The mapping φ is syntactically defined. If E contains no choice sequence variables then $\varphi(E)$ is E.*

The proof is similar to Troelstra's proof of the translation theorem for **LS** into the language without lawless sequence variables (cf. [8], 663ff), with a significant difference. Instead of the constant K_0 Troelstra used to represent the class of lawlike codes of continuous *total* functions, we can *define* the condition for e to be a lawlike code of a continuous partial function defined on all the R-lawless sequences:

$$J_0(e) \equiv \forall w(Seq(w)\ \&\ \forall n < lh(w)(e(\overline{w}(n)) = 0) \to \exists y(Seq(y)\ \&\ e(w * y) > 0)),$$
$$J_1(e) \equiv J_0(e)\ \&\ \forall w[e(w) > 0 \to Seq(w)\ \&\ \forall y(Seq(y) \to e(w * y) = e(w))].$$

By Lemma 1.9(i) and the next lemma, the conclusion of effective continuous choice for R-lawless sequences can be rewritten

$$\exists e \exists b(J_1(e) \,\&\, \forall n \forall w(e(w) = n+1 \to \forall \alpha[RLS(\alpha) \to A(w * \alpha, \lambda x.b(\langle n, x\rangle))])).$$

2.2 Lemma

(i) **IRS** + RLS1 *proves* $\forall e(J_0(e) \leftrightarrow \forall \alpha[RLS(\alpha) \to e(\alpha)\downarrow])$, *and*

(ii) **IRS** *proves* $\forall \alpha[RLS(\alpha) \leftrightarrow \forall e(J_j(e) \to e(\alpha)\downarrow)]$ *for* $j = 0, 1$,

where $e(\alpha)\downarrow$ *abbreviates* $\exists m\, (e(\overline{\alpha}(m)) > 0)$.

Proofs. (i) $\to$: Assume $J_0(e)$. Using $\mathrm{AC}^R_{00}!$ define a lawlike predictor g by

$$g(w) = \begin{cases} \langle\,\rangle & \text{if } \exists y \sqsubseteq w(e(y) > 0) \vee \neg Seq(w), \\ \mu y(Seq(y) \,\&\, e(w * y) > 0) & \text{otherwise.} \end{cases}$$

If $RLS(\alpha)$ then $\alpha \in \overline{\alpha}(n) * g(\overline{\alpha}(n))$ for some n, so $e(\alpha)\downarrow$.

(i) $\leftarrow$: Assume $\forall \alpha[RLS(\alpha) \to e(\alpha)\downarrow]$ and $Seq(w)$. By RLS1 there is an $\alpha \in w$ with $RLS(\alpha)$, so for some $y = \overline{\alpha}(m)$: $e(y) > 0 \,\&\, \forall n < m(e(\overline{\alpha}(n)) = 0)$. If also $\forall n < lh(w)(e(\overline{w}(n)) = 0)$ then $w \sqsubseteq y$, so $y = w * z$ where $e(w * z) > 0$.

(ii) $\to$ follows immediately from (i) $\to$ (with the fact that $\forall e(J_1(e) \to J_0(e))$). For (ii) $\leftarrow$: Assume $\forall e(J_1(e) \to e(\alpha)\downarrow)$ and let g be a lawlike predictor. Define e as follows:

$$e(w) = \begin{cases} 1 & \text{if } Seq(w) \,\&\, \exists y \sqsubseteq w(y * g(y) \sqsubseteq w), \\ 0 & \text{otherwise.} \end{cases}$$

Then $J_1(e)$ holds, so $e(\alpha)\downarrow$, so g correctly predicts α somewhere. □

The proof of the translation theorem depends on Lemmas 1.9, 2.2, and the following sequence of lemmas removing restricted existential quantifiers and reducing restricted R-formulas of the form $\forall \alpha[RLS(\alpha) \to A]$ and $\forall \alpha_i[RLS([\alpha_0, \ldots, \alpha_k]) \to A]$ to simpler formulas of the same kind. For the case that A is prime the reduction is complete in one step, even in **IRS** + RLS1.

2.3 Lemma

If $s(\alpha), t(\alpha)$ *are terms with no arbitrary choice variables but* α *free, and* a *is free for* α *in both, then* **IRS** + RLS1 *proves*

(i) $\forall \alpha[RLS(\alpha) \to s(\alpha) = t(\alpha)] \leftrightarrow \forall a[s(a) = t(a)]$ *and*
(ii) $\exists \alpha[RLS(\alpha) \,\&\, s(\alpha) = t(\alpha)] \leftrightarrow \exists a[s(a) = t(a)]$.

Proof. By induction on the complexity of the term $s(\alpha)$ (expressing the value of a primitive recursive function of α and the other free variables) **IRS** proves

$$\forall\alpha\exists x\exists y\forall\beta(\overline{\beta}(x) = \overline{\alpha}(x) \rightarrow s(\beta) = y).$$

Only the argument for (i) is completed here since the proof of (ii) is similar.

(i) $\leftarrow$: Assume $\forall a[s(a) = t(a)]$ and $RLS(\alpha)$. Let x, y, z satisfy $\forall\beta(\overline{\beta}(x) = \overline{\alpha}(x) \rightarrow s(\beta) = y \;\&\; t(\beta) = z]$ and let $w = \overline{\alpha}(x)$. Then $\forall a[\overline{a}(x) = w \rightarrow s(a) = y \;\&\; t(a) = z]$, so $y = z$ since $w * \lambda n.0$ is lawlike by $\mathrm{AC}^R_{00}!$, so $s(\alpha) = t(\alpha)$ since $s(\alpha) = y \;\&\; t(\alpha) = z$. So **IRS** proves $\forall a[s(a) = t(a)] \rightarrow \forall\alpha[RLS(\alpha) \rightarrow s(\alpha) = t(\alpha)]$.

(i) $\rightarrow$: Assume $\forall\alpha[RLS(\alpha) \rightarrow s(\alpha) = t(\alpha)]$. Let x, y, z satisfy $\forall\beta[\overline{\beta}(x) = \overline{a}(x) \rightarrow s(\beta) = y \;\&\; t(\beta) = z]$, so $s(a) = y \;\&\; t(a) = z$. By RLS1 there is a β such that $RLS(\beta) \;\&\; \overline{\beta}(x) = \overline{a}(x)$, so $s(\beta) = t(\beta)$ and $y = z$, so $s(a) = t(a)$. So **IRS** + RLS1 proves $\forall\alpha[RLS(\alpha) \rightarrow s(\alpha) = t(\alpha)] \rightarrow \forall a[s(a) = t(a)]$. □

2.4 Lemma

IRLS *proves*

$$\text{(i) } \forall\alpha[RLS(\alpha) \rightarrow A(\alpha) \;\&\; B(\alpha)] \leftrightarrow \forall\alpha[RLS(\alpha) \rightarrow A(\alpha)] \;\&\; \forall\alpha[RLS(\alpha) \rightarrow B(\alpha)],$$

$$\text{(ii) } \forall\alpha[RLS(\alpha) \rightarrow A(\alpha) \vee B(\alpha)] \leftrightarrow \exists e[J_1(e) \;\&\; \forall w(e(w) > 0 \rightarrow \forall\alpha[RLS(\alpha) \rightarrow A(w * \alpha)] \vee \forall\alpha[RLS(\alpha) \rightarrow B(w * \alpha)])],$$

$$\text{(iii) } \forall\alpha[RLS(\alpha) \rightarrow (A(\alpha) \rightarrow B(\alpha))] \leftrightarrow \forall w(Seq(w) \rightarrow (\forall\alpha[RLS(\alpha) \rightarrow A(w * \alpha)] \rightarrow \forall\alpha[RLS(\alpha) \rightarrow B(w * \alpha)])),$$

$$\text{(iv) } \forall\alpha[RLS(\alpha) \rightarrow \neg A(\alpha)] \leftrightarrow \forall w(Seq(w) \rightarrow \neg\forall\alpha[RLS(\alpha) \rightarrow A(w * \alpha)]),$$

for $A(\alpha), B(\alpha)$ *restricted, with no arbitrary choice sequence variables other than* α *occurring free.*

Proofs. (ii) follows from RLS4 using Lemmas 1.9 and 2.2 with the observation

$$\forall e(J_0(e) \rightarrow \exists g[J_1(g) \;\&\; \forall x\forall w(g(w) = x + 1 \leftrightarrow \exists y \sqsubseteq w\,(e(y) = x + 1 \;\&\; \forall z \sqsubset y\, e(z) = 0))].$$

Using Lemma 1.9 again: (iii) follows from RLS3, (iv) $\rightarrow$ follows from RLS1, and (iv) $\leftarrow$ is an easy consequence of RLS3. □

2.5 Lemma

IRLS *proves*

$$\exists\alpha[RLS(\alpha)\ \&\ A(\alpha)] \leftrightarrow \exists w(Seq(w)\ \&\ \forall\alpha[RLS(\alpha) \rightarrow A(w * \alpha)])$$

if $A(\alpha)$ *is restricted and contains free no arbitrary choice sequence variable but* α*.*

Proof. Immediate from RLS3. □

2.6 Lemma

For restricted formulas $A(\alpha, x)$*,* $A(\alpha, b)$ *containing free no arbitrary choice sequence variables other than* α*,* **IRLS** *proves*

(i) $\forall\alpha[RLS(\alpha) \rightarrow \exists x A(\alpha, x)] \leftrightarrow \exists e[J_1(e)\ \&$
$\forall w(e(w) > 0 \rightarrow \exists x \forall\alpha[RLS(\alpha) \rightarrow A(w * \alpha, x)])]$,

(ii) $\forall\alpha[RLS(\alpha) \rightarrow \exists b A(\alpha, b)] \leftrightarrow \exists e[J_1(e)\ \&$
$\forall w(e(w) > 0 \rightarrow \exists b \forall\alpha[RLS(\alpha) \rightarrow A(w * \alpha, b)])]$,

(iii) $\forall\alpha[RLS(\alpha) \rightarrow \forall x A(\alpha, x)] \leftrightarrow \forall x \forall\alpha[RLS(\alpha) \rightarrow A(\alpha, x)]$,

(iv) $\forall\alpha[RLS(\alpha) \rightarrow \forall b A(\alpha, b)] \leftrightarrow \forall b \forall\alpha[RLS(\alpha) \rightarrow A(\alpha, b)]$.

Proofs. (i) and (ii) are by RLS4; (iii) and (iv) are by predicate logic. □

2.7 Lemma

For $A(\alpha, \beta)$ *restricted with no arbitrary choice sequence variables free except the distinct variables* α, β*,* **IRLS** *proves*

(i) $\forall\alpha[RLS(\alpha) \rightarrow \forall\beta[RLS([\alpha, \beta]) \rightarrow A(\alpha, \beta)]] \leftrightarrow \forall\gamma[RLS(\gamma) \rightarrow A([\gamma]_0, [\gamma]_1)]$,

(ii) $\forall\alpha[RLS(\alpha) \rightarrow \exists\beta[RLS([\alpha, \beta])\ \&\ A(\alpha, \beta)]] \leftrightarrow$
$\exists e[J_1(e)\ \&\ \forall y(e(y) > 0 \rightarrow \exists w(Seq(w)\ \&\ \forall\gamma[RLS(\gamma) \rightarrow A(y * [\gamma]_0, w * [\gamma]_1)]))]$,

where in general $[\gamma]_0(n) = \gamma(2n)$ *and* $[\gamma]_1(n) = \gamma(2n + 1)$*.*

Proofs. (i) is immediate from the definitions (note that $\gamma = [\,[\gamma]_0, [\gamma]_1\,]$) and the fact that $RLS([\alpha, \beta]) \rightarrow RLS(\alpha)$ by Lemma 1.9(ii). (ii) follows from RLS2, RLS3 and the closure properties in Lemma 1.9. □

2.8 Proof of the translation theorem

Definition. The *index* of a restricted formula E of $\mathcal{L}$ is $I(E) = 2i + j + 2k$ where

1. i is the number of restricted existential sequence quantifiers occurring in E,
2. j is the number of restricted universal sequence quantifiers occurring in E and
3. k is the maximum number of logical symbols $(\&, \vee, \to, \neg, \forall, \exists)$ occurring in any part F of a subformula of E of any of the forms $\forall\alpha[RLS(\alpha) \to F]$, $\exists\alpha[RLS(\alpha) \,\&\, F]$, $\forall\alpha_i[RLS([\alpha_0, \ldots, \alpha_k]) \to F]$ or $\exists\alpha_i[RLS([\alpha_0, \ldots, \alpha_k]) \,\&\, F]$.

If $C \leftrightarrow D$ denotes any restricted R-formula of a type displayed in the statement of any of the Lemmas 2.3, 2.4, 2.5, 2.6 or 2.7, inspection shows that $I(C) > I(D)$.

The lemmas permit the reduction of a given restricted R-formula E to a formula F of the two-sorted language, with only number and lawlike sequence variables, such that **IRLS** proves $E \leftrightarrow F$. For a uniform translation, the sequence in which the lemmas are to be applied can be determined uniquely (modulo the renaming of variables) by the logical form of E, beginning with the leftmost occurrence of a restricted quantifier. The successive reductions produce a sequence $E_0, \ldots, E_q$ of restricted R-formulas with $I(E_i) > I(E_{i+1})$, where E_0 is E and $I(E_q) = 0$, so we can define $\varphi(E) = E_q$.

There are two wrinkles which are best illustrated by an example. Suppose E_i is a restricted R-formula of the form $\exists\alpha[RLS(\alpha) \,\&\, \forall\beta[RLS([\alpha, \beta]) \to A(\alpha, \beta)]]$. Lemma 2.5 reduces this to $\exists w[Seq(w) \,\&\, \forall\alpha[RLS(\alpha) \to \forall\beta[RLS([w*\alpha, \beta]) \to A(w * \alpha, \beta)]]]$, which is not restricted but can be simplified to $\exists w[Seq(w) \,\&\, \forall\alpha[RLS(\alpha) \to \forall\beta[RLS([\alpha, \beta]) \to A(w * \alpha, \beta)]]]$ using Lemma 1.9 once. If needed, repeated uses of Lemma 1.9 reduce $A(w * \alpha, \beta)$ to a restricted formula $A'(w, \alpha, \beta)$. Then E_{i+1} is $\exists w[Seq(w) \,\&\, \forall\alpha[RLS(\alpha) \to \forall\beta[RLS([\alpha, \beta]) \to A'(w, \alpha, \beta)]]]$, a restricted R-formula with $I(E_{i+1}) < I(E_i)$ since $A(\alpha, \beta)$ and $A'(w, \alpha, \beta)$ have the same number of logical symbols.

By Lemma 2.7(i), in the next step $\forall\alpha[RLS(\alpha) \to \forall\beta[RLS([\alpha, \beta]) \to A'(w, \alpha, \beta)]]$ is reduced to $\forall\gamma[RLS(\gamma) \to A'(w, [\gamma]_0, [\gamma]_1)]$, which may not be restricted. But if for example $A'(w, \alpha, \beta)$ is $\forall\delta[RLS([\alpha, \beta, \delta]) \to B(w, \alpha, \beta, \delta)]$, Lemma 1.9 reduces $A'(w, [\gamma]_0, [\gamma]_1)$ to $\forall\delta[RLS([\gamma, \delta]) \to B(w, [\gamma]_0, [\gamma]_1, \delta)]$ and eventually to a restricted formula $\forall\delta[RLS([\gamma, \delta]) \to B'(w, \gamma, \delta)]$. Then E_{i+2} is $\exists w[Seq(w) \,\&\, \forall\gamma[RLS(\gamma) \to \forall\delta[RLS([\gamma, \delta]) \to B'(w, \gamma, \delta)]]]$, a restricted R-formula with $I(E_{i+2}) < I(E_{i+1})$. □

2.9 Final remarks

Evidently the translation will be unique only up to congruence (renaming of bound variables). While technically RLS1′ is not restricted, it is equivalent over **IRS** to

$\exists\alpha[RLS(\alpha)\ \&\ 0=0]$, which reduces over **IRS** + RLS1 to $\exists w\forall\alpha[RLS(\alpha)\rightarrow 0=0]$ and then to $\exists w\forall a(0=0)$, which is equivalent in **IR** to $0=0$. RLS2′ permits a similar analysis over **IRS** + RLS2.

Note that the intuitionistic system **IRLS** proves AC^R_{01}; and if AC^R_{01} replaces $\mathrm{AC}^R_{00}!$ then RLS4 becomes provable from the other axioms of the semi-classical system **RLS**. It follows that if **IRS′** comes from **IRS** by strengthening $\mathrm{AC}^R_{00}!$ to AC^R_{01}, then

$$\mathbf{RLS} = \mathbf{IRS'} + \mathrm{RLS'}\ 1\text{-}3 + \mathrm{RLEM}.$$

Evidently **IRLS** is not a conservative extension of **IRS** or even **IRS′**, since $\exists\alpha\forall a\neg\forall x(\alpha(x)=a(x))$ is provable in **IRLS** but not in **IRS′**. Similarly, **RLS** is not a conservative extension of **IRS′** + RLEM. However, **IRS′** is a conservative extension of its two-sorted subsystem **B**. I am indebted to A. S. Troelstra (cf. [6]) for the hint that prompted these observations.

Two questions remain. Is **IRLS** a conservative extension of (two-sorted) constructive analysis **IR**? Is the semi-classical system **RLS** a conservative extension of two-sorted classical analysis **R**? Since **R** proves lawlike countable choice AC^R_{01} and a lawlike version $\mathrm{BI!}^R$ of the bar induction schema, the translation lemma suggests that **RLS** may be a conservative extension of **R**, but this is only a conjecture.

References

[1] S. C. Kleene. *Formalized recursive functionals and formalized realizability*. Number 89 in Memoirs. Amer. Math. Soc., 1969.

[2] S. C. Kleene and R. E. Vesley. *The Foundations of Intuitionistic Mathematics, Especially in Relation to Recursive Functions*. North Holland, 1965.

[3] G. Kreisel. Lawless sequences of natural numbers. *Comp. Math.*, 20:222–248, 1968.

[4] J. R. Moschovakis. Iterated definability, lawless sequences and Brouwer's continuum. In *Gödel's Disjunction: The Scope and Limits of Mathematical Knowledge*, pages 92–107, Oxford, 2016.

[5] J. R. Moschovakis. Embedding the constructive and the classical in the intuitionistic continuum, July 2014. Chiemsee summer school in Proof, Truth and Computation, slides from talk. `http:\www.math.ucla.edu~joan`.

[6] A. S. Troelstra. An addendum. *Annals of Math. Logic*, 3:437–439, 1971.

[7] A. S. Troelstra. *Choice Sequences, a Chapter of Intuitionistic Mathematics*. Oxford Logic Guides. Clarendon Press, 1977.

[8] A. S. Troelstra and D. van Dalen. *Constructivism in Mathematics: An Introduction, Volumes I and II*. North-Holland, 1988.

Received 22 April 2015

Classifying the provably total set functions of KP and KP($\mathcal{P}$)

Jacob Cook
Department of Pure Mathematics, University of Leeds, Leeds LS2 9JT, United Kingdom
jacob_knows@hotmail.co.uk

Michael Rathjen
Department of Pure Mathematics, University of Leeds, Leeds LS2 9JT, United Kingdom
rathjen@maths.leeds.ac.uk

Abstract

This article is concerned with classifying the provably total set-functions of Kripke-Platek set theory, **KP**, and Power Kripke-Platek set theory, **KP**($\mathcal{P}$), as well as proving several (partial) conservativity results. The main technical tool used in this paper is a relativisation technique where ordinal analysis is carried out relative to an arbitrary but fixed set x.

A classic result from ordinal analysis is the characterisation of the provably recursive functions of Peano Arithmetic, **PA**, by means of the fast growing hierarchy [10]. Whilst it is possible to formulate the natural numbers within **KP**, the theory speaks primarily about sets. For this reason it is desirable to obtain a characterisation of its provably total set functions. We will show that **KP** proves the totality of a set function precisely when it falls within a hierarchy of set functions based upon a relativised constructible hierarchy stretching up in length to any ordinal below the Bachmann-Howard ordinal. As a consequence of this result we obtain that **IKP** $+ \forall x \forall y\, (x \in y \vee x \notin y)$ is conservative over **KP** for Π_2-formulae, where **IKP** stands for intuitionistic Kripke-Platek set theory.

Part of the material is based upon research supported by the EPSRC of the UK through grant No. EP/K023128/1. This research was also supported by a Leverhulme Research Fellowship and a Marie Curie International Research Staff Exchange Scheme Fellowship within the 7th European Community Framework Programme. This publication was also made possible through the support of a grant from the John Templeton Foundation.

In a similar vein, utilising [56], it is shown that $\mathbf{KP}(\mathcal{P})$ proves the totality of a set function precisely when it falls within a hierarchy of set functions based upon a relativised von Neumann hierarchy of the same length. The relativisation technique applied to $\mathbf{KP}(\mathcal{P})$ with the global axiom of choice, $\mathbf{AC}_{global}$, also yields a parameterised extension of a result in [58], showing that $\mathbf{KP}(\mathcal{P}) + \mathbf{AC}_{global}$ is conservative over $\mathbf{KP}(\mathcal{P}) + \mathbf{AC}$ and $\mathbf{CZF} + \mathbf{AC}$ for $\Pi_2^{\mathcal{P}}$ statements. Here $\mathbf{AC}$ stands for the ordinary axiom of choice and $\mathbf{CZF}$ refers to constructive Zermelo-Fraenkel set theory.

1 Introduction

A major application of the techniques of ordinal analysis has been the classification of the provably total recursive functions of a theory. Usually the theories to which this methodology has been applied have been arithmetic theories, in that context it makes most sense to speak about arithmetic functions. The concept of a recursive function on natural numbers can be extended to a more general recursion theory on arbitrary sets. For more details see [38], [39] and [59]. Since **KP** speaks primarily about sets, it is perhaps desirable to obtain a classification of its provably total recursive set functions.

To provide some context we first state a classic result from proof theory, the classification of the *provably total recursive functions* of **PA**. A classification can be gleaned from Gentzen's 1938 [25] and 1943 [26] papers. The first explicit characterization of these functions as those definable by recursions on ordinals less than ε_0 was given by Kreisel [31, 32] in the early 1950s. Many people re-proved or provided variants of this classification result (see [64, Chap. 4] for the history). As to techniques for extracting numerical bounds from infinite proofs, Schwichtenberg's [63] and the considerably more elegant approach by Buchholz and Wainer in [10] and its generalization and simplification by Weiermann in [66] are worth mentioning. For the following definitions, suppose we have an ordinal representation system for ordinals below ε_0, together with an assignment of fundamental sequences to the limit ordinal terms. For an ordinal term α, let α_n denote the n-th element of the fundamental sequence for α, ie. $\alpha_{n+1} < \alpha_n$ and $\sup_{n<\omega}(\alpha_n) = \alpha$. There are certain technical properties that such an assignment must satisfy, these will not be gone into here, for a detailed presentation see [10].

Definition 1.1. For each $\alpha < \varepsilon_0$ we define the function $F_\alpha : \omega \to \omega$ by transfinite recursion as follows

$$F_0(n) := n + 1$$

$$F_{\alpha+1}(n) := F_\alpha^{n+1}(n) \ (:= \overbrace{F_\alpha \circ \ldots \circ F_\alpha}^{n+1}(n))$$

$$F_\alpha(n) := F_{\alpha_n}(n) \quad \text{if } \alpha \text{ is a limit.}$$

This hierarchy is known as the *fast growing hierarchy*. Given unary functions on the natural numbers f and g, we say that f majorises g if there is some n such that $(\forall m > n)(g(m) < f(m))$. For a recursive function f let $A_f(n, m)$ be the Σ formula expressing that on input n the Turing machine for computing f outputs m, to avoid frustrating counter examples let us suppose A_f does this in some 'natural' way.

Theorem 1.2. Suppose $f : \omega \to \omega$ is a recursive function. Then

i) If **PA** $\vdash \forall x \exists! y A_f(x, y)$ then f is majorised by F_α for some $\alpha < \varepsilon_0$.

ii) **PA** $\vdash \forall x \exists! y A_{F_\alpha}(x, y)$ for every $\alpha < \varepsilon_0$.

Proof. This classic result is proved in full in [10]. □

This chapter will be focused on obtaining a similar result for the provably total set functions of **KP**.[1] A similar role to the fast growing hierarchy in Theorem 1.2 will be played by the *relativised constructible hierarchy.*

Definition 1.3. Let X be any set. We may relativise the constructible hierarchy to X as follows:

$$L_0(X) := TC(\{X\}) \quad \text{the } \textit{transitive closure} \text{ of } \{X\}$$

$$L_{\alpha+1}(X) := \{B \subseteq L_\alpha(X) \ : \ B \text{ is definable over } \langle L_\alpha(X), \in \rangle\}$$

$$L_\theta(X) := \bigcup_{\xi<\theta} L_\xi(X) \quad \text{when } \theta \text{ is a limit.}$$

[1] There are many papers concerned with the provably recursive **number-theoretic** functions of **KP** and much stronger theories. The basic idea consists in adding another layer of control to the ordinal analysis that allows one to extract bounds for numerical witnesses. These techniques were initially engineered by Buchholz, Wainer [10] and Weiermann [66] and then got extended by Blankertz, Weiermann [5, 6, 7], Michelbrink [37], Pohlers and Stegert [42] to ever stronger theories. Another route for obtaining classifications of provably numerical functions proceeds as follows. The ordinal analysis of a set theory T shows that the arithmetic part of T can be reduced to **PA** plus transfinite induction for every ordinal below the proof-theoretic ordinal of T. Thus it suffices to characterize the provably numerical functions of the latter system. This leads to the descent recursive functions in the sense of [23]. That this method is perfectly general was first sketched in [23] and then proved rigorously in [12]. The latter approach has the advantage that the ordinal analysis of T needn't be burdened with the extra task of controlling numerical witnesses.

In section 2 we build an ordinal notation system relativised to an arbitrary set X, this will be used for the rest of the article. In section 3 we define the infinitary system $\mathbf{RS}_\Omega(X)$, based on the relativised constructible hierarchy and show that we can eliminate cuts for derivations of Σ formulae. In section 4 we embed **KP** into $\mathbf{RS}_\Omega(X)$, allowing us to obtain cut free infinitary derivations of **KP** provable Σ formulae. Technically we use Buchholz' operator controlled derivations (see [11]) which are also used in [41]. In section 5 we give a well ordering proof in **KP** for the ordinal notation system given in section 2. Finally we combine the results of this chapter to give a classification of the provably total set functions of **KP** in section 6. This result, whilst perhaps known to those who have thought hard about these things, has not appeared in the literature to date. Section 7 contains applications to semi-intuitionistic Kripke-Platek set theory. Section 8 carries out a relativised ordinal analysis of Power Kripke-Platek set theory, $\mathbf{KP}(\mathcal{P})$, from which ensues a classification of its provable set functions. This closely follows the treatment in [56]. In section 9, a further ingredient is added to the infinitary system by incorporating a global choice relation. Due to the relativisation one gets partial conservativity results for $\mathbf{KP}(\mathcal{P}) + \mathbf{AC}_{global}$ over $\mathbf{KP}(\mathcal{P}) + \mathbf{AC}$ and $\mathbf{CZF} + \mathbf{AC}$ that provide improvements on [58, Theorem 3.3] and [58, Corollary 5.2]. These theories can also be added to the list of theories [57, Theorem 15.1] with the same proof-theoretic strength.

2 A relativised ordinal notation system

The aim of this section is to relativise the construction of the Bachmann-Howard ordinal to contain an arbitrary set X or rather its rank θ. We will construct an ordinal representation system that will be set primitive recursive given access to an oracle for X. Here the notion of recursive and primitive recursive is extended to arbitrary sets, see [39] or [59] for more detail. The construction of an ordinal representation system for the Bachmann-Howard ordinal is now fairly standard in proof theory, carried out for example in [9]. Intuitively our system will appear similar, only the ordering W will be inserted as an initial segment before new ordinals start being 'named' via the collapsing function.

Before defining the formal terms and the procedure for computing their ordering, it is informative to give definitions for the corresponding ordinals and ordinal functions themselves. To this end we will begin working in **ZFC**, later it will become clear that the necessary ordinals can be expressed as formal terms and comparisons between these terms can be made primitive recursively relative to W.

In what follows **ON** will denote the class of all ordinals. First we require some information about the φ function on ordinals. These definitions and results are well known, see [62].

Definition 2.1. For each $\alpha \in \mathbf{ON}$ we define a class of ordinals $Cr(\alpha) \subseteq \mathbf{ON}$ and a class function

$$\varphi_\alpha : \mathbf{ON} \to \mathbf{ON}$$

by transfinite recursion.

i) $Cr(0) := \{\omega^\beta \mid \beta \in \mathbf{ON}\}$ and $\varphi_0(\beta) := \omega^\beta$.

ii) For $\alpha > 0$ $Cr(\alpha) := \{\beta \mid (\forall \gamma < \alpha)(\varphi_\gamma(\beta) = \beta)\}$.

iii) For each $\alpha \in \mathbf{ON}$ $\varphi_\alpha(\cdot)$ is the function enumerating $Cr(\alpha)$.

The convention is to write $\varphi\alpha\beta$ instead of $\varphi_\alpha(\beta)$. An ordinal $\beta \in Cr(0)$ is often referred to as *additive principal*, since for all $\beta_1, \beta_2 < \beta$ we have $\beta_1 + \beta_2 < \beta$.

Theorem 2.2.

i) $\varphi\alpha_1\beta_1 = \varphi\alpha_2\beta_2$ if and only if $\begin{cases} & \alpha_1 < \alpha_2 \text{ and } \beta_1 = \varphi\alpha_2\beta_2 \\ \text{or} & \alpha_1 = \alpha_2 \text{ and } \beta_1 = \beta_2 \\ \text{or} & \alpha_2 < \alpha_1 \text{ and } \varphi\alpha_1\beta_1 = \beta_2. \end{cases}$

ii) $\varphi\alpha_1\beta_1 < \varphi\alpha_2\beta_2$ if and only if $\begin{cases} & \alpha_1 < \alpha_2 \text{ and } \beta_1 < \varphi\alpha_2\beta_2 \\ \text{or} & \alpha_1 = \alpha_2 \text{ and } \beta_1 < \beta_2 \\ \text{or} & \alpha_2 < \alpha_1 \text{ and } \varphi\alpha_1\beta_1 < \beta_2. \end{cases}$

iii) For any additive principal β there are unique ordinals $\beta_1 \leq \beta$ and $\beta_2 < \beta$ such that $\beta = \varphi\beta_1\beta_2$.

Proof. This result is proved in full in [62]. □

Definition 2.3. We define $\Gamma_{(\cdot)} : \mathbf{ON} \to \mathbf{ON}$ to be the class function enumerating the ordinals $\beta > 0$ such that for all $\beta_1, \beta_2 < \beta$ we have $\varphi\beta_1\beta_2 < \beta$. Ordinals of the form Γ_β will be referred to as *strongly critical.*

Now let $\theta \in \mathbf{ON}$ be the unique ordinal that is the set-theoretic rank of X.

Definition 2.4. Let Ω be the least uncountable cardinal greater than θ. The sets $B_\theta(\alpha) \subseteq \mathbf{ON}$ and ordinals $\psi_\theta(\alpha)$ are defined by transfinite recursion on α as follows:

$$B_\theta(\alpha) := \text{Closure of } \{0, \Omega\} \cup \{\Gamma_\beta \,:\, \beta \leq \theta\} \text{ under } +,\ \varphi \text{ and } \psi_\theta|_\alpha$$
$$\psi_\theta(\alpha) := \min\{\beta \,:\, \beta \notin B_\theta(\alpha)\}$$

For the remainder of this section, since θ remains fixed, the subscripts will be dropped from B_θ and ψ_θ to improve readability. At first glance it may appear strange having the elements from θ inserted into the Γ-numbers. Ultimately we aim to have $+$ and φ as primitive symbols in our notation system, simply having θ as an initial segment here would cause problems with unique representation. Some ordinals could get a name directly from θ and other names by applying $+$ and φ to smaller elements.

Lemma 2.5. For each $\alpha \in \mathbf{ON}$:

i) The cardinality of $B(\alpha)$ is $\max\{\aleph_0, |\theta|\}$, where $|\theta|$ denotes the cardinality of θ.

ii) $\psi\alpha < \Omega$.

Proof. i) Let

$$\begin{aligned} B^0(\alpha) :=& \{0, \Omega\} \cup \{\Gamma_\beta : \beta \leq \theta\} \\ B^{n+1}(\alpha) :=& B^n(\alpha) \cup \{\xi + \eta : \xi, \eta \in B^n(\alpha)\} \\ & \cup \{\varphi\xi\eta : \xi, \eta \in B^n(\alpha)\} \\ & \cup \{\psi\xi : \xi \in B^n(\alpha) \cap \alpha\}. \end{aligned}$$

Observe that $B(\alpha) = \bigcup_{n<\omega} B^n(\alpha)$, this can be proved by a straightforward induction on n.

If θ is finite then, again by induction on n, we can show that each $B^n(\alpha)$ is also finite. Since $B(\alpha)$ is a countable union of finite sets and $\omega \subseteq B(\alpha)$ it follows that it must have cardinality $\aleph_0$.

Now suppose θ is infinite, so $B(\alpha)$ is the countable union of sets of cardinality $|\theta|$ and thus also has cardinality $|\theta|$.

ii) If $\psi\alpha \geq \Omega$ then $\Omega \subset B(\alpha)$ contradicting i). □

Lemma 2.6.

i) If $\gamma \leq \delta$ then $B(\gamma) \subseteq B(\delta)$ and $\psi\gamma \leq \psi\delta$.

ii) If $\gamma \in B(\delta) \cap \delta$ then $\psi\gamma < \psi\delta$.

iii) If $\gamma \leq \delta$ and $[\gamma, \delta) \cap B(\gamma) = \emptyset$ then $B(\gamma) = B(\delta)$.

iv) If ξ is a limit then $B(\xi) = \bigcup_{\eta<\xi} B(\eta)$.

v) $\psi\gamma$ is a strongly critical and $\psi\gamma \geq \Gamma_{\theta+1}$.

vi) $B(\gamma) \cap \Omega = \psi\gamma$.

vii) If ξ is a limit then $\psi\xi = \sup_{\eta<\xi}\psi\eta$.

viii) $\psi(\gamma+1) \leq (\psi\gamma)^\Gamma$, where δ^Γ denotes the smallest strongly critical ordinal above δ.

ix) If $\alpha \in B(\alpha)$ then $\psi(\alpha+1) = (\psi\alpha)^\Gamma$.

x) If $\alpha \notin B(\alpha)$ then $\psi(\alpha+1) = \psi\alpha$ and $B(\alpha+1) = B(\alpha)$.

xi) If $\gamma \in B(\gamma)$ and $\delta \in B(\delta)$ then [$\gamma < \delta$ if and only if $\psi\gamma < \psi\delta$].

Proof. i) Suppose $\gamma \leq \delta$, now note that $B(\delta)$ is closed under $\psi|_\delta$ which includes $\psi|_\gamma$ so $B(\gamma) \subseteq B(\delta)$. From this it immediately follows from the definition that $\psi\gamma \leq \psi\delta$.

ii) From $\gamma \in B(\delta) \cap \delta$ we get $\psi\gamma \in B(\delta)$, thus $\psi\gamma < \psi\delta$ b the definition of $\psi\delta$.

iii) It is enough to show that $B(\gamma)$ is closed under $\psi|_\delta$. Let $\beta \in B(\gamma)$ and $\beta < \delta$, then by assumption $\beta < \gamma$, thus $\psi\beta \in B(\gamma)$.

iv) By i) we have $\bigcup_{\eta<\xi} B(\eta) \subseteq B(\xi)$. It remains to verify that $Y := \bigcup_{\eta<\xi} B(\eta)$ is closed under $\psi|_\xi$. So let $\delta \in Y \cap \xi$, since ξ is a limit there is some $\xi_0 < \xi$ such that $\delta \in Y \cap \xi_0$ and there is some $\xi_1 < \xi$ such that $\delta \in B(\xi_1)$. Therefore $\delta \in B(\xi^*) \cap \xi^*$ where $\xi^* = \max\{\xi_0, \xi_1\}$, thus $\psi\delta \in B(\xi^*) \subseteq Y$.

v) We may write the ordinal $\psi\alpha$ in Cantor normal form, so that $\psi\alpha = \omega^{\alpha_1}+\ldots+\omega^{\alpha_n}$ with $\alpha_1 \geq \ldots \geq \alpha_n$. If $n > 1$ then $\alpha_1, \ldots, \alpha_n < \psi\alpha$ which implies by the definition of $\psi\alpha$ that $\alpha_1, \ldots, \alpha_n \in B(\alpha)$. But by closure of $B(\alpha)$ under $+$ and φ we get $\varphi 0\alpha_1+\ldots+\varphi 0\alpha_n = \omega^{\alpha_1}+\ldots+\omega^{\alpha_n} \in B(\alpha)$ contradicting $\psi\alpha \notin B(\alpha)$. Thus $\psi\alpha$ is additive principal and it follows from Theorem 2.2iii) that we may find ordinals $\gamma \leq \psi\alpha$ and $\delta < \psi\alpha$ such that $\psi\alpha = \varphi\gamma\delta$. If $\delta > 0$ then $\gamma < \psi\alpha$ since $\gamma \leq \varphi\gamma 0 < \varphi\gamma\delta$, but if $\delta, \gamma < \psi\alpha$ then we have $\delta, \gamma \in B(\alpha)$ and hence $\varphi\gamma\delta \in B(\alpha)$ contradicting $\psi\alpha \notin B(\alpha)$. Thus $\psi\alpha = \varphi\gamma 0$, but if $\gamma < \psi\alpha$ then again we get $\varphi\gamma 0 \in B(\alpha)$; a contradiction. So it must be the case that $\psi\alpha = \gamma$, ie. $\psi\alpha$ is strongly critical.

For the second part note that $\psi\alpha \neq \Gamma_\beta$ for any $\beta \leq \theta$ since by definition each such $\Gamma_\beta \in B(\alpha)$.

vi) By 2.5ii) and the definition of ψ it is clear that $\psi\alpha \subseteq B(\alpha) \cap \Omega$. Now let

$$Y := \psi\alpha \cup \{\delta \geq \Omega \mid \delta \in B(\alpha)\}.$$

By v) Y contains $0, \Omega$ and Γ_β for $\beta \leq \theta$, moreover it is closed under $+$ and φ. It remains to show that Y is closed under $\psi|_\alpha$, this follows immediately from ii).

vii) Let ξ be a limit ordinal. Using parts vi), iv) and i) we have

$$\psi\xi = B(\xi) \cap \Omega = (\bigcup_{\eta<\xi} B(\eta)) \cap \Omega = \bigcup_{\eta<\xi}(B(\eta) \cap \Omega) = \bigcup_{\eta<\xi} \psi\eta = \sup_{\eta<\xi}\psi\eta.$$

viii) Let

$$Y := (\psi\alpha)^\Gamma \cup \{\delta \geq \Omega \mid \delta \in B(\alpha)\}.$$

Y is closed under $+$ and φ, also it contans Γ_β for any $\beta \leq \theta$ by v). Moreover it contains $\psi\gamma$ for any $\gamma \leq \alpha$ by i), so it is closed under $\psi|_{(\alpha+1)}$. Therefore Y must contain $B(\alpha+1)$, and so $\psi(\alpha+1) \leq (\psi\alpha)^\Gamma$.

ix) From $\alpha \in B(\alpha)$ we get $\alpha \in B(\alpha+1)$, it then follows from ii) that $\psi\alpha < \psi(\alpha+1)$. Thus $\psi(\alpha+1) \leq (\psi\alpha)^\Gamma$ by viii) and $\psi(\alpha+1) \geq (\psi\alpha)^\Gamma$ from v), so it must be the case that $\psi(\alpha+1) = (\psi\alpha)^\Gamma$.

x) Suppose $\alpha \notin B(\alpha)$, then $[\alpha, \alpha+1) \cap B(\alpha) = \emptyset$ so we may apply iii) to give $B(\alpha+1) = B(\alpha)$ from which $\psi(\alpha+1) = \psi\alpha$ follows immediately.

xi) Suppose $\gamma \in B(\gamma)$ and $\delta \in B(\delta)$. If $\gamma < \delta$ then from ix) we get $\psi(\gamma+1) = (\psi\gamma)^\Gamma > \psi\gamma$, but by i) $\psi(\gamma+1) \leq \psi\delta$.

Now if $\psi\gamma < \psi\delta$ then from the contraposition of i) we get $\gamma < \delta$. □

Definition 2.7. We write

i) $\alpha =_{NF} \alpha_1 + \ldots + \alpha_n$ if $\alpha = \alpha_1 + \ldots + \alpha_n$, $n > 1$, $\alpha_1, \ldots, \alpha_n$ are additive principal numbers and $\alpha_1 \geq \ldots \geq \alpha_n$.

ii) $\alpha =_{NF} \varphi\gamma\delta$ if $\alpha = \varphi\gamma\delta$ and $\gamma, \delta < \varphi\gamma\delta$.

iii) $\alpha =_{NF} \psi\gamma$ if $\alpha = \psi\gamma$ and $\gamma \in B(\gamma)$

Lemma 2.8.

i) If $\alpha =_{NF} \alpha_1 + \ldots + \alpha_n$ then for any $\eta \in \mathbf{ON}$

$$\alpha \in B(\eta) \quad \text{if and only if} \quad \alpha_1, \ldots, \alpha_n \in B(\eta).$$

ii) If $\alpha =_{NF} \varphi\gamma\delta$ then for any $\eta \in \mathbf{ON}$

$$\alpha \in B(\eta) \quad \text{if and only if} \quad \gamma, \delta \in B(\eta).$$

iii) If $\alpha =_{NF} \psi\gamma$ then for any $\eta \in \mathbf{ON}$

$$\alpha \in B(\eta) \quad \text{if and only if} \quad \gamma \in B(\eta) \cap \eta.$$

Proof. i) Suppose $\alpha =_{NF} \alpha_1 + \ldots + \alpha_n$, the $\Leftarrow$ direction is clear from the closure of $B(\eta)$ under $+$. For the other direction let

$$AP(\alpha) := \begin{cases} \emptyset & \text{if } \alpha = 0 \\ \{\alpha\} & \text{if } \alpha \text{ is additive principal} \\ \{\alpha_1, \ldots, \alpha_n\} & \text{if } \alpha =_{NF} \alpha_1 + \ldots + \alpha_n. \end{cases}$$

$AP(\alpha)$ stands for the *additive predecessors* of α. Now let

$$Y := \{\gamma \in B(\eta) \mid AP(\gamma) \subseteq B(\eta)\}.$$

Observe that $0, \Omega \in Y$ and $\{\Gamma_\beta \mid \beta \leq \theta\} \subseteq Y$. Now choose any $\gamma, \delta \in Y$, we have $AP(\gamma + \delta) \subseteq AP(\gamma) \cup AP(\delta) \subseteq B(\eta)$, thus Y is closed under $+$. Now $AP(\varphi\gamma\delta) = \{\varphi\gamma\delta\}$ since the range of φ is the additive principal numbers thus Y is closed under φ. Finally $AP(\psi\gamma) = \{\psi\gamma\}$ for any $\gamma \in Y \cap \eta$ so Y is closed under $\psi|_\eta$. It follows that $B(\eta) \subseteq Y$ and thus the other direction is proved.

ii) Again the $\Leftarrow$ direction follows immediately from the closure of $B(\eta)$ under φ. For the other direction we let

$$PP(\alpha) := \begin{cases} \emptyset & \text{if } \alpha = 0 \\ \{\alpha\} & \text{if } \alpha \text{ is strongly critical} \\ \{\gamma, \delta\} & \text{if } \alpha =_{NF} \varphi\gamma\delta \\ \{\alpha_1, \ldots, \alpha_n\} & \text{if } \alpha =_{NF} \alpha_1 + \ldots + \alpha_n. \end{cases}$$

For want of a better phrase $PP(\alpha)$ stands for the *predicative predecessors* of α. Now set

$$Y := \{\gamma \in B(\eta) \mid PP(\gamma) \subseteq B(\eta)\}$$

It is easily seen that Y contains $0,\Omega$ and Γ_β for any $\beta \leq \theta$. $PP(\gamma+\delta) \subseteq PP(\gamma) \cup PP(\delta)$ so Y is closed under $+$. $PP(\varphi\gamma\delta) \subseteq \{\gamma,\delta\}$ so Y is closed under φ. Finally $PP(\psi\gamma) = \{\psi\gamma\}$ for any $\gamma < \eta$ by 2.6v). It follows that Y must contain $B(\eta)$, which proves the $\Rightarrow$ direction.

iii) Suppose $\alpha =_{NF} \psi\gamma$, the $\Leftarrow$ direction is clear by the closure of $B(\eta)$ under $\psi|_\eta$. For the other direction suppose $\alpha \in B(\eta)$, from this we get $\psi\gamma < \psi\eta$ which gives us $\gamma < \eta$. Now by assumption $\gamma \in B(\gamma)$, and $B(\gamma) \subseteq B(\eta)$ so $\gamma \in B(\eta) \cap \eta$. □

In order to create an ordinal notation system from the ordinal functions described above, we single out a set $R(\theta)$ of ordinals which have a unique canonical description.

Definition 2.9. We give an inductive definition of the set $R(\theta)$, and the complexity $G\alpha < \omega$ for every $\alpha \in R(\theta)$.

(R1) $0,\Omega \in R(\theta)$ and $G0 := G\Omega := 0$.

(R2) For each $\beta \leq \theta$, $\Gamma_\beta \in R(\theta)$ and $G\Gamma_\beta := 0$.

(R3) If $\alpha =_{NF} \alpha_1 + \ldots + \alpha_n$ and $\alpha_1,\ldots,\alpha_n \in R(\theta)$ then $\alpha \in R(\theta)$ and $G\alpha := \max\{G\alpha_1,\ldots,G\alpha_n\}+1$.

(R4) If $\gamma,\delta < \Omega$, $\alpha =_{NF} \varphi\gamma\delta$ and $\gamma,\delta \in R(\theta)$ then $\alpha \in R(\theta)$ and $G\alpha := \max\{G\gamma, G\delta\}+1$.

(R5) If $\gamma \geq \Omega$, $\alpha =_{NF} \varphi 0\gamma$ and $\gamma \in R(\theta)$ then $\alpha \in R(\theta)$ and $G\alpha := G\gamma + 1$.

(R6) If $\alpha =_{NF} \psi\gamma$ and $\gamma \in R(\theta)$ then $\alpha \in R(\theta)$ and $G\alpha := G\gamma + 1$.

Lemma 2.10. Every element $\alpha \in R(\theta)$ is included due to precisely one of the rules (R1)-(R6) and thus the complexity $G\alpha$ is uniquely defined.

Proof. This follows immediately from 2.8. □

Our goal is to turn $R(\theta)$ into a formal representation system, the main obstacle to this is that it is not immediately clear how to deal with the constraint $\gamma \in B(\gamma)$ in a computable way. This problem leads to the following definition.

Definition 2.11. To each $\alpha \in R(\theta)$ we assign a set $K\alpha$ of ordinal terms by induction on the complexity $G\alpha$:

(K1) $K0 := K\Omega := K\Gamma_\beta := \emptyset$ for all $\beta \leq \theta$.

(K2) If $\alpha =_{NF} \alpha_1 + \ldots + \alpha_n$ then $K\alpha := K\alpha_1 \cup \ldots \cup K\alpha_n$.

(K3) If $\alpha =_{NF} \varphi\gamma\delta$ then $K\alpha := K\gamma \cup K\delta$.

(K4) If $\alpha =_{NF} \psi\gamma$ then $K\alpha := \{\gamma\} \cup K\gamma$.

$K\alpha$ consists of the ordinals that occur as arguments of the ψ function in the normal form representation of α. Note that each ordinal in $K\alpha$ belongs to $R(\theta)$ itself and has complexity lower than $G\alpha$.

Lemma 2.12. For any $\alpha, \eta \in R(\theta)$

$$\alpha \in B(\eta) \quad \text{if and only if} \quad (\forall \xi \in K\alpha)(\xi < \eta).$$

Proof. The proof is by induction on $G\alpha$. If $G\alpha = 0$ then $\alpha \in B(\eta)$ for any η, and $K\alpha = \emptyset$ by (K1) so the result holds.

Case 1. If $\alpha =_{NF} \alpha_1 + \ldots + \alpha_n$ then $\alpha \in B(\eta)$ iff $\alpha_1, \ldots, \alpha_n \in B(\eta)$ by 2.8i). Now inductively $\alpha_1, \ldots, \alpha_n \in B(\eta)$ iff $(\forall \xi \in K\alpha_1 \cup \ldots \cup K\alpha_n)(\xi < \eta)$, but by (K2) $K\alpha = K\alpha_1 \cup \ldots \cup K\alpha_n$.

Case 2. If $\alpha =_{NF} \varphi\gamma\delta$ we may argue in a similar fashion to Case 1, using 2.8ii) and (K3) instead.

Case 3. If $\alpha =_{NF} \psi\gamma$ then $\alpha \in B(\eta)$ iff $\gamma \in B(\eta) \cap \eta$ by 2.8iii). Now by induction hypothesis $\gamma \in B(\eta) \cap \eta$ iff $(\forall \xi \in K\gamma)(\xi < \eta)$ and $\gamma < \eta$, and by (K4) this occurs precisely when $(\forall \xi \in K\alpha)(\xi < \eta)$. □

Recall that θ is the rank of X. Let

$$\begin{aligned} \mathcal{L}_\theta &:= \{0, \Omega, +, \varphi, \psi\} \cup \{\Gamma_\xi \ : \ \xi \leq \theta\} \quad \text{and} \\ \mathcal{L}_\theta^* &:= \{s \mid s \text{ is a finite string of symbols from } \mathcal{L}_\theta\}. \end{aligned}$$

Now let $T(\theta) \subseteq \mathcal{L}_\theta^*$ be the set of strings that correspond to ordinals in $R(\theta)$ expressed in normal form. Owing to Lemma 2.10 there is a one to one correspondence between $T(\theta)$ and $R(\theta)$. The ordering on $T(\theta)$ induced from the ordering of the ordinals in $R(\theta)$ will be denoted $\prec$. To differentiate between elements of the two sets, Greek letters $\alpha, \beta, \gamma, \eta, \xi, \ldots$ range over ordinals and Roman letters $a, b, c, d, e, \ldots$ range over finite strings from $\mathcal{L}_\theta^*$.

Theorem 2.13. The set $T(\theta)$ and the relation $\prec$ on $T(\theta)$ are set primitive recursive in θ.

Proof. Below a θ-primitive recursive procedure means a procedure that is primitive recursive in the two parameters θ and the ordering $<_\theta$ on the ordinals $\xi \leq \theta$. We need to provide the following two procedures:

A) A θ-primitive recursive procedure which decides for $a \in \mathcal{L}^*_\theta$ whether $a \in T(\theta)$.

B) A θ-primitive recursive procedure which decides for non-identical $a, b \in T(\theta)$ whether $a \prec b$ or $b \prec a$.

We define **A)** and **B)** simultaneously by induction on the term complexity Ga.

For the base stage of **A)** we have $0, \Omega \in T(\theta)$ and $\Gamma_\xi \in T(\theta)$ for all $\xi \leq \theta$.

For the base stage of **B)** we have $0 \prec \Gamma_\xi \prec \Omega$ for all $\xi \leq \theta$ and the terms Γ_ξ inherit the ordering from θ, for which we have access to an oracle.

For the inductive stage of **A)** we require the following 3 things:

A1) A θ-primitive recursive procedure that on input $a_1, \ldots, a_n \in T(\theta)$ decides whether $a_1 + \ldots + a_n \in T(\theta)$.

A2) A θ-primitive recursive procedure that on input $a_1, a_2 \in T(\theta)$ decides whether $\varphi a_1 a_2 \in T(\theta)$.

A3) A θ-primitive recursive procedure that on input $a \in T(\theta)$ decides whether $\psi a \in T(\theta)$.

For **A1)** we need to decide if $n > 1$ and if $a_1 \succeq \ldots \succeq a_n$, which we can do by the induction hypothesis. We also need to decide if $a_1, \ldots, a_n$ are additive principal; all terms other than those of the form $b_1 + \ldots + b_m$ $(m > 1)$ and 0 are additive principal.

For **A2)**, first let ORD_θ denote the set of $\mathcal{L}_\theta$ strings which represent an ordinal (not necessarily in normal form), ie. each function symbol has the correct arity. Next we define the set of strings which correspond to the strongly critical ordinals, where $\equiv$ signifies identity of strings.

$$SC_\theta := \{\Omega\} \cup \{\Gamma_\xi \ : \ \xi \leq \theta\} \cup \{a \in ORD_\theta \ : \ a \equiv \psi b\}.$$

We may decide membership of SC_θ in a θ-primitive recursive fashion. For the decision procedure we split into cases based upon the form of a_2:

i) If $a_2 \equiv 0$ then $\varphi a_1 a_2 \in T(\theta)$ whenever $a_1 \notin SC_\theta$.

ii) If $a_2 \in SC_\theta$ then $\varphi a_1 a_2 \in T(\theta)$ whenever $a_1 \succeq a_2$ and $a_2 \neq \Omega$.

iii) If $a_2 \succ \Omega$ then $\varphi a_1 a_2 \in T(\theta)$ exactly when $a_1 = 0$.

iv) If $a_2 \equiv b_1 + \ldots + b_n \prec \Omega$, with $n > 1$ then $\varphi a_1 a_2 \in T(\theta)$ regardless of the form of a_1.

iv) If $a_2 \equiv \varphi b_1 b_2 \prec \Omega$ then $\varphi a_1 a_2 \in T(\theta)$ whenever $a_1 \succeq b_1$.

For a rigorous treatment of the φ function see [62].

The function K from Definition 2.11 lifts to a θ-primitive recursive function on $T(\theta)$. Moreover every $b \in Ka$ is a member of $T(\theta)$ of lower complexity than a. Owing to Lemma 2.12, for the decision procedure **A3)** we may first compute Ka, then check whether $(\forall b \in Ka)(b \prec a)$, which we may do by the induction hypothesis.

Finally for the inductive stage of **B)**, given two elements of $T(\theta)$ we may decide their ordering using the following procedure.

B1) $0 \prec a$ for every $a \neq 0$.

B2) $\Gamma_\xi \prec \Omega$ for every $\xi \leq \theta$.

B3) The elements Γ_ξ inherit the ordering from θ.

B4) If $a \in SC_\theta$ or $a \equiv \varphi bc$ then $a_1 + \ldots + a_n \prec a$ if $a_1 \prec a$.

B5) If $a \in SC_\theta$ then $\varphi bc \prec a$ if $b, c \prec a$.

B6) $\psi b \prec \Omega$ for all b.

B7) $\psi a \succ \Gamma_\xi$ for all $\xi \leq \theta$.

B8) $a_1 + \ldots + a_n \prec b_1 + \ldots + b_m$ if $n < m$ and $(\forall i \leq n)[a_i \equiv b_i]$
or $\exists i \leq \min(n, m)[\forall j < i(a_j = b_j)$ and $a_i \prec b_i]$.

B9) $\varphi a_1 b_1 \prec \varphi a_2 b_2$ if $a_1 \prec a_2 \ \wedge \ b_1 \prec \varphi a_2 b_2$
or $a_1 = a_2 \ \wedge \ b_1 \prec b_2$
or $a_2 \prec a_1 \ \wedge \ \varphi a_1 b_1 \prec b_2$.

B10) $\psi a \prec \psi b$ if $a \prec b$.

□

3 The proof theory of $\mathbf{RS}_\Omega(X)$

3.1 A Tait-style sequent calculus formulation of KP

Definition 3.1. The language of **KP** consists of free variables $a_0, a_1, \ldots$, bound variables $x_0, x_1, \ldots$, the binary predicate symbols $\in$, $\notin$ and the logical symbols $\vee, \wedge, \forall, \exists$ as well as parentheses), (.

The atomic formulas are those of the form

$$(a \in b) \quad , \quad (a \notin b).$$

The formulas of **KP** are defined inductively by:

i) Atomic formulas are formulas.

ii) If A and B are formulas then so are $A \vee B$ and $A \wedge B$.

iii) If $A(b)$ is a formula in which the bound variable x does not occur, then $\forall x A(x)$, $\exists x A(x)$, $(\forall x \in a)A(x)$ and $(\exists x \in a)A(x)$ are all formulas.

Quantifiers of the form $\exists x$ and $\forall x$ will be called unbounded and those of the form $(\exists x \in a)$ and $(\forall x \in a)$ will be referred to as bounded quantifiers.

A formula is said to be Δ_0 if it contains no unbounded quantifiers. A formula is said to be Σ (Π) if it contains no unbounded universal (existential) quantifiers.

The negation $\neg A$ of a formula A is obtained from A by undergoing the following operations:

i) Replacing every occurrence of $\in$,$\notin$ with $\notin$,$\in$ respectively.

ii) Replacing any occurrence of $\wedge, \vee, \forall x, \exists x, (\forall x \in a), (\exists x \in a)$ with $\vee, \wedge, \exists x, \forall x$, $(\exists x \in a), (\forall x \in a)$ respectively.

Thus the negation of a formula A is in negation normal form. The expression $A \to B$ will be considered shorthand for $\neg A \vee B$.

The expression $a = b$ is to be treated as an abbreviation for $(\forall x \in a)(x \in b) \wedge (\forall x \in b)(x \in a)$.

The derivations of **KP** take place in a Tait-style sequent calculus, finite sets of formulae denoted by Greek capital letters are derived. Intuitively the sequent Γ may be read as the disjunction of formulae occuring in Γ.

The axioms of **KP** are:

Logical axioms:	$\Gamma, A, \neg A$ for any formula A.
Extensionality:	$\Gamma, a = b \wedge B(a) \rightarrow B(b)$ for any formula $B(a)$.
Pair:	$\Gamma, \exists z(a \in z \wedge b \in z)$.
Union:	$\Gamma, \exists z(\forall y \in a)(\forall x \in y)(x \in z)$.
Δ_0-Separation:	$\Gamma, \exists y[(\forall x \in y)(x \in a \wedge B(x)) \wedge (\forall x \in a)(B(x) \rightarrow x \in y)]$ for any Δ_0-formula $B(a)$.
Set Induction:	$\Gamma, \forall x[(\forall y \in x F(y) \rightarrow F(x)] \rightarrow \forall x F(x)$ for any formula $F(a)$.
Infinity:	$\Gamma, \exists x[(\exists z \in x)(z \in x) \wedge (\forall y \in x)(\exists z \in x)(y \in z)]$.
Δ_0-Collection:	$\Gamma, (\forall x \in a)\exists y G(x, y) \rightarrow \exists z(\forall x \in a)(\exists y \in z)G(x, y)$ for any Δ_0-formula G.

The rules of inference are

$$(\wedge)\ \frac{\Gamma, A \qquad \Gamma, B}{\Gamma, A \wedge B}$$

$$(\vee)\ \frac{\Gamma, A}{\Gamma, A \vee B} \qquad \frac{\Gamma, B}{\Gamma, A \vee B}$$

$$(b\exists)\ \frac{\Gamma, a \in b \wedge F(a)}{\Gamma, (\exists x \in b)F(x)} \qquad (\exists)\ \frac{\Gamma, F(a)}{\Gamma, \exists x F(x)}$$

$$(b\forall)\ \frac{\Gamma, a \in b \rightarrow F(a)}{\Gamma, (\forall x \in b)F(x)} \qquad (\forall)\ \frac{\Gamma, F(a)}{\Gamma, \forall x F(x)}$$

$$(\text{Cut})\ \frac{\Gamma, A \qquad \Gamma, \neg A}{\Gamma}$$

In both ($b\forall$) and ($\forall$), the variable a must not be present in the conclusion, such a variable is referred to as the *eigenvariable* of the inference.

The *minor formulae* of an inference are those rendered prominently in the premises, the other formulae in the premises will be referred to as *side formulae.* The *principal* formula of an inference is the one rendered prominently in the conclusion. Note that the principal formula can also be a side formula of that inference, when this happens we say that there has been a *contraction.* The rule (Cut) has no principal formula.

Formally, bounded and unbounded quantifiers are treated as logically separate operations. However, it is important to know and ensure that they interact with one another as expected.

Lemma 3.2. The following are derivable within **KP**:

i) $(\forall x \in b)F(x) \leftrightarrow \forall x(x \in b \to F(x))$.

ii) $(\exists x \in b)F(x) \leftrightarrow \exists x(x \in b \wedge F(x))$.

Proof. We verify only i) as the proof of ii) is very similar. First note that $a \in b \wedge \neg F(a), a \in b \to F(a)$ is a logical axiom of **KP**, we have the following derivation in **KP**.

$$\dfrac{(\vee)\text{ twice}\,\dfrac{(\forall)\,\dfrac{(b\exists)\,\dfrac{a \in b \wedge \neg F(a), a \in b \to F(a)}{(\exists x \in b)\neg F(x), a \in b \to F(a)}}{(\exists x \in b)\neg F(x), \forall x(x \in b \to F(x))}}{(\forall x \in b)F(x) \to \forall x(x \in b \to F(x))} \qquad (\vee)\text{ twice}\,\dfrac{(b\forall)\,\dfrac{(\exists)\,\dfrac{a \in b \wedge \neg F(a), a \in b \to F(a)}{\exists x(x \in b \wedge \neg F(x)), a \in b \to F(a)}}{\exists x(x \in b \wedge \neg F(x)), (\forall x \in b)F(x)}}{\forall x(x \in b \to F(x)) \to (\forall x \in b)F(x)}}{(\forall x \in b)F(x) \leftrightarrow \forall x(x \in b \to F(x))}\,(\wedge)$$

□

3.2 The infinitary system $\mathbf{RS}_\Omega(X)$

Let X be an arbitrary (well founded) set and let θ be the set-theoretic rank of X (hereby referred to as the $\in$-rank). Henceforth all ordinals are assumed to belong to the ordinal notation system $T(\theta)$ developed in the previous section. The system $\mathbf{RS}_\Omega(X)$ will be an infinitary proof system based on $L_\Omega(X)$; the relativised constructible hierarchy up to Ω.

Definition 3.3. We give an inductive definition of the set $\mathcal{T}$ of $\mathbf{RS}_\Omega(X)$ terms, to each term $t \in \mathcal{T}$ we assign an ordinal level $|\,t\,|$

i) For every $u \in TC(\{X\})$, $\bar{u} \in \mathcal{T}$ and $|\,\bar{u}\,| := \Gamma_{\text{rank}(u)}$ [here rank(u) is the $\in$-rank of u and TC denotes the *transitive closure* operator.] Note that rank$(u) \leq \theta$.

ii) For every $\alpha < \Omega$, $\mathbb{L}_\alpha(X) \in \mathcal{T}$ and $|\,\mathbb{L}_\alpha(X)\,| := \Gamma_{\theta+1} + \alpha$.

iii) If $\alpha < \Omega$, $A(a, b_1, \ldots, b_n)$ is a formula of **KP** with all free variables displayed and $s_1, \ldots, s_n$ are terms with levels less than $\Gamma_{\theta+1} + \alpha$ then

$$[x \in \mathbb{L}_\alpha(X) | A(x, s_1, \ldots, s_n)^{\mathbb{L}_\alpha(X)}]$$

is a term of level $\Gamma_{\theta+1} + \alpha$. Here the superscript $\mathbb{L}_\alpha(X)$ indicates that all unbounded quantifiers occuring in A are replaced by quantifiers bounded by $\mathbb{L}_\alpha(X)$.

The terms of $\mathbf{RS}_\Omega(X)$ are to be viewed as purely formal, syntactic objects. However their names are highly suggestive of the intended interpretation in the relativised constructible hierarchy up to Ω.

Definition 3.4. The formulae of $\mathbf{RS}_\Omega(X)$ are of the form $A(s_1, \ldots, s_n)$, where $A(a_1, \ldots, a_n)$ is a formula of **KP** with all free variables displayed and $s_1, \ldots, s_n$ are $\mathbf{RS}_\Omega(X)$ terms.

Formulae of the form $\bar{u} \in \bar{v}$ and $\bar{u} \notin \bar{v}$ will be referred to as *basic*. The properties Δ_0, Σ and Π are inherited from **KP** formulae.

Note that the system $\mathbf{RS}_\Omega(X)$ does not contain free variables.

For the remainder of this section we shall refer to $\mathbf{RS}_\Omega(X)$ terms and formulae simply as terms and formulae.

For any formula A we define

$$\begin{aligned} k(A) :=& \{|\,t\,| \mid t \text{ occurs in } A, \text{ subterms included}\} \\ &\cup \{\Omega \mid \text{if } A \text{ contains an unbounded quantifier}\}. \end{aligned}$$

If Γ is a finite set of the $\mathbf{RS}_\Omega(X)$ formulae $A_1, \ldots, A_n$ then we define

$$k(\Gamma) := k(A_1) \cup \ldots \cup k(A_n).$$

Abbreviations 3.5.

i) For $\mathbf{RS}_\Omega(X)$ terms s and t, the expression $s = t$ will be considered as shorthand for

$$(\forall x \in s)(x \in t) \wedge (\forall x \in t)(x \in s).$$

ii) If $|\,s\,| < |\,t\,|$, $A(s,t)$ is an $\mathbf{RS}_\Omega(X)$ formula and $\diamond$ is a propositional connective we define:

$$s \,\dot{\in}\, t \diamond A(s,t) := \begin{cases} s \in t \diamond A(s,t) & \text{if } t \equiv \bar{u} \\ A(s,t) & \text{if } t \equiv \mathbb{L}_\alpha(X) \\ B(s) \diamond A(s,t) & \text{if } t \equiv [x \in \mathbb{L}_\alpha(X) \mid B(x)]. \end{cases}$$

Our aim will be to remove cuts from certain $\mathbf{RS}_\Omega(X)$ derivations of Σ sentences. In order to do this we need to express a certain kind of uniformity in infinite derivations. The right tool for expressing this uniformity was developed by Buchholz in [11] and is termed *operator control*.

Definition 3.6. Let $\mathcal{P}(\mathbf{ON}) := \{Y : \text{ Y is a set of ordinals}\}$. A class function

$$\mathcal{H} : \mathcal{P}(\mathbf{ON}) \to \mathcal{P}(\mathbf{ON})$$

is called an Operator if the following conditions are satisfied for $Y, Y' \in \mathcal{P}(\mathbf{ON})$.

(H1) $0 \in \mathcal{H}(Y)$ and $\Gamma_\beta \in \mathcal{H}(Y)$ for any $\beta \leq \theta + 1$.

(H2) If $\alpha =_{\text{NF}} \alpha_1 + \ldots + \alpha_n$ then $\alpha \in \mathcal{H}(Y)$ iff $\alpha_1, \ldots, \alpha_n \in \mathcal{H}(Y)$.

(H3) If $\alpha =_{\text{NF}} \varphi\alpha_1\alpha_2$ then $\alpha \in \mathcal{H}(Y)$ iff $\alpha_1, \alpha_2 \in \mathcal{H}(Y)$.

(H4) $Y \subseteq \mathcal{H}(Y)$.

(H5) $Y' \subseteq \mathcal{H}(Y) \Rightarrow \mathcal{H}(Y') \subseteq \mathcal{H}(Y)$.

Note that this definition of operator, as with the infinitary system $\mathbf{RS}_\Omega(X)$ is dependent on the set X and its $\in$-rank θ.

Abbreviations 3.7. For an operator $\mathcal{H}$:

i) We write $\alpha \in \mathcal{H}$ instead of $\alpha \in \mathcal{H}(\emptyset)$.

ii) Likewise $Y \subseteq \mathcal{H}$ is shorthand for $Y \subseteq \mathcal{H}(\emptyset)$.

iii) For any $\mathbf{RS}_\Omega(X)$ term t, $\mathcal{H}[t](Y) := \mathcal{H}(Y \cup |\, t \,|)$.

iv) If $\mathfrak{X}$ is an $\mathbf{RS}_\Omega(X)$ formula or set of formulae then $\mathcal{H}[\mathfrak{X}](Y) := \mathcal{H}(Y \cup k(\mathfrak{X}))$.

Lemma 3.8. Let $\mathcal{H}$ be an operator s an $\mathbf{RS}_\Omega(X)$ term and $\mathfrak{X}$ an $\mathbf{RS}_\Omega(X)$ formula or set of formulae.

i) If $Y \subseteq Y'$ then $\mathcal{H}(Y) \subseteq \mathcal{H}(Y')$.

ii) $\mathcal{H}[s]$ and $\mathcal{H}[\mathfrak{X}]$ are operators.

iii) If $|\, s \,| \in \mathcal{H}$ then $\mathcal{H}[s] = \mathcal{H}$.

iv) If $k(\mathfrak{X}) \subseteq \mathcal{H}$ then $\mathcal{H}[\mathfrak{X}] = \mathcal{H}$.

Proof. These results are easily checked, they are proved in full in [50]. □

Definition 3.9. If $\mathcal{H}$ is an operator, α an ordinal and Γ a finite set of $\mathbf{RS}_\Omega(X)$-formulae, we give an inductive definition of the relation $\mathcal{H} \overset{\alpha}{\vdash} \Gamma$ by recursion on α. ($\mathcal{H}$*-controlled derivability in* $\mathbf{RS}_\Omega(X)$.) We require always that

$$\{\alpha\} \cup k(\Gamma) \subseteq \mathcal{H}$$

this condition will not be repeated in the inductive clauses pertaining to the axioms and inference rules below. We have the following axioms:

$$\mathcal{H} \overset{\alpha}{\vdash} \Gamma, \bar{u} \in \bar{v} \quad \text{if} \quad u, v \in TC(X) \text{ and } u \in v$$
$$\mathcal{H} \overset{\alpha}{\vdash} \Gamma, \bar{u} \notin \bar{v} \quad \text{if} \quad u, v \in TC(X) \text{ and } u \notin v.$$

The following are the inference rules of $\mathbf{RS}_\Omega(X)$, the column on the right gives the requirements on the ordinals, terms and formulae for each rule.

$$(\wedge) \quad \frac{\mathcal{H} \overset{\alpha_0}{\vdash} \Gamma, A \qquad \mathcal{H} \overset{\alpha_1}{\vdash} \Gamma, B}{\mathcal{H} \overset{\alpha}{\vdash} \Gamma, A \wedge B} \qquad \alpha_0, \alpha_1 < \alpha$$

$$(\vee) \quad \frac{\mathcal{H} \overset{\alpha_0}{\vdash} \Gamma, C \quad \text{for some } C \in \{A, B\}}{\mathcal{H} \overset{\alpha}{\vdash} \Gamma, A \vee B} \qquad \alpha_0 < \alpha$$

$$(\notin) \quad \frac{\mathcal{H}[s] \overset{\alpha_s}{\vdash} \Gamma, s \,\dot{\in}\, t \to r \neq s \quad \text{for all } |s| < |t|}{\mathcal{H} \overset{\alpha}{\vdash} \Gamma, r \notin t} \qquad \begin{array}{r} \alpha_s < \alpha \\ r \in t \text{ is not basic} \end{array}$$

$$(\in) \quad \frac{\mathcal{H} \overset{\alpha_0}{\vdash} \Gamma, s \,\dot{\in}\, t \wedge r = s}{\mathcal{H} \overset{\alpha}{\vdash} \Gamma, r \in t} \qquad \begin{array}{r} \alpha_0 < \alpha \\ |s| < |t| \\ |s| < \Gamma_{\theta+1} + \alpha \\ r \in t \text{ is not basic} \end{array}$$

$$(b\forall) \quad \frac{\mathcal{H}[s] \overset{\alpha_s}{\vdash} \Gamma, s \,\dot{\in}\, t \to A(s) \quad \text{for all } |s| < |t|}{\mathcal{H} \overset{\alpha}{\vdash} \Gamma, (\forall x \in t) A(x)} \qquad \alpha_s < \alpha$$

$$(b\exists) \quad \frac{\mathcal{H} \overset{\alpha_0}{\vdash} \Gamma, s \,\dot{\in}\, t \wedge A(s)}{\mathcal{H} \overset{\alpha}{\vdash} \Gamma, (\exists x \in t) A(x)} \qquad \begin{array}{r} \alpha_0 < \alpha \\ |s| < |t| \\ |s| < \Gamma_{\theta+1} + \alpha \end{array}$$

$$(\forall)\quad \frac{\mathcal{H}[s] \overset{\alpha_s}{\vdash} \Gamma, A(s) \quad \text{for all } s}{\mathcal{H} \overset{\alpha}{\vdash} \Gamma, \forall x A(x)} \qquad \alpha_s < \alpha$$

$$(\exists)\quad \frac{\mathcal{H} \overset{\alpha_0}{\vdash} \Gamma, A(s)}{\mathcal{H} \overset{\alpha}{\vdash} \Gamma, \exists x A(x)} \qquad \begin{array}{c}\alpha_0 < \alpha \\ |\, s \,| < \Gamma_{\theta+1} + \alpha\end{array}$$

$$(\mathrm{Cut})\quad \frac{\mathcal{H} \overset{\alpha_0}{\vdash} \Gamma, A \qquad \mathcal{H} \overset{\alpha_0}{\vdash} \Gamma, \neg A}{\mathcal{H} \overset{\alpha}{\vdash} \Gamma} \qquad \alpha_0 < \alpha$$

$$(\Sigma\text{-Ref}_\Omega(X))\quad \frac{\mathcal{H} \overset{\alpha_0}{\vdash} \Gamma, A}{\mathcal{H} \overset{\alpha}{\vdash} \Gamma, \exists z A^z} \qquad \begin{array}{c}\alpha_0, \Omega < \alpha \\ A \text{ is a } \Sigma \text{ formula}\end{array}$$

A^z results from A by restricting all unbounded quantifiers in A to z. The reason for the condition preventing the derivation of basic formulas in the rules $(\in)$ and $(\notin)$ is to prevent derivations of sequents which are already axioms, as this would cause a hindrance to cut-elimination. The condition that $|\, s \,| < \Gamma_{\theta+1} + \alpha$ in $(\in)$ and $(\exists)$ inferences will allow us to place bounds on the location of witnesses in derivable Σ formulas.

3.3 Cut elimination for $\mathbf{RS}_\Omega(X)$

We need to keep track of the complexity of cuts appearing in a derivation, to this end we define the *rank* of an $\mathbf{RS}_\Omega(X)$ formula.

Definition 3.10. The rank of a term or formula is defined by recursion on the construction as follows:

1. $rk(\bar{u}) := \Gamma_{rank(u)}$
2. $rk(\mathbb{L}_\alpha(X)) := \Gamma_{\theta+1} + \omega \cdot \alpha$
3. $rk([x \in \mathbb{L}_\alpha(X) | F(x)]) := \max(\Gamma_{\theta+1} + \omega \cdot \alpha + 1, rk(F(\bar{\emptyset})) + 2)$
4. $rk(s \in t) := rk(s \notin t) := \max(rk(t) + 1, rk(s) + 6)$
5. $rk((\exists x \in \bar{u}) F(x)) := rk((\forall x \in \bar{u}) F(x)) := \max(rk(\bar{u}) + 3, rk(F(\bar{\emptyset})) + 2)$.

6. $rk((\exists x \in t)F(x)) := rk((\forall x \in t)F(x)) := \max(rk(t), rk(F(\bar{\emptyset})) + 2)$ if t is not of the form $\bar{u}$.

7. $rk(\exists x F(x)) := rk(\forall x F(x)) := \max(\Omega, rk(F(\bar{\emptyset})) + 1)$

8. $rk(A \wedge B) := rk(A \vee B) := \max(rk(A), rk(B)) + 1$

$\mathcal{H} \overset{\alpha}{\underset{\rho}{\vdash}} \Gamma$ will be used to denote that $\mathcal{H} \overset{\alpha}{\vdash} \Gamma$ and all cut formulas appearing in the derivation have rank $< \rho$.

Observation 3.11. i) For each term t, $rk(t) = \omega \cdot |\, t \,| + n$ for some $n < \omega$.

ii) For each formula A, $rk(A) = \omega \cdot \max(k(A)) + n$ for some $n < \omega$.

iii) $rk(A) < \Omega$ if and only if A is Δ_0.

The next Lemma shows that the rank of a formula A is determined *only* by $\max(k(A))$ and the logical structure of A.

Lemma 3.12. For each formula $A(s)$, if $|\, s \,| < \max(k(A(s)))$ then $rk(A(s)) = rk(A(\bar{\emptyset}))$.

Proof. The proof is by induction on the complexity of A.

Case 1. If $A(s) \equiv s \in t$ then by assumption $|\, s \,| < |\, t \,|$, so $rk(A(s)) = rk(t) + 1 = rk(A(\bar{\emptyset}))$.

Case 2. If $A(s) \equiv t \in s$ we may argue in a similar fashion to Case 1.

Case 3. It cannot be the case that $A(s) \equiv s \in s$.

Case 4. If $A(s) \equiv (\exists y \in \bar{u})B(y, s)$ then

$$rk(A(s)) = \max(rk(\bar{u}) + 3, rk(B(\bar{\emptyset}, s)) + 2)$$

and

$$rk(A(\bar{\emptyset})) = \max(rk(\bar{u}) + 3, rk(B(\bar{\emptyset}, \bar{\emptyset})) + 2).$$

4.1 If $|\, \bar{u} \,| > \max(k(B(\bar{\emptyset}, \bar{\emptyset})))$ then $|\, s \,| < |\, \bar{u} \,|$ by assumption, so using observation 3.11ii) gives us

$$rk(A(s)) = rk(\bar{u}) + 3 = rk(A(\bar{\emptyset})).$$

4.2 If $|\, \bar{u} \,| \leq \max(k(B(\bar{\emptyset}, \bar{\emptyset}))$ then $|\, s \,| < \max(k(B(\bar{\emptyset}, \bar{\emptyset})))$ by assumption, so by induction hypothesis

$$rk(B(\bar{\emptyset}, s)) = rk(B(\bar{\emptyset}, \bar{\emptyset}))$$

and hence using Observation 3.11ii) gives us

$$rk(A(s)) = rk(B(\bar{\emptyset}, \bar{\emptyset})) + 2 = rk(A(\bar{\emptyset})).$$

Case 5. If $A(s) \equiv (\exists y \in t)B(y, s)$ for some t not of the form $\bar{u}$, we may argue in a similar way to case 4.

Case 6. $A(s) \equiv (\exists y \in s)B(y, s)$, now $|\, s \,| < \max(k(A(\bar{\emptyset}))) = \max(k(B(\bar{\emptyset}, \bar{\emptyset})))$, so by induction hypothesis

$$rk(B(\bar{\emptyset}, s)) = rk(B(\bar{\emptyset}, \bar{\emptyset}))$$

and hence using observation 3.11 we see that

$$\begin{aligned} rk(A(s)) &= rk(B(\bar{\emptyset}, s)) + 2 \\ &= rk(B(\bar{\emptyset}, \bar{\emptyset})) + 2 \\ &= rk(A(\bar{\emptyset})). \end{aligned}$$

Case 7. If $A(s) \equiv \exists x B(x, s)$ then $|\, s \,| < \max(k(A(s))) = \max(k(B(\emptyset, s)))$ by assumption, so we may apply the induction hypothesis to see that $rk(A(s)) = \max(\Omega, rk(B(\emptyset, s)) + 1) = \max(\Omega, rk(B(\emptyset, \emptyset)) + 1) = rk(A(\emptyset))$.

Case 8. All other cases are either propositional in which case we may just use the induction hypothesis directly or are dual to cases already considered. □

Definition 3.13. To each non-basic formula A we assign an infinitary disjunction $(\bigvee A_i)_{i \in y}$ or conjunction $(\bigwedge A_i)_{i \in y}$ as follows:

1. $r \in t :\simeq \bigvee(s \,\dot{\in}\, t \wedge r = s)_{|\, s \,| < |\, t \,|}$ provided $r \in t$ is not a basic formula.

2. $(\exists x \in t)B(x) :\simeq \bigvee(s \,\dot{\in}\, t \wedge B(s))_{|\, s \,| < |\, t \,|}$

3. $\exists x B(x) :\simeq \bigvee(B(s))_{s \in \mathcal{T}}$

4. $B_0 \vee B_1 :\simeq \bigvee(B_i)_{i \in \{0,1\}}$

5. $\neg B :\simeq \bigwedge(\neg B_i)_{i \in y}$ if B is of the form considered in 1.-4.

The idea is that the infinitary conjunction or disjunction lists the premises required to derive A as the principal formula of an $\mathbf{RS}_\Omega(X)$-inference different from (Σ-$\mathrm{Ref}_\Omega(X)$) or (Cut).

Lemma 3.14. *If* $A \simeq (\bigvee A_i)_{i\in y}$ *or* $A \simeq (\bigwedge A_i)_{i\in y}$ *then*

$$\forall i \in y(rk(A_i) < rk(A)).$$

Proof. We need only treat the case where $A \simeq (\bigvee A_i)_{i\in y}$ since the other case is dual to this one. We proceed by induction on the complexity of A.

Case 1. Suppose $A \equiv r \in t$ then by assumption either r or t is not of the form $\bar{u}$, we split cases based on the form of t.

1.1 If $t \equiv \bar{u}$ then r is not of the form $\bar{v}$ and $rk(A) = rk(r) + 6$. In this case $A_i \equiv \bar{v} \in \bar{u} \wedge \bar{v} = r$ for some $|\,\bar{v}\,| < |\,\bar{u}\,|$ and we have

$$\begin{aligned} rk(A_i) &= \max(rk(\bar{v} \in \bar{u}), rk(\bar{v} = r)) + 1 \\ &= rk(\bar{v} = r) + 1 \\ &= \max(rk((\forall x \in \bar{v})(x \in r)), rk((\forall x \in r)(x \in \bar{v}))) + 2 \\ &= rk(r) + 5 < rk(r) + 6 = rk(A). \end{aligned}$$

1.2 If $t \equiv \mathbb{L}_\alpha(X)$ then $A_i \equiv s = r$ for some $|\,s\,| < |\,t\,|$. So we have

$$\begin{aligned} rk(A_i) &= rk((\forall x \in s)(x \in r) \wedge (\forall x \in r)(x \in s)) \\ &= \max(rk(s) + 4, rk(r) + 4) \\ &< \max(rk(r) + 1, rk(t) + 6) = rk(A). \end{aligned}$$

1.3 If $t \equiv [x \in \mathbb{L}_\alpha(X) | B(x)]$ then $A_i \equiv B(s) \wedge s = r$ for some $|\,s\,| < |\,t\,|$. So we have

$$rk(A_i) = \max(rk(B(s)) + 1, rk(r = s) + 1).$$

First note that $rk(r = s) + 1 = \max(rk(s) + 5, rk(r) + 5) < rk(A)$. So it remains to verify that $rk(B(s)) + 1 < rk(A)$, for this it is enough to show that $rk(B(s)) < rk(t)$.

1.3.1 If $\max(k(B(s))) \leq |\,s\,|$ then by Observation 3.11ii) we have $rk(B(s)) + 1 < \omega \cdot |\,s\,| + \omega \leq rk(t)$.

1.3.2 Otherwise $\max(k(B(s))) > |\,s\,|$ then by Lemma 3.12 we have

$$\begin{aligned} rk(B(s)) + 1 &= rk(B(\bar{\emptyset})) + 1 \\ &< \max(\Gamma_{\theta+1} + \omega \cdot \alpha + 1, rk(B(\bar{\emptyset})) + 2) = rk(t). \end{aligned}$$

Case 2. Suppose $A \equiv (\exists x \in t)B(x)$, we split into cases based on the form of t.

2.1 If $t \equiv \bar{u}$ then $rk(A) := \max(rk(\bar{u})+3, rk(B(\bar{\emptyset}))+2)$. In this case $A_i \equiv \bar{v} \in \bar{u} \wedge B(\bar{v})$ for some $|\,\bar{v}\,| < |\,\bar{u}\,|$, so we have

$$rk(A_i) = \max(rk(\bar{u}) + 2, rk(B(\bar{v})) + 1).$$

Clearly $rk(\bar{u}) + 2 < rk(\bar{u}) + 3$ so it remains to verify that $rk(B(\bar{v})) + 1 < rk(A)$.

2.1.1 If $|\bar{v}| \geq \max(k(B(\bar{v})))$ then by Observation 3.11i) $rk(B(\bar{v})) + 1 < rk(\bar{u}) < rk(\bar{u}) + 3$.

2.1.2 If $|\bar{v}| < \max(k(B(\bar{v})))$ then by Lemma 3.12 $rk(B(\bar{v})) + 1 = rk(B(\bar{\emptyset})) + 1 < rk(B(\bar{\emptyset})) + 2$.

2.2 Now suppose $t \equiv \mathbb{L}_\alpha(X)$, so $rk(A) = \max(rk(t), rk(B(\bar{\emptyset})) + 2)$. In this case $A_i = B(s)$ for some $|\,s\,| < |\,t\,|$.

2.2.1 If $|\,s\,| \geq \max(k(B(s)))$ then $rk(B(s)) < rk(t)$ by Observation 3.11.

2.2.2 If $|\,s\,| < \max(k(B(s)))$ then by Lemma 3.12 $rk(B(s)) = rk(B(\bar{\emptyset})) < rk(A)$.

2.3. Now suppose $t \equiv [y \in \mathbb{L}_\alpha(X) \mid C(y)]$, so we have

$$\begin{aligned} rk(A) &:= \max(rk(t), rk(B(\bar{\emptyset})) + 2) \\ &= \max(\Gamma_{\theta+1} + \omega \cdot \alpha + 1, rk(C(\bar{\emptyset})) + 2, rk(B(\bar{\emptyset})) + 2). \end{aligned}$$

In this case $A_i \equiv C(s) \wedge B(s)$ for some $|\,s\,| < |\,t\,|$.

2.3.1 If $|\,s\,| < \max(k(B(s)))$ then $rk(B(s)) + 1 = rk(B(\bar{\emptyset})) + 1 < rk(B(\bar{\emptyset})) + 2$. It remains to show that $rk(C(s)) < rk(A)$.

2.3.1.1 If $\max(k(C(s))) < |\,t\,|$ then $rk(C(s)) + 1 < rk(t)$ by Observation 3.11.

2.3.1.2 Now if $\max(k(C(s))) \geq |\,t\,|$ then we may apply Lemma 3.12 to give

$$rk(C(s)) + 1 = rk(C(\bar{\emptyset})) + 1 < rk(C(\bar{\emptyset})) + 2 \leq rk(A).$$

2.3.2 If $|\,s\,| \geq \max(k(B(s)))$ then $rk(B(s)) < \Gamma_{\theta+1} + \omega \cdot \alpha$ by Observation 3.11. Now we may apply the same argument as in 2.3.1.1 and 2.3.1.2 to yield $rk(C(s)) + 1 < rk(A)$.

Case 3. If $A \equiv \exists x B(x)$ then $rk(A) := \max(\Omega, rk(B(\bar{\emptyset})) + 1)$. In this case $A_i \equiv B(s)$ for some term s.

3.1 If B contains an unbounded quantifier then by Lemma 3.12 $rk(B(s)) = rk(B(\bar{\emptyset})) < rk(A)$.

3.2 If B does not contain an unbounded quantifier then $rk(B(s)) < \Omega$ by Observation 3.11iii).

Case 4. If $A \equiv B \vee C$ then the result is clear immediately from the definition of $rk(A)$. □

Lemma 3.15. Let $\mathcal{H}$ be an arbitrary operator.

i) If $\alpha \leq \alpha' \in \mathcal{H}$, $\rho \leq \rho'$, $k(\Gamma') \subseteq \mathcal{H}$ and $\mathcal{H} \overset{\alpha}{\underset{\rho}{\vdash}} \Gamma$ then $\mathcal{H} \overset{\alpha'}{\underset{\rho'}{\vdash}} \Gamma, \Gamma'$.

ii) If C is a basic formula which holds true in the set X and $\mathcal{H} \overset{\alpha}{\underset{\rho}{\vdash}} \Gamma, \neg C$ then $\mathcal{H} \overset{\alpha}{\underset{\rho}{\vdash}} \Gamma$.

iii) If $\mathcal{H} \overset{\alpha}{\underset{\rho}{\vdash}} \Gamma, A \vee B$ then $\mathcal{H} \overset{\alpha}{\underset{\rho}{\vdash}} \Gamma, A, B$.

iv) If $A \simeq \bigwedge(A_i)_{i \in y}$ and $\mathcal{H} \overset{\alpha}{\underset{\rho}{\vdash}} \Gamma, A$ then $(\forall i \in y)\, \mathcal{H}[i] \overset{\alpha}{\underset{\rho}{\vdash}} \Gamma, A_i$.

v) If $\gamma \in \mathcal{H}$ and $\mathcal{H} \overset{\alpha}{\underset{\rho}{\vdash}} \Gamma, \forall x F(x)$ then $\mathcal{H} \overset{\alpha}{\underset{\rho}{\vdash}} \Gamma, (\forall x \in \mathbb{L}_\gamma(X))F(x)$.

Proof. All proofs are by induction on α.

i) If Γ is an axiom then Γ, Γ' is also an axiom, and since $\{\alpha'\} \cup k(\Gamma') \subseteq \mathcal{H}$ there is nothing to show.

Now suppose Γ is the result of an inference

$$\text{(I)}\ \frac{\ldots \mathcal{H}_i \overset{\alpha_i}{\underset{\rho}{\vdash}} \Gamma_i \ldots}{\mathcal{H} \overset{\alpha}{\underset{\rho}{\vdash}} \Gamma}\ (i \in y) \quad \alpha_i < \alpha.$$

Using the induction hypothesis we have

$$\ldots \mathcal{H}_i \overset{\alpha_i}{\underset{\rho'}{\vdash}} \Gamma_i, \Gamma' \ldots \quad (i \in y) \quad \alpha_i < \alpha.$$

It's worth noting that $k(\Gamma') \subseteq \mathcal{H}_i$, since $\mathcal{H}_i(\emptyset) \supseteq \mathcal{H}(\emptyset)$, this can be observed by looking at each inference rule.

Finally we may apply the inference (I) again to obtain

$$\mathcal{H} \overset{\alpha'}{\underset{\rho'}{\vdash}} \Gamma, \Gamma'$$

as required.

ii) If $\Gamma, \neg C$ is an axiom then so is Γ so there is nothing to show.

Now suppose $\Gamma, \neg C$ was derived as the result of an inference rule (I), then $\neg C$ cannot have been the principal formula since it is basic so we have the premise(s)

$$\mathcal{H}_i \overset{\alpha_i}{\underset{\rho}{\vdash}} \Gamma_i, \neg C \quad \alpha_i < \alpha.$$

Now by induction hypothesis we obtain

$$\mathcal{H}_i \overset{\alpha_i}{\underset{\rho}{\vdash}} \Gamma_i \quad \alpha_i < \alpha$$

to which we may apply the inference rule (I) to complete the proof.

iii) If $\Gamma, A \vee B$ is an axiom then Γ, A, B is also an axiom. If $A \vee B$ was not the principal formula of the last inference then we can apply the induction hypothesis to its premises and then the same inference again.

Now suppose that $A \vee B$ was the principal formula of the last inference. So we have

$$\mathcal{H} \overset{\alpha_0}{\underset{\rho}{\vdash}} \Gamma, C \quad \text{or} \quad \mathcal{H} \overset{\alpha_0}{\underset{\rho}{\vdash}} \Gamma, C, A \vee B \quad \text{where } C \in \{A, B\} \text{ and } \alpha_0 < \alpha.$$

By i) we may assume that we are in the latter case. By the induction hypothesis, and a contraction, we obtain

$$\mathcal{H} \overset{\alpha_0}{\underset{\rho}{\vdash}} \Gamma, A, B$$

Finally using i) yields

$$\mathcal{H} \overset{\alpha}{\underset{\rho}{\vdash}} \Gamma, A, B\,.$$

iv) If Γ, A is an axiom, then Γ is also an axiom since A cannot be the *active part* of an axiom, so Γ, A_i is an axiom for any $i \in y$. If A was not the principal formula of the last inference then we may apply the induction hypothesis to its premises and then use that inference again.

Now suppose A was the principal formula of the last inference. With the possible use of part i), we may assume we are in the following situation:

$$\mathcal{H}[i] \overset{\alpha_i}{\underset{\rho}{\vdash}} \Gamma, A, A_i \quad (\forall i \in y) \quad \alpha_i < \alpha.$$

Inductively and via a contraction we obtain

$$\mathcal{H}[i] \mid\frac{\alpha_i}{\rho} \Gamma, A_i\,.$$

Here it is important to note that $\mathcal{H}[i][i] \equiv \mathcal{H}[i]$. To which we may apply part i) to obtain

$$\mathcal{H}[i] \mid\frac{\alpha}{\rho} \Gamma, A_i$$

as required.

v) The interesting case is where $\forall x F(x)$ was the principal formula of the last inference. In this case we may assume we are in the following situation:

(1) $$\mathcal{H}[s] \mid\frac{\alpha_s}{\rho} \Gamma, \forall x F(x), F(s) \quad \text{for all terms } s, \text{ with } \alpha_s < \alpha.$$

Using the induction hypothesis yields

(2) $$\mathcal{H}[s] \mid\frac{\alpha_s}{\rho} \Gamma, (\forall x \in \mathbb{L}_\gamma(X))F(x), F(s)\,.$$

Note that for $|\,s\,| < \Gamma_{\theta+1} + \gamma$ we have $s \mathrel{\dot\in} \mathbb{L}_\gamma(X) \to F(s) \equiv F(s)$. So as a subset of (2) we have

$$\mathcal{H}[s] \mid\frac{\alpha_s}{\rho} \Gamma, (\forall x \in \mathbb{L}_\gamma(X))F(x), s \mathrel{\dot\in} \mathbb{L}_\gamma(X) \to F(s)$$

for all $|\,s\,| < \Gamma_{\theta+1} + \gamma$, with $\alpha_s < \alpha$. From which one application of $(b\forall)$ gives us the desired result. □

Lemma 3.16 (Reduction for $\mathbf{RS}_\Omega(X)$)**.** Suppose $C \equiv \bar{u} \in \bar{v}$ or $C \simeq \bigvee(C_i)_{i \in y}$ and $rk(C) := \rho \neq \Omega$.

$$\text{If} \quad [\mathcal{H} \mid\frac{\alpha}{\rho} \Lambda, \neg C \quad \& \quad \mathcal{H} \mid\frac{\beta}{\rho} \Gamma, C] \quad \text{then} \quad \mathcal{H} \mid\frac{\alpha+\beta}{\rho} \Lambda, \Gamma\,.$$

Proof. If $C \equiv \bar{u} \in \bar{v}$ then by 3.15ii) we have either $\mathcal{H} \mid\frac{\alpha}{\rho} \Lambda$ or $\mathcal{H} \mid\frac{\beta}{\rho} \Gamma$. Hence using 3.15i) we obtain $\mathcal{H} \mid\frac{\alpha+\beta}{\rho} \Lambda, \Gamma$ as required.

Now suppose $C \simeq \bigvee(C_i)_{i \in y}$, we proceed by induction on β. We have

(1) $$\mathcal{H} \mid\frac{\alpha}{\rho} \Lambda, \neg C$$

(2) $$\mathcal{H} \mid\frac{\beta}{\rho} \Gamma, C\,.$$

If C was not the principal formula of the last inference in (2), then we may apply the induction hypothesis to the premises of that inference and then the same inference again. Now suppose C was the principal formula of the last inference in (2). If B was the principal formula of the inference (Σ-Ref$_\Omega(X)$), then B is of the form $\exists z F(s_1, \ldots, s_n)^z$, which implies $rk(B) = \Omega$, therefore the last inference in (2) was not (Σ-Ref$_\Omega(X)$). So we have

$$(3) \qquad \mathcal{H} \overset{\beta_0}{\underset{\rho}{\vdash}} \Gamma, C, C_{i_0} \quad \text{for some } i_0 \in y,\ \beta_0 < \beta \text{ with } |\, i_0 \,| < \Gamma_{\theta+1} + \beta.$$

The induction hypothesis applied to (2) and (3) yields

$$(4) \qquad \mathcal{H} \overset{\alpha+\beta_0}{\underset{\rho}{\vdash}} \Lambda, \Gamma, C_{i_0}\,.$$

Now applying Lemma 3.15iv) to (1) provides

$$(5) \qquad \mathcal{H}[i_0] \overset{\alpha}{\underset{\rho}{\vdash}} \Lambda, \neg C_{i_0}\,.$$

But $|\, i_0 \,| \in \mathcal{H}$ by (4), which means $\mathcal{H}[i_0] = \mathcal{H}$ by Lemma 3.8iv), so in fact we have

$$(6) \qquad \mathcal{H} \overset{\alpha}{\underset{\rho}{\vdash}} \Lambda, \neg C_{i_0}\,.$$

Thus we may apply (Cut) to (4) and (6) (noting that $rk(C_{i_0}) < rk(C) := \rho$ by Lemma 3.14) to obtain

$$\mathcal{H} \overset{\alpha+\beta}{\underset{\rho}{\vdash}} \Lambda, \Gamma$$

as required. □

Theorem 3.17 (Predicative cut elimination for $\mathbf{RS}_\Omega(X)$).
If $\mathcal{H} \overset{\beta}{\underset{\rho+\omega^\alpha}{\vdash}} \Gamma$ and $\Omega \notin [\rho, \rho + \omega^\alpha)$ and $\alpha \in \mathcal{H}$ then $\mathcal{H} \overset{\varphi\alpha\beta}{\underset{\rho}{\vdash}} \Gamma$.

Proof. The proof is by main induction on α and subsidiary induction on β. If Γ is an axiom then the result is immediate. If the last inference was anything other that (Cut) we may apply the subsidiary induction hypothesis to its premises and then the same inference again. The crucial case is where the last inference was (Cut), so suppose there is a formula C with $rk(C) < \rho + \omega^\alpha$ such that

$$(1) \qquad \mathcal{H} \overset{\beta_0}{\underset{\rho+\omega^\alpha}{\vdash}} \Gamma, C \quad \text{with } \beta_0 < \beta.$$

$$(2) \qquad \mathcal{H} \overset{\beta_0}{\underset{\rho+\omega^\alpha}{\vdash}} \Gamma, \neg C \quad \text{with } \beta_0 < \beta.$$

Applying the subsidiary induction hypothesis to (1) and (2) yields

$$(3) \qquad \mathcal{H} \overset{\varphi\alpha\beta_0}{\underset{\rho}{\vdash}} \Gamma, C\,.$$

$$(4) \qquad \mathcal{H} \overset{\varphi\alpha\beta_0}{\underset{\rho}{\vdash}} \Gamma, \neg C\,.$$

Case 1. If $rk(C) < \rho$ then we may apply (Cut) to (3) and (4), noting that $\varphi\alpha\beta_0 + 1 < \varphi\alpha\beta \in \mathcal{H}$, to give the desired result.

Case 2. Now suppose $rk(C) \in [\rho, \rho + \omega^\alpha)$, so we may write $rk(C)$ in the following form:

(5) $$rk(C) = \rho + \omega^{\alpha_1} + \ldots + \omega^{\alpha_n} \quad \text{with } \alpha > \alpha_1 \geq \ldots \geq \alpha_n.$$

If $n = 0$, this means that $rk(C) = \rho$. From (3) we know that $k(C) \subseteq \mathcal{H}$ and thus $rk(C) \in \mathcal{H}$. Now (5) and (H2) and (H3) from Definition 3.6 give us $\alpha_1, \ldots, \alpha_n \in \mathcal{H}$. Since $rk(C) \neq \Omega$ we may apply the Reduction Lemma 3.16 to (3) and (4) to obtain

(6) $$\mathcal{H} \mid\frac{\varphi\alpha\beta_0+\varphi\alpha\beta_0}{\rho+\omega^{\alpha_1}+\ldots+\omega^{\alpha_n}} \Gamma\,.$$

Now $\varphi\alpha\beta_0 + \varphi\alpha\beta_0 < \varphi\alpha\beta$, so by Lemma 3.15i) we have

(7) $$\mathcal{H} \mid\frac{\varphi\alpha\beta}{\rho+\omega^{\alpha_1}+\ldots+\omega^{\alpha_n}} \Gamma\,.$$

Applying the main induction hypothesis (since $\alpha_n < \alpha$) to (7) gives

$$\mathcal{H} \mid\frac{\varphi\alpha_n(\varphi\alpha\beta)}{\rho+\omega^{\alpha_1}+\ldots+\omega^{\alpha_{n-1}}} \Gamma\,.$$

But since $\varphi\alpha\beta$ is a fixed point of the function $\varphi\alpha_n(\cdot)$ we have

$$\mathcal{H} \mid\frac{\varphi\alpha\beta}{\rho+\omega^{\alpha_1}+\ldots+\omega^{\alpha_{n-1}}} \Gamma\,.$$

Now since $\alpha_1, \ldots, \alpha_{n-1} < \alpha$ we may repeat this application of the main induction hypothesis a further $n - 1$ times to obtain

$$\mathcal{H} \mid\frac{\varphi\alpha\beta}{\rho} \Gamma$$

as required. □

Lemma 3.18 (Boundedness for $\mathbf{RS}_\Omega(X)$)**.** If C is a Σ formula, $\alpha \leq \beta < \Omega$, $\beta \in \mathcal{H}$ and $\mathcal{H} \mid\frac{\alpha}{\rho} \Gamma, C$ then $\mathcal{H} \mid\frac{\alpha}{\rho} \Gamma, C^{\mathbb{L}_\beta(X)}$.

Proof. The proof is by induction on α. If C is basic then $C \equiv C^{\mathbb{L}_\beta(X)}$ so there is nothing to show. If C was not the principal formula of the last inference then we may apply the induction hypothesis to its premises and then the same inference again. Now suppose C was the principal formula of the last inference. The last inference cannot have been (Σ-Ref$_\Omega(X)$) since $\alpha < \Omega$.

Case 1. Suppose $C \simeq \bigwedge(C_i)_{i\in y}$ and $\mathcal{H}[i] \overset{\alpha_i}{\underset{\rho}{\vdash}} \Gamma, C, C_i$ with $\alpha_i < \alpha$. Since C is a Σ formula, there must be some $\eta \in \mathcal{H}(\emptyset) \cap \Omega$ such that $(\forall s \in y)(|\, s\,| < \eta)$. Therefore $C^{\mathbb{L}_\beta(X)} \simeq \bigwedge(C_i^{\mathbb{L}_\beta(X)})_{i\in y}$. Now two applications of the induction hypothesis gives

$$\mathcal{H}[i] \overset{\alpha_i}{\underset{\rho}{\vdash}} \Gamma, C^{\mathbb{L}_\beta(X)}, C_i^{\mathbb{L}_\beta(X)}$$

to which we may apply the appropriate inference to gain the desired result.

Case 2. Now suppose $C \simeq \bigvee(C_i)_{i\in y}$ and $\mathcal{H} \overset{\alpha_0}{\underset{\rho}{\vdash}} \Gamma, C, C_{i_0}$, with $i_0 \in y$, $|\, i_0\,| < \Gamma_{\theta+1}+\alpha$ and $\alpha_0 < \alpha$. In this case $C^{\mathbb{L}_\beta(X)} \simeq \bigvee(C_i)_{i\in y'}$ where either $y' = y$ or $y' = \{i \in y \,|\, |\, i\,| < \Gamma_{\theta+1} + \beta\}$. Now by assumption $|\, i_0\,| < \Gamma_{\theta+1} + \alpha < \Gamma_{\theta+1} + \beta$, so $i_0 \in y'$. Thus using the same inference again, or $(b\exists)$ in the case that the last inference was $(\exists)$, we obtain

$$\mathcal{H} \overset{\alpha}{\underset{\rho}{\vdash}} \Gamma, C^{\mathbb{L}_\beta(X)}$$

as required. □

Definition 3.19. For each $\eta \in T(\theta)$ we define

$$\mathcal{H}_\eta : \mathcal{P}(\mathbf{ON}) \mapsto \mathcal{P}(\mathbf{ON})$$
$$\mathcal{H}_\eta(Y) := \bigcap\{B(\alpha) \,|\, Y \subseteq B(\alpha) \text{ and } \eta < \alpha\}$$

Lemma 3.20. For any η, $\mathcal{H}_\eta$ is an operator.

Proof. We must verify the conditions (H1) - (H5) from Definition 3.6.

(H1) Clearly $0 \in \mathcal{H}_\eta(Y)$ and $\{\Gamma_\beta \,|\, \beta \leq \theta\} \subseteq \mathcal{H}_\eta(Y)$ since these belong in any of the sets $B(\alpha)$. It remains to note that $\mathcal{H}_\eta(Y) \supseteq B(1)$ and since $\Gamma_{\theta+1} = \psi 0 \in B(1)$ we have $\Gamma_{\theta+1} \in \mathcal{H}_\eta(Y)$.

(H2) and (H3) follow immediately from Lemma 2.8i) and ii) respectively.

(H4) is clear from the definition. Now for (H5) suppose $Y' \subseteq \mathcal{H}_\eta(Y)$, then $Y' \subseteq B(\alpha)$ for every α such that $\eta < \alpha$ and $Y \subseteq B(\alpha)$. It follows that $\mathcal{H}_\eta(Y') \subseteq \mathcal{H}_\eta(Y)$. □

Lemma 3.21. **i)** $\mathcal{H}_\eta(Y)$ is closed under φ and $\psi|_{\eta+1}$.

ii) If $\delta < \eta$ then $\mathcal{H}_\delta(Y) \subseteq \mathcal{H}_\eta(Y)$.

iii) If $\delta < \eta$ and $\mathcal{H}_\delta \overset{\alpha}{\underset{\rho}{\vdash}} \Gamma$ then $\mathcal{H}_\eta \overset{\alpha}{\underset{\rho}{\vdash}} \Gamma$

Proof. i) Note that for any X, $\mathcal{H}_\eta(X) = B(\alpha)$ for some $\alpha \geq \eta + 1$.

ii) follows immediately from the definition of $\mathcal{H}_\eta$ and iii) follows easily from ii). □

Lemma 3.22. Suppose $\eta \in B(\eta)$ and for any ordinal β let $\hat{\beta} := \eta + \omega^{\Omega+\beta}$.

i) If $\alpha \in \mathcal{H}_\eta$ then $\hat{\alpha}, \psi\hat{\alpha} \in \mathcal{H}_{\hat{\alpha}}$.

ii) If $\alpha_0 \in \mathcal{H}_\eta$ and $\alpha_0 < \alpha$ then $\psi\hat{\alpha_0} < \psi\hat{\alpha}$

Proof. i) First note that $\mathcal{H}_\eta(\emptyset) = B(\eta + 1)$. Now from $\alpha, \eta \in B(\eta + 1)$ we get $\hat{\alpha} \in B(\eta + 1)$ and thus $\hat{\alpha} \in B(\hat{\alpha})$. It follows that $\psi\hat{\alpha} \in B(\hat{\alpha} + 1) = \mathcal{H}_{\hat{\alpha}}(\emptyset)$.

ii) Suppose that $\alpha_0 \in \mathcal{H}_\eta$ and $\alpha_0 < \alpha$, using the preceding argument we get that $\psi\hat{\alpha_0} \in B(\hat{\alpha_0} + 1) \subseteq B(\hat{\alpha})$, thus $\psi\hat{\alpha_0} < \psi\hat{\alpha}$. □

Theorem 3.23 (Collapsing for $\mathbf{RS}_\Omega(X)$)**.** Suppose Γ is a set of Σ formulae and $\eta \in B(\eta)$.

$$\text{If} \quad \mathcal{H}_\eta \overset{\alpha}{\underset{\Omega+1}{\vdash}} \Gamma \quad \text{then} \quad \mathcal{H}_{\hat{\alpha}} \overset{\psi\hat{\alpha}}{\underset{\psi\hat{\alpha}}{\vdash}} \Gamma\,.$$

Proof. We proceed by induction on α. First note that from $\alpha \in \mathcal{H}_\eta$ we get $\hat{\alpha}, \psi\hat{\alpha} \in \mathcal{H}_{\hat{\alpha}}$ from Lemma 3.22i).

If Γ is an axiom then the result follows by Lemma 3.15i). So suppose Γ arose as the result of an inference, we shall distinguish cases according to the last inference of $\mathcal{H}_\eta \overset{\alpha}{\underset{\Omega+1}{\vdash}} \Gamma$.

Case 1. Suppose $A \simeq \bigwedge(A_i)_{i \in y} \in \Gamma$ and $\mathcal{H}_\eta[i] \overset{\alpha_i}{\underset{\Omega+1}{\vdash}} \Gamma, A_i$ with $\alpha_i < \alpha$ for each $i \in y$. Since A is a Σ formula, we must have $\sup\{|i| \mid i \in y\} < \Omega$, therefore as $k(A) \subseteq \mathcal{H}_\eta = B(\eta + 1)$ we must have $\sup\{|i| \mid i \in y\} < \psi(\eta + 1)$. It follows that for any $i \in y$ $|i| \in \mathcal{H}_\eta$ and thus $\mathcal{H}_\eta[i] = \mathcal{H}_\eta$. This means that we may use the induction hypothesis to give

$$\mathcal{H}_{\hat{\alpha_i}} \overset{\psi\hat{\alpha_i}}{\underset{\psi\hat{\alpha_i}}{\vdash}} \Gamma, A_i \quad \text{for all } i \in y.$$

Now applying Lemma 3.21ii) we get

$$\mathcal{H}_{\hat{\alpha}} \overset{\psi\hat{\alpha_i}}{\underset{\psi\hat{\alpha_i}}{\vdash}} \Gamma, A_i \quad \text{for all } i \in y.$$

Upon noting that $\psi\hat{\alpha}_i < \psi\hat{\alpha}$ by 3.22ii) we may apply the appropriate inference to obtain

$$\mathcal{H}_{\hat{\alpha}} \mathrel{\substack{\psi\hat{\alpha} \\ \overline{} \\ \psi\hat{\alpha}}}\!\!\vdash \Gamma\,.$$

Case 2. Now suppose that $A \simeq \bigvee(A_i)_{i\in y} \in \Gamma$ and $\mathcal{H}_\eta \vdash^{\alpha_0}_{\Omega+1} \Gamma, A_{i_0}$ with $i_0 \in y$, $|i_0| \in \mathcal{H}_\eta$ and $\alpha_0 < \alpha$. We may immediately apply the induction hypothesis to obtain

$$\mathcal{H}_{\hat{\alpha}} \vdash^{\psi\hat{\alpha_0}}_{\psi\hat{\alpha_0}} \Gamma, A_{i_0}\,.$$

Now we want to be able to apply the appropriate inference to derive Γ but first we must check that $|\, i_0\,| < \Gamma_{\theta+1} + \psi\hat{\alpha}$. Since $|\, i_0\,| \in \mathcal{H}_\eta = B(\eta+1)$ we have

$$|\, i_0\,| < \psi(\eta+1) < \psi\hat{\alpha} \leq \Gamma_{\theta+1} + \psi\hat{\alpha}.$$

Therefore we may apply the appropriate inference to yield

$$\mathcal{H}_{\hat{\alpha}} \vdash^{\psi\hat{\alpha}}_{\psi\hat{\alpha}} \Gamma\,.$$

Case 3. Now suppose the last inference was (Σ-Ref$_\Omega(X)$) so we have $\exists z F^z \in \Gamma$ and $\mathcal{H}_\eta \vdash^{\alpha_0}_{\Omega+1} \Gamma, F$ with $\alpha_0 < \alpha$ and F a Σ formula. Applying the induction hypothesis we have

$$\mathcal{H}_{\hat{\alpha}} \vdash^{\psi\hat{\alpha_0}}_{\psi\hat{\alpha_0}} \Gamma, F\,.$$

Applying Boundedness 3.18 we obtain

$$\mathcal{H}_{\hat{\alpha}} \vdash^{\psi\hat{\alpha_0}}_{\psi\hat{\alpha_0}} \Gamma, F^{\mathbb{L}_{\psi\hat{\alpha_0}}(X)}\,.$$

Now by Lemma 3.22 $|\,\mathbb{L}_{\psi\hat{\alpha_0}}(X)\,| = \Gamma_{\theta+1} + \psi\hat{\alpha_0} < \Gamma_{\theta+1} + \psi\hat{\alpha}$, so we may apply ($\exists$) to obtain

$$\mathcal{H}_{\hat{\alpha}} \vdash^{\psi\hat{\alpha}}_{\psi\hat{\alpha}} \Gamma, \exists z F^z$$

as required.

Case 4. Finally suppose the last inference was (Cut), so for some A with $rk(A) \leq \Omega$ we have

(1) $$\mathcal{H}_\eta \vdash^{\alpha_0}_{\Omega+1} \Gamma, A \quad \text{with } \alpha_0 < \alpha.$$

(2) $$\mathcal{H}_\eta \vdash^{\alpha_0}_{\Omega+1} \Gamma, \neg A \quad \text{with } \alpha_0 < \alpha.$$

4.1 If $rk(A) < \Omega$ then A is Δ_0. In this case both A and $\neg A$ are Σ formulae so we may immediately apply the induction hypothesis to both (1) and (2) giving

$$\mathcal{H}_{\hat{\alpha_0}} \overset{\psi\hat{\alpha_0}}{\underset{\psi\hat{\alpha_0}}{\vdash}} \Gamma, A \tag{3}$$

$$\mathcal{H}_{\hat{\alpha_0}} \overset{\psi\hat{\alpha_0}}{\underset{\psi\hat{\alpha_0}}{\vdash}} \Gamma, \neg A\,. \tag{4}$$

Since $k(A) \subseteq \mathcal{H}_\eta(\emptyset) = B(\eta+1)$ and A is Δ_0 it follows from Observation 3.11 that $rk(A) \in B(\eta+1) \cap \Omega$. Thus $rk(A) < \psi(\eta+1) < \psi\hat{\alpha}$, so we may apply (Cut) to complete this case.

4.2 Finally suppose $rk(A) = \Omega$. Without loss of generality we may assume that $A \equiv \exists z F(z)$ with F a Δ_0 formula. We may immediately apply the induction hypothesis to (1) giving

$$\mathcal{H}_{\hat{\alpha_0}} \overset{\psi\hat{\alpha_0}}{\underset{\psi\hat{\alpha_0}}{\vdash}} \Gamma, A\,. \tag{5}$$

Applying Boundedness 3.18 to (5) yields

$$\mathcal{H}_{\hat{\alpha_0}} \overset{\psi\hat{\alpha_0}}{\underset{\psi\hat{\alpha_0}}{\vdash}} \Gamma, A^{\mathbb{L}_{\psi\hat{\alpha_0}}(X)}\,. \tag{6}$$

Now using Lemma 3.15v) on (2) yields

$$\mathcal{H}_{\hat{\alpha_0}} \overset{\alpha_0}{\underset{\Omega+1}{\vdash}} \Gamma, \neg A^{\mathbb{L}_{\psi\hat{\alpha_0}}(X)}\,. \tag{7}$$

Observe that since $\eta, \alpha_0 \in \mathcal{H}_\eta$ we have $\hat{\alpha_0} \in B(\eta+1) \subseteq B(\hat{\alpha_0})$. So since $\Gamma, \neg A^{\mathbb{L}_{\psi\hat{\alpha_0}}(X)}$ is a set of Σ-formulae we may apply the induction hypothesis to (7) giving

$$\mathcal{H}_{\alpha_1} \overset{\psi\alpha_1}{\underset{\psi\alpha_1}{\vdash}} \Gamma, \neg A^{\mathbb{L}_{\psi\hat{\alpha_0}}} \quad \text{where } \alpha_1 := \hat{\alpha_0} + \omega^{\Omega+\alpha_0}. \tag{8}$$

Now

$$\alpha_1 = \hat{\alpha_0} + \omega^{\Omega+\alpha_0} = \eta + \omega^{\Omega+\alpha_0} + \omega^{\Omega+\alpha_0} < \eta + \omega^{\Omega+\alpha} := \hat{\alpha}.$$

Owing to Lemma 3.22ii) we have $\psi\hat{\alpha_0}, \psi\alpha_1 < \psi\hat{\alpha}$, thus we may apply (Cut) to (6) and (8) giving

$$\mathcal{H}_{\hat{\alpha}} \overset{\psi\hat{\alpha}}{\underset{\psi\hat{\alpha}}{\vdash}} \Gamma$$

as required. □

4 Embedding KP into $\mathbf{RS}_\Omega(X)$

Definition 4.1. i) Given ordinals $\alpha_1, \ldots, \alpha_n$. The expression $\omega^{\alpha_1}\#\ldots\#\omega^{\alpha_n}$ denotes the ordinal $\omega^{\alpha_{p(1)}} + \ldots + \omega^{\alpha_{p(n)}}$, where $p : \{1, \ldots, n\} \mapsto \{1, \ldots, n\}$ such that $\alpha_{p(1)} \geq \ldots \geq \alpha_{p(n)}$. More generally $\alpha\#0 := 0\#\alpha := 0$ and $\alpha\#\beta := \omega^{\alpha_1}\#\ldots\#\omega^{\alpha_n}\#\omega^{\beta_1}\#\ldots\#\omega^{\beta_m}$ for $\alpha =_{NF} \omega^{\alpha_1} + \ldots + \omega^{\alpha_n}$ and $\beta =_{NF} \omega^{\beta_1} + \ldots + \omega^{\beta_m}$.

ii) If A is any $\mathbf{RS}_\Omega(X)$-formula then $no(A) := \omega^{rk(A)}$.

iii) If $\Gamma = \{A_1, \ldots, A_n\}$ is a set of $\mathbf{RS}_\Omega(X)$-formulae then

$$no(\Gamma) := no(A_1)\#\ldots\#no(A_n).$$

iv) $\Vdash \Gamma$ will be used to abbreviate that

$$\mathcal{H}[\Gamma] \overset{no(\Gamma)}{\underset{0}{\vdash}} \Gamma \quad \text{holds for any operator } \mathcal{H}.$$

v) $\Vdash^\alpha_\rho \Gamma$ will be used to abbreviate that

$$\mathcal{H}[\Gamma] \overset{no(\Gamma)\#\alpha}{\underset{\rho}{\vdash}} \Gamma \quad \text{holds for any operator } \mathcal{H}.$$

As might be expected $\Vdash^\alpha \Gamma$ and $\Vdash_\rho \Gamma$ stand for $\Vdash^\alpha_0 \Gamma$ and $\Vdash^0_\rho \Gamma$ respectively.

The following lemma shows that under certain conditions we may use $\Vdash$ as a calculus.

Lemma 4.2. i) If Γ follows from premises Γ_i by an $\mathbf{RS}_\Omega(X)$ inference other than (Cut) or (Σ-Ref$_\Omega(X)$) and without contractions then

$$\text{if } \Vdash^\alpha_\rho \Gamma_i \quad \text{then} \quad \Vdash^\alpha_\rho \Gamma.$$

ii) If $\Vdash^\alpha_\rho \Gamma, A, B$ then $\Vdash^\alpha_\rho \Gamma, A \vee B$.

Proof. Part i) follows from Lemma 3.14. It also needs to be noted that if the last inference was universal with premises $\{\Gamma_i\}_{i \in Y}$, then $\mathcal{H}[\Gamma_i] \subseteq \mathcal{H}[i]$.

For part ii) suppose $\Vdash^\alpha_\rho \Gamma, A, B$, so we have

$$\mathcal{H}[\Gamma] \overset{no(\Gamma,A,B)\#\alpha}{\underset{\rho}{\vdash}} \Gamma, A, B\,.$$

Two applications of $(\vee)$ and a contraction yields

$$\mathcal{H}[\Gamma] \overset{no(\Gamma,A,B)\#\alpha+2}{\underset{\rho}{\vdash}} \Gamma, A \vee B\,.$$

It remains to note that since $\omega^{rk(A\vee B)}$ is additive principal, Lemma 3.14 gives us

$$\begin{aligned} no(\Gamma, A, B)\#\alpha + 2 &= no(\Gamma)\#\alpha\#\omega^{rk(A)}\#\omega^{rk(B)} + 2 < no(\Gamma)\#\alpha\#\omega^{rk(A\vee B)} \\ &= no(\Gamma, A \vee B)\#\alpha. \end{aligned}$$

So we may complete the proof with an application of Lemma 3.15i). □

Lemma 4.3. Let A be an $\mathbf{RS}_\Omega(X)$ formula and s, t be $\mathbf{RS}_\Omega(X)$ terms.

i) $\Vdash A, \neg A$.

ii) $\Vdash s \notin s$.

iii) $\Vdash s \subseteq s$ where $s \subseteq s :\equiv (\forall x \in s)(x \in s)$.

iv) If $|s| < |t|$ then $\Vdash s \dot{\in} t \to s \in t$ and $\Vdash \neg(s \dot{\in} t), s \in t$.

v) $\Vdash s \neq t, t = s$.

vi) If $|s| < |t|$ and $\Vdash \Gamma, A, B$ then $\Vdash \Gamma, s \dot{\in} t \to A, s \dot{\in} t \wedge B$.

vii) If $|s| < \Gamma_{\theta+1} + \alpha$ then $\Vdash s \in \mathbb{L}_\alpha(X)$.

Proof. i) We use induction of $rk(A)$, and split into cases based upon the form of A:

Case 1. Suppose $A \equiv \bar{u} \in \bar{v}$. In this case either A or $\neg A$ is an axiom so there is nothing to show.

Case 2. Suppose $A \equiv r \in t$ where $\max(|r|, |t|) \geq \Gamma_{\theta+1}$. By Lemma 3.14 and the induction hypothesis we have $\Vdash s \dot{\in} t \wedge r = s, s \dot{\in} t \to r \neq s$ for all $|s| < |t|$. Thus we have the following template for derivations in $\mathbf{RS}_\Omega(X)$:

$$\begin{array}{c} (\in)\dfrac{\Vdash s \dot{\in} t \wedge r = s, s \dot{\in} t \to r \neq s}{\Vdash r \in t, s \dot{\in} t \to r \neq s} \\ (\notin)\dfrac{\phantom{\Vdash r \in t, s \dot{\in} t \to r \neq s}}{\Vdash r \in t, r \notin t} \end{array}$$

Case 3. Suppose $A \equiv (\exists x \in t)F(x)$. By Lemma 3.14 and the induction hypothesis we have $\Vdash s \dot{\in} t \wedge F(s), s \dot{\in} t \to \neg F(s)$ for all $|s| < |t|$. We have the following template for derivations in $\mathbf{RS}_\Omega(X)$:

$$\begin{array}{c} (b\exists)\dfrac{\Vdash s \dot{\in} t \wedge F(s), s \dot{\in} t \to \neg F(s) \quad \text{for all } |s| < |t|}{\Vdash (\exists x \in t)F(x), s \dot{\in} t \to \neg F(s)} \\ (b\forall)\dfrac{\phantom{\Vdash (\exists x \in t)F(x), s \dot{\in} t \to \neg F(s)}}{\Vdash (\exists x \in t)F(x), (\forall x \in t)\neg F(x)} \end{array}$$

Case 4. $A \equiv A_0 \vee A_1$. We have the following template for derivations in $\mathbf{RS}_\Omega(X)$:

$$(\wedge)\,\dfrac{(\vee)\,\dfrac{\Vdash A_0, \neg A_0}{\Vdash A_0 \vee A_1, \neg A_0} \qquad (\vee)\,\dfrac{\Vdash A_1, \neg A_1}{\Vdash A_0 \vee A_1, \neg A_1}}{\Vdash A_0 \vee A_1, \neg A_0 \wedge \neg A_1}$$

All other cases may be seen as variations of those above.

ii) We proceed by induction on $rk(s)$. If s is of the form $\bar{u}$ then $s \notin s$ is already an axiom. Inductively we have $\Vdash r \notin r$ for all $|r| < |s|$. Now suppose s is of the form $\mathbb{L}_\alpha(X)$, in this case $r \notin r \equiv r \dot{\in} s \wedge r \notin r$ so we have the following template for derivations in $\mathbf{RS}_\Omega(X)$:

$$(\notin)\,\dfrac{3.5\mathrm{ii})\,\dfrac{(\vee)\,\dfrac{(b\exists)\,\dfrac{\Vdash r \dot{\in} s \wedge r \notin r}{\Vdash (\exists x \in s)(x \notin r)}}{\Vdash s \neq r}}{\Vdash r \dot{\in} s \rightarrow s \neq r}}{\Vdash s \notin s}$$

Now suppose s is of the form $[x \in \mathbb{L}_\alpha(X) \mid B(x)]$, by i) we have $\Vdash B(r), \neg B(r)$ for any $|r| < |s|$. We have the following template for derivations in $\mathbf{RS}_\Omega(X)$:

$$(\notin)\,\dfrac{\text{Lemma 4.2ii})\,\dfrac{(\vee)\,\dfrac{(b\exists)\,\dfrac{(\wedge)\,\dfrac{\Vdash r \notin r \quad \Vdash B(r), \neg B(r) \quad \text{for any } |r| < |s|}{\Vdash B(r) \wedge r \notin r, \neg B(r)}}{\Vdash (\exists x \in s)(x \notin r), \neg B(r)}}{\Vdash s \neq r, \neg B(r)}}{\Vdash B(r) \rightarrow s \neq r}}{\Vdash s \notin s}$$

iii) Again we proceed by induction on $rk(s)$. If $s \equiv \bar{u}$ then $\Vdash \bar{v} \notin \bar{u}, \bar{v} \in \bar{u}$ for any $|\bar{v}| < |\bar{u}|$ by part i), so we have the following template for derivations in $\mathbf{RS}_\Omega(X)$:

$$(b\forall)\,\dfrac{\text{Lemma 4.2ii})\,\dfrac{\Vdash \bar{v} \notin \bar{u}, \bar{v} \in \bar{u}}{\Vdash \bar{v} \in \bar{u} \rightarrow \bar{v} \in \bar{u}}}{\Vdash (\forall x \in s)(x \in s)}$$

Suppose $s \equiv \mathbb{L}_\alpha(X)$, by the induction hypothesis we have $\Vdash r \subseteq r$ for all $|r| < |s|$. We have the following template for derivations in $\mathbf{RS}_\Omega(X)$:

$$(b\forall)\,\dfrac{\text{3.5ii)}\,\dfrac{(\in)\,\dfrac{(\wedge)\,\dfrac{\Vdash r \subseteq r \qquad \Vdash r \subseteq r}{\Vdash r = r}}{\Vdash r \in s}}{\Vdash r \mathrel{\dot{\in}} s \to r \in s}}{\Vdash (\forall x \in s)(x \in s)}$$

Finally suppose $s \equiv [x \in \mathbb{L}_\alpha(X) \mid B(x)]$, again by the induction hypothesis we have $\Vdash r \subseteq r$ for all $|\, r \,| < |\, s \,|$. Also by part i) we have $\Vdash \neg B(r), B(r)$ for all such r. We have the following template for derivations in $\mathbf{RS}_\Omega(X)$:

$$(b\forall)\,\dfrac{\text{Lemma 4.2ii)}\,\dfrac{(\in)\,\dfrac{(\wedge)\,\dfrac{(\wedge)\,\dfrac{\Vdash \neg B(r), r \subseteq r}{\Vdash \neg B(r), r = r} \qquad \Vdash \neg B(r), B(r)}{\Vdash \neg B(r), B(r) \wedge r = r}}{\Vdash \neg B(r), r \in s}}{\Vdash B(r) \to r \in s}}{\Vdash (\forall x \in s)(x \in s)}$$

iv) Was shown whilst proving iii).

v) By part i) we have $\Vdash \neg(s \subseteq t), s \subseteq t$ and $\Vdash \neg(t \subseteq s), t \subseteq s$ for all $|\, s \,| < |\, t \,|$. We have the following template for derivations in $\mathbf{RS}_\Omega(X)$.

$$\text{3.5i)}\,\dfrac{(\wedge)\,\dfrac{(\vee)\,\dfrac{\Vdash \neg(s \subseteq t), s \subseteq t}{\Vdash \neg(s \subseteq t) \vee \neg(t \subseteq s), s \subseteq t} \qquad (\vee)\,\dfrac{\Vdash \neg(t \subseteq s), t \subseteq s}{\Vdash \neg(t \subseteq s) \vee \neg(s \subseteq t), t \subseteq s}}{\Vdash \neg(s \subseteq t) \vee \neg(t \subseteq s), s \subseteq t \wedge t \subseteq s}}{\Vdash s \neq t, t = s}$$

vi) If $t \equiv \mathbb{L}_\alpha(X)$ then this result is trivial since $s \mathrel{\dot{\in}} t \to A := A$ and $s \mathrel{\dot{\in}} t \wedge B := B$.

Now if $t \equiv \bar{u}$ then $s \mathrel{\dot{\in}} t := s \in t$ and if $t \equiv [x \in \mathbb{L}_\alpha(X) \mid C(x)]$ then $s \mathrel{\dot{\in}} t := C(s)$. In either case we have the following template for derivations in $\mathbf{RS}_\Omega(X)$:

$$(\wedge)\,\dfrac{(\vee)\,\dfrac{\Vdash \Gamma, A, B}{\Vdash \Gamma, s \mathrel{\dot{\in}} t \to A, B} \qquad (\vee)\,\dfrac{\Vdash \Gamma, \neg(s \mathrel{\dot{\in}} t), s \mathrel{\dot{\in}} t \quad \text{by i)}}{\Vdash \Gamma, s \mathrel{\dot{\in}} t \to A, s \mathrel{\dot{\in}} t}}{\Vdash \Gamma, s \mathrel{\dot{\in}} t \to A, s \mathrel{\dot{\in}} t \wedge B}$$

vii) By part iii) we have $\Vdash s = s$ for all $|\, s \,| < \Gamma_{\theta+1} + \alpha$ which means we have $\Vdash s \mathrel{\dot{\in}} \mathbb{L}_\alpha(X) \wedge s = s$ for all such s. From which one application of $(\in)$ gives the desired result. □

Lemma 4.4 (Extensionality). For any $\mathbf{RS}_\Omega(X)$ formula $A(s_1, \dots, s_n)$,

$$\Vdash [s_1 \neq t_1], \dots, [s_n \neq t_n], \neg A(s_1, \dots, s_n), A(t_1, \dots, t_n).$$

Where $[s_i \neq t_i] := \neg(s_i \subseteq t_i), \neg(t_i \subseteq s_i)$.

Proof. The proof is by induction on $rk(A(s_1, \dots, s_n))\#rk(A(t_1, \dots, t_n))$.

Case 1. Suppose $A(s_1, s_2) \equiv s_1 \in s_2$. By the induction hypothesis we have $\Vdash [s_1 \neq t_1], [s \neq t], s_1 \neq s, t_1 = t$ for all $|\, s \,| < |\, s_2 \,|$ and all $|\, t \,| < |\, t_2 \,|$. What follows is a template for derivations in $\mathbf{RS}_\Omega(X)$, for ease of reading the principal formula of each inference is underlined (some lines do not necessarily represent single inferences, but in these cases it is clear how to extend the concept of "principal formula" in a sensible way).

$$\begin{array}{rl}
 & \Vdash [s_1 \neq t_1], [s \neq t], s_1 \neq s, t_1 = t \\
\hline
(\vee) & \Vdash [s_1 \neq t_1], \underline{s \neq t}, s_1 \neq s, t_1 = t \\
\hline
\text{Lemma 4.3 vi)} & \Vdash [s_1 \neq t_1], \underline{t \dot{\in} t_2 \to s \neq t}, s_1 \neq s, \underline{t \dot{\in} t_2 \wedge t_1 = t} \\
\hline
(\in) & \Vdash [s_1 \neq t_1], t \dot{\in} t_2 \to s \neq t, s_1 \neq s, \underline{t_1 \in t_2} \\
\hline
(\notin) & \Vdash [s_1 \neq t_1], \underline{s \notin t_2}, s_1 \neq s, t_1 \in t_2 \\
\hline
\text{Lemma 4.3 vi)} & \Vdash [s_1 \neq t_1], \underline{s \dot{\in} s_2 \wedge s \notin t_2}, \underline{s \dot{\in} s_2 \to s_1 \neq s}, t_1 \in t_2 \\
\hline
(b\exists) & \Vdash [s_1 \neq t_1], \underline{(\exists x \in s_2)(x \notin t_2)}, s \dot{\in} s_2 \to s_1 \neq s, t_1 \in t_2 \\
\hline
(\notin) & \Vdash [s_1 \neq t_1], (\exists x \in s_2)(x \notin t_2), \underline{s_1 \notin s_2}, t_1 \in t_2 \\
\hline
\text{Lemma 3.15i)} & \Vdash [s_1 \neq t_1], \underline{s_2 \neq t_2}, s_1 \notin s_2, t_1 \in t_2
\end{array}$$

Case 2. Suppose $A(s_1) \equiv s_1 \in s_1$. In this case $\neg A(s_1) \equiv s_1 \notin s_1$ so the result follows from Lemma 4.3ii).

Case 3. Suppose $A(s_1, \dots, s_n) \equiv (\exists y \in s_i)(B(y, s_1, \dots, s_n))$ for some $1 \leq i \leq n$. Inductively we have

$$\Vdash [s_1 \neq t_1], \dots, [s_n \neq t_n], \neg B(r, s_1, \dots, s_n), B(r, t_1, \dots, t_n)$$

for all $|\, r \,| < |\, s_i \,|$. Now by applying 4.3iv) we obtain

$$\Vdash [s_1 \neq t_1], \dots, [s_n \neq t_n], r \dot{\in} s_i \to \neg B(r, s_1, \dots, s_n), r \dot{\in} s_i \wedge B(r, t_1, \dots, t_n).$$

To which we may apply $(b\exists)$ followed by $(b\forall)$ to arrive at the desired conclusion.

Case 4. Suppose $A(s_1,\ldots,s_n) \equiv (\exists x \in r)B(x,s_1,\ldots,s_n)$ for some r not present in $s_1,\ldots,s_n$. From the induction hypothesis we have, for all $|p|<|r|$:

$$\Vdash [s_1 \neq t_1],\ldots,[s_n \neq t_n], p \dot{\in} r \to \neg B(p,s_1,\ldots,s_n), p \dot{\in} r \wedge B(p,t_1,\ldots,t_n)$$

Applying ($b\exists$) followed by ($b\forall$) gives us the desired result.

The cases where $A(s_1,\ldots,s_n) \equiv \exists x B(x,s_1,\ldots,s_n)$ or $A(s_1,\ldots,s_n) \equiv B \vee C$ may be treated in a similar manner to case 4. All other cases are dual to one of the ones considered above. □

Lemma 4.5 (Set Induction). For any $\mathbf{RS}_\Omega(X)$-formula F:

$$\Vdash^{\omega^{rk(A)}} \forall x[(\forall y \in x)F(y) \to F(x)] \to \forall x F(x)$$

where $A := \forall x[(\forall y \in x)F(y) \to F(x)]$.

Proof. Claim:

$$(*) \qquad \mathcal{H}[A,s] \,\Big|\!\frac{\omega^{rk(A)}\#\omega^{|s|+1}}{0}\, \neg A, F(s) \quad \text{for any term } s.$$

We begin by verifying (*) using induction on $|s|$. From the induction hypothesis we know that

$$(1) \qquad \mathcal{H}[A,t] \,\Big|\!\frac{\omega^{rk(A)}\#\omega^{|t|+1}}{0}\, \neg A, F(t) \quad \text{for all } |t|<|s|.$$

By applying ($\vee$) if necessary to (1) we obtain

$$(2) \qquad \mathcal{H}[A,t,s] \,\Big|\!\frac{\omega^{rk(A)}\#\omega^{|t|+1}+1}{0}\, \neg A, t \dot{\in} s \to F(t) \quad \text{for all } |t|<|s|.$$

To which we may apply ($b\forall$) yielding

$$(3) \qquad \mathcal{H}[A,s] \,\Big|\!\frac{\eta+2}{0}\, \neg A, (\forall y \in s)F(y) \quad \text{where } \eta := \omega^{rk(A)}\#\omega^{|s|}.$$

Observe that $no(\neg F(s), F(s)) < \omega^{rk(A)}$, so by Lemma 4.3i) we have

$$(4) \qquad \mathcal{H}[A,s] \,\Big|\!\frac{\eta+2}{0}\, \neg F(s), F(s)\,.$$

Applying ($\wedge$) to (3) and (4) yields

$$(5) \qquad \mathcal{H}[A,s] \,\Big|\!\frac{\eta+3}{0}\, \neg A, (\forall y \in s)F(y) \wedge \neg F(s), F(s)\,.$$

To which we may apply (∃) to otain

$$\text{(6)} \qquad \mathcal{H}[A,s] \overset{\eta+4}{\underset{0}{\vdash}} \neg A, \exists x[(\forall y \in x)F(y) \wedge \neg F(x)], F(s)\,.$$

It remains to observe that $\neg A \equiv \exists x[(\forall y \in x)F(y) \wedge \neg F(x)]$ and that $\eta + 4 < \omega^{rk(A)}\#\omega^{|s|+1}$, and hence we may apply Lemma 3.15i) to provide

$$\text{(7)} \qquad \mathcal{H}[A,s] \overset{\omega^{rk(A)}\#\omega^{|s|+1}}{\underset{0}{\vdash}} \neg A, F(s)$$

so the claim is verified.

Applying (∀) to (*) gives

$$\mathcal{H}[A] \overset{\omega^{rk(A)}\#\Omega}{\underset{0}{\vdash}} \neg A, \forall x F(x)\,.$$

Now by two applications of (∨) we may conclude

$$\mathcal{H}[A] \overset{\omega^{rk(A)}\#\Omega+2}{\underset{0}{\vdash}} A \to \forall x F(x)\,.$$

It remains to note that $no(A \to \forall x F(x)) \geq \omega^{\Omega+1} > \Omega + 2$, so we have

$$\text{(1)} \qquad \overset{\omega^{rk(A)}}{\underset{0}{\Vdash}} A \to (\forall x \in \mathbb{L}_\alpha(X))F(x)$$

as required. □

Lemma 4.6 (Infinity). Suppose $\omega < \mu < \Omega$, then

$$\Vdash (\exists x \in \mathbb{L}_\mu(X))[(\exists z \in x)(z \in x) \wedge (\forall y \in x)(\exists z \in x)(y \in z)].$$

Proof. The following gives a template for derivations in $\mathbf{RS}_\Omega(X)$, the idea is that $\mathbb{L}_\omega(X)$ serves as a witness inside $\mathbb{L}_\mu(X)$.

$$\dfrac{\dfrac{\dfrac{\dfrac{\dfrac{\dfrac{\text{Lemma 4.3vii)}}{\Vdash s \in \mathbb{L}_k(X) \quad \text{for any } |s| < |\mathbb{L}_k(X)| \text{ and } k < \omega.}}{\Vdash \mathbb{L}_k(X) \,\dot{\in}\, \mathbb{L}_\omega(X) \wedge s \in \mathbb{L}_k(X)}\,\text{3.5ii)}}{\Vdash (\exists z \in \mathbb{L}_\omega(X))(s \in \mathbb{L}_k(X))}\,(b\exists)}{\Vdash s \,\dot{\in}\, \mathbb{L}_\omega(X) \to (\exists z \in \mathbb{L}_\omega(X))(s \in z)}\,\text{3.5ii)}}{\Vdash (\forall y \in \mathbb{L}_\omega(X))(\exists z \in \mathbb{L}_\omega(X))(y \in z)}\,(b\forall) \qquad \dfrac{\dfrac{\Vdash \mathbb{L}_0(X) \in \mathbb{L}_\omega(X)}{\Vdash \mathbb{L}_0(X) \,\dot{\in}\, \mathbb{L}_\omega(X) \wedge \mathbb{L}_0(X) \in \mathbb{L}_\omega(X)}\,\text{3.5ii)}}{\Vdash (\exists z \in \mathbb{L}_\omega(X))(z \in \mathbb{L}_\omega(X))}\,(b\exists)}{\Vdash (\forall y \in \mathbb{L}_\omega(X))(\exists z \in \mathbb{L}_\omega(X))(y \in z) \wedge (\exists z \in \mathbb{L}_\omega(X))(z \in \mathbb{L}_\omega(X))}\,(\wedge)$$

$$\dfrac{\dfrac{\Vdash (\forall y \in \mathbb{L}_\omega(X))(\exists z \in \mathbb{L}_\omega(X))(y \in z) \wedge (\exists z \in \mathbb{L}_\omega(X))(z \in \mathbb{L}_\omega(X))}{\Vdash \mathbb{L}_\omega(X) \,\dot{\in}\, \mathbb{L}_\mu(X) \wedge [(\forall y \in \mathbb{L}_\omega(X))(\exists z \in \mathbb{L}_\omega(X))(y \in z) \wedge (\exists z \in \mathbb{L}_\omega(X))(z \in \mathbb{L}_\omega(X))]}\,\text{3.5ii)}}{\Vdash (\exists x \in \mathbb{L}_\mu(X))[(\exists z \in x)(z \in x) \wedge (\forall y \in x)(\exists z \in x)(y \in z)]}\,(b\exists)$$

□

Lemma 4.7 (Δ_0-Separation). Suppose $A(a, b_1, \dots, b_n)$ be a Δ_0-formula of **KP** with all free variables indicated, μ a limit ordinal and $|s|, |t_0|, \dots, |t_n| < \Gamma_{\theta+1} + \mu$.

$$\Vdash (\exists y \in \mathbb{L}_\mu(X))\Big[(\forall x \in y)(x \in s \wedge A(x, t_1, \dots, t_n)) \\ \wedge (\forall x \in s)(A(x, t_1, \dots, t_n) \to x \in y)\Big]$$

Proof. Let $\alpha := \max\{|s|, |t_0|, \dots, |t_n|\} + 1$ and note that $\alpha < \Gamma_{\theta+1} + \mu$ since μ ia a limit. Now let β be the unique ordinal such that $\alpha = \Gamma_{\theta+1} + \beta$ if such an ordinal exists, if not set $\beta := 0$. Now define

$$t := [z \in \mathbb{L}_\beta(X) \mid z \in s \wedge B(z)]$$

where $B(z) := A(z, t_1, \dots, t_n)$. We have the following templates for derivations in $\mathbf{RS}_\Omega(X)$:

$$(b\forall)\ \dfrac{3.5\text{ii)}\ \dfrac{\text{Lemma 4.2ii)}\ \dfrac{\dfrac{\text{Lemma 4.3 i)}}{\Vdash \neg(r \in s \wedge B(r)), r \in s \wedge B(r) \quad \text{for all } |r| < \alpha}}{\Vdash (r \in s \wedge B(r)) \to r \in s \wedge B(r)}}{\Vdash r \mathrel{\dot\in} t \to r \in s \wedge B(r)}}{\Vdash (\forall x \in t)(x \in s \wedge B(r))}$$

In the following derivation r ranges over terms $|r| < |s|$.

$$(b\forall)\ \dfrac{\text{Lemma 4.2ii)}\ \dfrac{\text{Lemma 4.2ii)}\ \dfrac{(\in)\ \dfrac{3.5\text{ii)}\ \dfrac{(\wedge)\ \dfrac{(\wedge)\ \dfrac{\dfrac{\text{Lemma 4.3 iv)}}{\Vdash \neg(r \mathrel{\dot\in} s), r \in s} \quad \dfrac{\text{Lemma 4.3 i)}}{\Vdash \neg B(r), B(r)}}{\Vdash \neg(r \mathrel{\dot\in} s), \neg B(r), r \in s \wedge B(r)} \quad \dfrac{\text{Lemma 4.3 iii)}}{\Vdash r = r}}{\Vdash \neg(r \mathrel{\dot\in} s), \neg B(r), (r \in s \wedge B(r)) \wedge r = r}}{\Vdash \neg(r \mathrel{\dot\in} s), \neg B(r), r \mathrel{\dot\in} t \wedge r = r}}{\Vdash \neg(r \mathrel{\dot\in} s), \neg B(r), r \in t}}{\Vdash \neg(r \mathrel{\dot\in} s), (B(r) \to r \in t)}}{\Vdash r \mathrel{\dot\in} s \to (B(r) \to r \in t)}}{\Vdash (\forall x \in s)(B(x) \to x \in t)}$$

Now applying $(\wedge)$ to the two preceding derivations and noting that $|t| < \Gamma_{\theta+1} + \mu$ gives us

$$\Vdash t \mathrel{\dot\in} \mathbb{L}_\mu(X) \wedge [(\forall x \in t)(x \in s \wedge B(r)) \wedge (\forall x \in s)(B(x) \to x \in t)]$$

to which we may apply $(b\exists)$ to obtain

$$\Vdash (\exists y \in \mathbb{L}_\mu(X))[(\forall x \in y)(x \in s \wedge B(x)) \wedge (\forall x \in s)(B(x) \to x \in y)].$$

It should also be checked that

$$t \in \mathcal{H}[(\exists y \in \mathbb{L}_\mu(X))[(\forall x \in y)(x \in s \wedge B(x)) \wedge (\forall x \in s)(B(x) \rightarrow x \in y)]]$$

but this is the case since

$$|s|, |t_0|, \ldots, |t_n| \in k((\exists y \in \mathbb{L}_\mu(X))[(\forall x \in y)(x \in s \wedge B(x)) \wedge (\forall x \in s)(B(x) \rightarrow x \in y)])$$

and $|t| = \max\{\max\{|s|, |t_0|, \ldots, |t_n|\} + 1, \Gamma_{\theta+1}\}$. □

Lemma 4.8 (Pair and Union). Let μ be a limit ordinal and let s, t be $\mathbf{RS}_\Omega(X)$-terms such that $|s|, |t| < \Gamma_{\theta+1} + \mu$, then

i) $\Vdash (\exists z \in \mathbb{L}_\mu(X))(s \in z \wedge t \in z)$

ii) $(\exists z \in \mathbb{L}_\mu(X))(\forall y \in s)(\forall x \in y)(x \in z)$.

Proof. Let $\alpha := \max\{|s|, |t|\} + 1$, now let β be the unique ordinal such that $\alpha = \Gamma_{\theta+1} + \beta$ if such an ordinal exists, otherwise set $\beta := 0$. Now by Lemma 4.3vii) we have

$$\Vdash s \in \mathbb{L}_\beta(X) \quad \text{and} \quad \Vdash t \in \mathbb{L}_\beta(X).$$

Now by $(\wedge)$ and noticing that $\beta < \mu$ since μ is a limit, we have

$$\Vdash \mathbb{L}_\beta(X) \mathrel{\dot{\in}} \mathbb{L}_\mu(X) \wedge (s \in \mathbb{L}_\beta(X) \wedge t \in \mathbb{L}_\beta(X)).$$

To which we may apply $(b\exists)$ to obtain the desired result.

ii) Let β be the unique ordinal such that $|s| = \Gamma_{\theta+1} + \beta$ if such an ordinal exists, otherwise let $\beta = 0$. By Lemma 4.3vii) we have $\Vdash r \in \mathbb{L}_\beta(X)$ for any $|r| < |s|$. In the following template for derivations in $\mathbf{RS}_\Omega(X)$, r and t range over terms such that $|r| < |t| < |s|$:

$$\cfrac{\cfrac{\cfrac{\cfrac{\cfrac{\cfrac{\Vdash r \in \mathbb{L}_\beta(X)}{\Vdash r \mathrel{\dot{\in}} t \rightarrow r \in \mathbb{L}_\beta(X)}\ (\vee)\text{ if necessary}}{\Vdash (\forall x \in t)(x \in \mathbb{L}_\beta(X))}\ (b\forall)}{\Vdash t \mathrel{\dot{\in}} s \rightarrow (\forall x \in t)(x \in \mathbb{L}_\beta(X))}\ (\vee)\text{ if necessary}}{\Vdash (\forall y \in s)(\forall x \in y)(x \in \mathbb{L}_\beta(X))}\ (b\forall)}{\Vdash \mathbb{L}_\beta(X) \mathrel{\dot{\in}} \mathbb{L}_\mu(X) \wedge (\forall y \in s)(\forall x \in y)(x \in \mathbb{L}_\beta(X)) \quad \text{since } \beta < \mu}\ 3.5\text{ii)}}{\Vdash (\exists z \in \mathbb{L}_\mu(X))(\forall y \in s)(\forall x \in y)(x \in z)}\ (b\exists)$$

□

Lemma 4.9 (Δ_0-Collection)**.** Suppose $F(a,b)$ is any Δ_0 formula of **KP**.

$$\Vdash (\forall x \in s)\exists y F(x,y) \to \exists z(\forall x \in s)(\exists y \in z)F(x,y)$$

Proof. By Lemma 4.3i) we have

$$\Vdash \neg(\forall x \in s)\exists y F(x,y), (\forall x \in s)\exists y F(x,y).$$

Applying (Σ-Ref$_\Omega(X)$) yields

$$\mathcal{H}[(\forall x \in s)\exists y F(x,y)] \mid\!\frac{\alpha+1}{0} \neg(\forall x \in s)\exists y F(x,y), \exists z(\forall x \in s)(\exists y \in z)F(x,y)$$

where $\alpha := \omega^{rk((\forall x \in s)\exists y F(x,y))} \# \omega^{rk((\forall x \in s)\exists y F(x,y))}$. Now two applications of ($\vee$) provides

$$\mathcal{H}[(\forall x \in s)\exists y F(x,y)] \mid\!\frac{\alpha+3}{0} (\forall x \in s)\exists y F(x,y) \to \exists z(\forall x \in s)(\exists y \in z)F(x,y)\,.$$

It remains to note that

$$\alpha + 3 < \omega^{rk(\forall x \in s)\exists y F(x,y)+1} = no((\forall x \in s)\exists y F(x,y) \to \exists z(\forall x \in s)(\exists y \in z)F(x,y))$$

so the proof is complete. □

Theorem 4.10. If $\mathbf{KP} \vdash \Gamma(a_1,\ldots,a_n)$ where $\Gamma(a_1,\ldots,a_n)$ is a finite set of formulae whose free variables are amongst $a_1,\ldots,a_n$, then there is some $m < \omega$ (which we may compute from the derivation) such that

$$\mathcal{H}[s_1,\ldots,s_n] \mid\!\frac{\Omega\cdot\omega^m}{\Omega+m} \Gamma(s_1,\ldots,s_n)$$

for any operator $\mathcal{H}$ and any $\mathbf{RS}_\Omega(X)$ terms $s_1,\ldots,s_n$.

Proof. Suppose $\Gamma(a_1,\ldots,a_n) \equiv \{A_1(a_1,\ldots,a_n),\ldots,A_k(a_1,\ldots,a_n)\}$. Note that for any choice of terms $s_1,\ldots,s_n$ and each $1 \le i \le k$

$$\begin{aligned} rk(A_i(s_1,\ldots,s_n)) &= \omega\cdot\max(k(A_i(s_1,\ldots,s_n))) + m_i \quad \text{for some } m_i < \omega \\ &\le \omega\cdot\Omega + m_i = \Omega + m_i. \end{aligned}$$

Therefore

$$no(A_i(s_1,\ldots,s_n)) = \omega^{rk(A_i(s_1,\ldots,s_n))} \le \omega^{\Omega+m_i} = \omega^\Omega \cdot \omega^{m_i} = \Omega\cdot\omega^{m_i}.$$

So letting $m = \max(m_1, \dots, m_k) + 1$ we have

$$\begin{aligned} no(\Gamma(s_1, \dots, s_n)) &\leq \Omega \cdot \omega^{m_1} \# \dots \# \Omega \cdot \omega^{m_n} \\ &= \Omega \cdot (\omega^{m_1} \# \dots \# \omega^{m_n}) \\ &\leq \Omega \cdot \omega^m \end{aligned}$$

The proof now proceeds by induction on the **KP** derivation. If $\Gamma(a_1, \dots, a_n)$ is an axiom of **KP** then the result follows from 4.3i), 4.4, 4.5, 4.6, 4.7, 4.8 or 4.9.

Now suppose that $\Gamma(a_1, \dots, a_n)$ arises as the result of an inference rule.

Case 1. Suppose the last inference was $(b\forall)$, so $(\forall x \in a_i)F(x, \bar{a}) \in \Gamma(\bar{a})$ and we are in the following situation in **KP**

$$(b\forall)\ \frac{\Gamma(\bar{a}), c \in a_i \to F(c, \bar{a})}{\Gamma(\bar{a})}$$

where c is different from $a_1, \dots, a_n$. Inductively we have some $m < \omega$ such that

$$(1) \qquad \mathcal{H}[\bar{s}, r] \overset{\Omega \cdot \omega^m}{\underset{\Omega + m}{\vdash}} \Gamma(\bar{s}), r \in s_i \to F(r, \bar{s}) \quad \text{for all } |\, r \,| < |\, s_i \,|.$$

1.1 If s_i is of the form $\bar{u}$ we may immediately apply $(b\forall)$ to complete this case.

Suppose $s_i \equiv \mathbb{L}_\alpha(X)$ for some α. Applying Lemma 3.15iii) to (1) gives

$$(2) \qquad \mathcal{H}[\bar{s}, r] \overset{\Omega \cdot \omega^m}{\underset{\Omega + m}{\vdash}} \Gamma(\bar{s}), \neg(r \in s_i), F(r, \bar{s})\,.$$

Since $|\, r \,| < |\, s \,|$, by Lemma 4.3vii) we have

$$(3) \qquad \Vdash r \in s.$$

Applying (Cut) to (1) and (2) yields

$$(4) \qquad \mathcal{H}[\bar{s}, r] \overset{\Omega \cdot \omega^m + 1}{\underset{\Omega + m}{\vdash}} \Gamma(\bar{s}), F(r, \bar{s})\,.$$

To which we may apply $(b\forall)$ to complete this case.

Suppose $s_i \equiv [x \in \mathbb{L}_\alpha(X) \mid B(x)]$, again we may apply Lemma 3.15iii) to (1) to obtain

$$(5) \qquad \mathcal{H}[\bar{s}, r] \overset{\Omega \cdot \omega^m}{\underset{\Omega + m}{\vdash}} \Gamma(\bar{s}), \neg(r \in s_i), F(r, \bar{s})\,.$$

Since $|\, r \,| < |\, s \,|$ by Lemma 4.3iv) we have

$$\Vdash \neg(r \dot{\in} s), r \in s. \tag{6}$$

Applying (Cut) to (5) and (6) yields

$$\mathcal{H}[\bar{s}, r] \overset{\Omega\cdot\omega^m+1}{\underset{\Omega+m}{\vdash}} \Gamma(\bar{s}), \neg(r \dot{\in} s_i), F(r, \bar{s})\,. \tag{7}$$

Now two applications of ($\vee$) provide

$$\mathcal{H}[\bar{s}, r] \overset{\Omega\cdot\omega^m+3}{\underset{\Omega+m}{\vdash}} \Gamma(\bar{s}), r \dot{\in} s_i \to F(r, \bar{s})\,. \tag{8}$$

To which we may apply ($b\forall$) to complete this case.

Case 2. Suppose the last inference was ($\forall$) so $\forall x A(x, \bar{a}) \in \Gamma(\bar{a})$ and we are in the following situation in **KP**

$$(\forall)\ \frac{\Gamma(\bar{a}), F(c, \bar{a})}{\Gamma(\bar{a})}$$

where c is different from $a_1, \ldots a_n$. Inductively we have some $m < \omega$ such that

$$\mathcal{H}[\bar{s}, r] \overset{\Omega\cdot\omega^m}{\underset{\Omega+m}{\vdash}} \Gamma(\bar{s}), F(r, \bar{s}) \quad \text{for all terms } r.$$

We may immediately apply ($\forall$) to complete this case.

Case 3. Suppose the last inference was ($b\exists$) so $(\exists x \in s_i)F(x, \bar{s}) \in \Gamma(\bar{s})$ and we are in the following situation in **KP**

$$(b\exists)\ \frac{\Gamma(\bar{a}), c \in a_i \wedge F(c, \bar{a})}{\Gamma(\bar{a})}$$

3.1 Suppose c is different from $a_1, \ldots, a_n$. Using the induction hypothesis we find some $m < \omega$ such that

$$\mathcal{H}[\bar{s}] \overset{\Omega\cdot\omega^m}{\underset{\Omega+m}{\vdash}} \Gamma(\bar{s}), \bar{\emptyset} \in s_i \wedge F(\bar{\emptyset}, \bar{s})\,. \tag{9}$$

3.1.1 If s_i is of the form $\bar{u}$ we may immediately apply ($b\exists$) to complete the case.

3.1.2 Suppose s_i is of the form $\mathbb{L}_\alpha(X)$. Applying Lemma 3.15iv) to (1) yields

$$\mathcal{H}[\bar{s}] \overset{\Omega\cdot\omega^m}{\underset{\Omega+m}{\vdash}} \Gamma(\bar{s}), F(\bar{\emptyset}, \bar{s})\,. \tag{10}$$

Noting that in this case $\bar{\emptyset} \mathrel{\dot\in} s \wedge F(\bar{\emptyset}, \bar{s}) \equiv F(\bar{\emptyset}, \bar{s})$, we may apply $(b\exists)$ to complete this case.

3.1.3 Suppose s_i is of the form $[x \in \mathbb{L}_\alpha(X) \mid B(x)]$. First we must verify the following claim

$$(*) \qquad \Vdash \neg(\bar{\emptyset} \in s_i \wedge F(\bar{\emptyset}, \bar{s})), \bar{\emptyset} \mathrel{\dot\in} s_i \wedge F(\bar{\emptyset}, \bar{s}).$$

Note that owing to Lemma 4.4 we have $\Vdash [r \neq \bar{\emptyset}], \neg B(r), B(\bar{\emptyset})$ for all $|r| < |s_i|$. In the following template for derivations in $\mathbf{RS}_\Omega(X)$ r ranges over terms $|r| < |s_i|$.

$$\text{Lemma 4.2ii)}\ \dfrac{\text{(}\wedge\text{)}\ \dfrac{(\notin)\ \dfrac{\text{Lemma 4.2ii)}\ \dfrac{\text{Lemma 4.2ii)}\ \dfrac{\Vdash [r \neq \bar{\emptyset}], \neg B(r), B(\bar{\emptyset})}{\Vdash r \neq \bar{\emptyset}, \neg B(r), B(\bar{\emptyset})}}{\Vdash B(r) \rightarrow r \neq \bar{\emptyset}, B(\bar{\emptyset})}}{\Vdash \neg(\bar{\emptyset} \in s_i), B(\bar{\emptyset})} \qquad \dfrac{\text{Lemma 4.3i)}}{\Vdash \neg F(\bar{\emptyset}, \bar{s}), F(\bar{\emptyset}, \bar{s})}}{\Vdash \neg(\bar{\emptyset} \in s_i), \neg F(\bar{\emptyset}, \bar{s}), B(\bar{\emptyset}) \wedge F(\bar{\emptyset}, \bar{s})}}{\Vdash \neg(\bar{\emptyset} \in s_i) \vee \neg F(\bar{\emptyset}, \bar{s}), B(\bar{\emptyset}) \wedge F(\bar{\emptyset}, \bar{s})}$$

Now applying (Cut) to (9) and (*) we get

$$(11) \qquad \mathcal{H}[\bar{s}] \, \frac{\ \Omega \cdot \omega^m + 1\ }{\Omega + m'} \, \Gamma(\bar{s}), \bar{\emptyset} \mathrel{\dot\in} s_i \wedge F(\bar{\emptyset}, \bar{s})\,.$$

Note the possible increase in cut rank. We may apply $(b\exists R)$ to (11) to complete this case.

3.2 Suppose c is one of $a_1, \ldots, a_n$, without loss of generality let us assume $c = a_1$. Applying the induction hypothesis we can compute some $m < \omega$ such that

$$(12) \qquad \mathcal{H}[\bar{s}] \, \frac{\ \Omega \cdot \omega^m\ }{\Omega + m} \, \Gamma(\bar{s}), s_1 \in s_i \wedge F(s_1, \bar{s}).$$

Note that in fact 3.2 subsumes 3.1 since we can conclude (12) from the induction hypothesis regardless of whether or not c is a member of $\bar{a}$. To help with clarity 3.1 is left in the proof above, but in later embeddings we shall dispense with such cases.

If s_1 and s_i are of the form $\bar{u}$ and $\bar{v}$ with $|s_1| < |s_i|$ then we may immediately apply $(b\exists)$ to complete this case. If this is not the case then we verify the following claim

$$(**) \qquad \Vdash \neg(s_1 \in s_i \wedge F(s_1, \bar{s})), (\exists x \in s_i) F(x, \bar{s}).$$

To prove (**) we split into cases based on the form of s_i.

3.2.1 Suppose s_i is of the form $\bar{u}$.

3.2.1.1 If s_1 is also of the form $\bar{v}$ [remember that by assumption $|\,s_1\,| \geq |\,s_i\,|$] then $\neg(s_1 \in s_i), F(s_1, \bar{s}), (\exists x \in s_i)F(x, \bar{s})$ is an axiom so we may apply $(\vee)$ twice to complete this case.

3.2.1.2 Now suppose s_1 is not of the form $\bar{v}$. We have following template for derivations in $\mathbf{RS}_\Omega(X)$, here r ranges over terms with $|\,r\,| < |\,s_i\,|$.

$$
\begin{array}{rl}
 & \text{Lemma 4.3i)} \qquad\qquad \text{Lemma 4.4} \\
 & \Vdash \neg(r \in s_i), r \in s_i \qquad \Vdash r \neq s_1, \neg F(s_1, \bar{s}), F(r, \bar{s}) \\ \hline
(\wedge) & \Vdash \neg(r \in s_i), r \neq s_1, \neg F(s_1, \bar{s}), r \in s_i \wedge F(r, \bar{s}) \\ \hline
(b\exists) & \Vdash \neg(r \in s_i), r \neq s_1, \neg F(s_1, \bar{s}), (\exists x \in s_i)F(x, \bar{s}) \\ \hline
\text{Lemma 4.2ii)} & \Vdash r \in s_i \rightarrow r \neq s_1, \neg F(s_1, \bar{s}), (\exists x \in s_i)F(x, \bar{s}) \\ \hline
(\notin) & \Vdash \neg(s_1 \in s_i), \neg F(s_1, \bar{s}), (\exists x \in s_i)F(x, \bar{s}) \\ \hline
\text{Lemma 4.2ii)} & \Vdash \neg(s_1 \in s_i) \vee \neg F(s_1, \bar{s}), (\exists x \in s_i)F(x, \bar{s})
\end{array}
$$

3.2.2 Now suppose s_i is of the form $\mathbb{L}_\alpha(X)$. In the following template for derivations in $\mathbf{RS}_\Omega(X)$ r ranges over terms with $|\,r\,| < |\,s_i\,|$.

$$
\begin{array}{rl}
 & \text{Lemma 4.4} \\
 & \Vdash r \neq s_1, \neg F(s_1, \bar{s}), F(x, \bar{s}) \\ \hline
\text{3.5ii)} & \Vdash r \neq s_1, \neg F(s_1, \bar{s}), r \,\dot{\in}\, s_i \wedge F(x, \bar{s}) \\ \hline
(b\exists) & \Vdash r \neq s_1, \neg F(s_1, \bar{s}), (\exists x \in s_i)F(x, \bar{s}) \\ \hline
\text{3.5ii)} & \Vdash r \,\dot{\in}\, s_i \rightarrow r \neq s_1, \neg F(s_1, \bar{s}), (\exists x \in s_i)F(x, \bar{s}) \\ \hline
(\notin) & \Vdash \neg(s_1 \in s_i), \neg F(s_1, \bar{s}), (\exists x \in s_i)F(x, \bar{s}) \\ \hline
\text{Lemma 4.2ii)} & \Vdash \neg(s_1 \in s_i) \vee \neg F(s_1, \bar{s}), (\exists x \in s_i)F(x, \bar{s})
\end{array}
$$

3.2.3 Finally suppose s_i is of the form $[x \in \mathbb{L}_\alpha \mid B(x)]$. In the following template for derivations in $\mathbf{RS}_\Omega(X)$ r ranges over terms with $|\,r\,| < |\,s_i\,|$.

$$
\begin{array}{rl}
 & \text{Lemma 4.3i)} \qquad\qquad \text{Lemma 4.4} \\
 & \Vdash \neg B(r), B(r) \qquad \Vdash r \neq s, \neg F(s_1, \bar{s}), F(r, \bar{s}) \\ \hline
(\wedge) & \Vdash \neg B(r), r \neq s_1, \neg F(s_1, \bar{s}), B(r) \wedge F(r, \bar{s}) \\ \hline
(b\exists) & \Vdash \neg B(r), r \neq s_1, \neg F(s_1, \bar{s}), (\exists x \in s_i)F(x, \bar{s}) \\ \hline
\text{Lemma 4.2ii)} & \Vdash B(r) \rightarrow r \neq s_1, \neg F(s_1, \bar{s}), (\exists x \in s_i)F(x, \bar{s}) \\ \hline
(\notin) & \Vdash \neg(s_1 \in s_i), \neg F(s_1, \bar{s}), (\exists x \in s_i)F(x, \bar{s}) \\ \hline
\text{Lemma 4.2ii)} & \Vdash \neg(s_1 \in s_i) \vee \neg F(s_1, \bar{s}), (\exists x \in s_i)F(x, \bar{s})
\end{array}
$$

This completes the proof of the claim (**). It remains to note that we may apply (Cut) to (**) and (12) to complete Case 3.

Case 4. Suppose the last inference was ($\exists$) so $\exists x F(x, \bar{s}) \in \Gamma(\bar{s})$ and we are in the following situation in **KP**:

$$(\exists)\ \frac{\Gamma(\bar{a}), F(c, \bar{a})}{\Gamma(\bar{a})}$$

Let $p = s_j$ if $c = a_j$ otherwise let $p = \bar{\emptyset}$, from the induction hypothesis we can compute some $m < \omega$ such that

$$\mathcal{H}[\bar{s}] \overset{\Omega \cdot \omega^m}{\underset{\Omega + m}{\vdash}} \Gamma(\bar{s}), F(p, \bar{s}) \,.$$

Applying ($\exists$) completes this case.

Case 5. If the last inference was ($\wedge$) or ($\vee$) the result follows immediately by applying the corresponding $\mathbf{RS}_\Omega(X)$ inference to the induction hypotheses.

Case 6. Finally suppose the last inference was (Cut). So we are in the following situation in **KP**

$$(\mathrm{Cut})\ \frac{\Gamma(\bar{a}), B(\bar{a}, \bar{b}) \qquad \Gamma(\bar{a}), \neg B(\bar{a}, \bar{b})}{\Gamma(\bar{a})}$$

Here $\bar{b} := b_1, \ldots, b_l$ denotes the free variables occurring in B that are different from $a_1, \ldots, a_n$. Let $\bar{\bar{\emptyset}}$ denote the sequence of l occurrences of $\bar{\emptyset}$. From the induction hypothesis we find m_1 and m_2 such that

$$\mathcal{H}[\bar{s}] \overset{\Omega \cdot \omega^{m_1}}{\underset{\Omega + m_1}{\vdash}} \Gamma(\bar{s}), B(\bar{s}, \bar{\bar{\emptyset}})$$
$$\mathcal{H}[\bar{s}] \overset{\Omega \cdot \omega^{m_1}}{\underset{\Omega + m_2}{\vdash}} \Gamma(\bar{s}), \neg B(\bar{s}, \bar{\bar{\emptyset}})$$

To which we may apply (Cut) to complete the proof. $\square$

5 A well ordering proof in KP

The aim of this section is to give a well ordering proof in KP for initial segments of formal ordinal terms from $T(\theta)$. First let

(2)
$$\begin{aligned} e_0 &:= \Omega + 1 \\ e_{n+1} &:= \omega^{e_n}. \end{aligned}$$

Each e_n is a formal term belonging to every representation system $T(\theta)$ from 2.13. Although the term is the same, the order type of terms in $T(\theta)$ below e_n will be dependent upon θ. We aim to verify that for every $n < \omega$

$$\begin{aligned}\mathbf{KP} \vdash A_n(\theta) :=& \exists\alpha\exists f[\mathrm{dom}(f) = \alpha \wedge \mathrm{range}(f) = \{a \in T(\theta) \mid a \prec \psi_\theta(e_n))\} \\ & \wedge \forall\gamma, \delta \in \mathrm{dom}(f)(\gamma < \delta \rightarrow f(\gamma) \prec f(\delta))]\end{aligned}$$

where in the above formula $\prec$ denotes the ordering on $T(\theta)$. Formally $A_n(\theta)$ is a Σ-formula of **KP** in which θ is a parameter (free variable) ranging over ordinals. For the remainder of this section we argue informally in **KP**. The symbols $\alpha, \beta, \gamma, \ldots$ are to be **KP**-variables ranging over ordinals and are ordered by $<$, the symbols $a, b, c, \ldots$ are seen as **KP**-variables ranging over codes of formal terms from $T(\theta)$, these are ordered by $\prec$. For the remainder of this section the variable θ will remain free as we argue in **KP**, for ease of reading we shall simply write Ω and ψ instead of Ω_θ and ψ_θ. This proof is an adaptation to the relativised case of a well ordering proof in [50] or [54].

Definition 5.1. The set Acc_θ is defined by

$$\begin{aligned}\mathrm{Acc}_\theta :=& \{a \in T(\theta) \mid a \prec \Omega \wedge \exists\alpha\exists f[\mathrm{dom}(f) = \alpha \wedge \mathrm{range}(f) = \{b : b \preceq a\} \\ & \wedge \forall\gamma, \delta \in \mathrm{dom}(f)(\gamma < \delta \rightarrow f(\gamma) \prec f(\delta))]\}.\end{aligned}$$

Lemma 5.2 (Acc_θ-induction)**.** For any **KP**-formula $F(a)$ we have

$$(\forall a \in \mathrm{Acc}_\theta)[(\forall b \prec a)F(b) \rightarrow F(a)] \rightarrow (\forall a \in \mathrm{Acc}_\theta)F(a).$$

Proof. For $a \in \mathrm{Acc}_\theta$ let $o(a)$ and f_a be the unique ordinal and function such that $o(a) = \mathrm{dom}(f_a)$, $\{b : b \preceq a\} = \mathrm{range}(f_a)$ and $\forall\gamma, \delta \in o(a)(\gamma < \delta \rightarrow f_a(\gamma) \prec f_a(\delta))$. Now for a contradiction let us assume that

$$(\forall a \in \mathrm{Acc}_\theta)[(\forall b \prec a)F(b) \rightarrow F(a)] \quad \text{but} \quad \neg F(a_0) \text{ for some } a_0 \in \mathrm{Acc}_\theta$$

Using set induction/foundation we may pick a_0 such that $o(a_0)$ is minimal. (Note that here we must make use of the full set induction schema of **KP** since the formula F is of unbounded complexity.) Now for any $b \prec a_0$ we have $o(b) < o(a_0)$, thus by our choice of a_0 we get $F(b)$, thus we have

$$(\forall b \prec a_0)F(b).$$

So by assumption we have $F(a_0)$, contradiction. □

Lemma 5.3. Acc_θ has the following closure properties:

i) $b \in \mathrm{Acc}_\theta \wedge a \prec b \quad \rightarrow \quad a \in \mathrm{Acc}_\theta$

ii) $(\forall a \prec b)(a \in \mathrm{Acc}_\theta) \quad \rightarrow \quad b \in \mathrm{Acc}_\theta$

iii) $a, b \in \mathrm{Acc}_\theta \quad \rightarrow \quad a + b \in \mathrm{Acc}_\theta$

iv) $a, b \in \mathrm{Acc}_\theta \quad \rightarrow \quad \varphi ab \in \mathrm{Acc}_\theta$

v) $(\forall \beta \leq \theta)\ \Gamma_\beta \in \mathrm{Acc}_\theta$

Proof. i) Using the notation defined at the start of the proof of Lemma 5.2 we may define

$$o(a) := \{\delta \in o(b) \mid f_b(\delta) \preceq a\} \quad \text{and} \quad f_a := f_b|_{o(a)+1}$$

thus witnessing that $a \in \mathrm{Acc}_\theta$.

ii) Let us assume that $(\forall a \prec b)(a \in \mathrm{Acc}_\theta)$, we must verify that $b \in \mathrm{Acc}_\theta$. Using Δ_0-Separation and Infinity we may form the set $\{a \mid a \prec b\}$, therefore $f := \bigcup_{a \prec b} f_a$ is a set by Δ_0-Collection and Union. Let $\beta := \mathrm{dom}(f)$. Setting $o(b) := \beta + 1$ and $f_b := f \cup \{(\beta, b)\}$ furnishes us with the correct witnesses to confirm that $b \in \mathrm{Acc}_\theta$.

iii) Firstly we must specify what $a + b$ means, since it may not be the case that the string $a + b$ is a term in $T(\theta)$. However, we may define a θ-primitive recursive function $+ : T(\theta) \times T(\theta) \rightarrow T(\theta)$ which corresponds to ordinal addition.

Let us assume that $(\forall c \prec b)(a + c \in \mathrm{Acc}_\theta)$, now if we can show that $a + b \in \mathrm{Acc}_\theta$ then the desired result will follow from Acc_θ-induction (5.2). Now let $d \prec a + b$, either $d \preceq a$ in which case $d \in \mathrm{Acc}_\theta$ by i) or $d \succ a$ and thus $d = a + c$ for some unique $c \prec b$. Such a c may be determined in a θ-primitive recursive fashion, hence $d \in \mathrm{Acc}_\theta$ by assumption. Thus we have

$$(\forall d \prec a + b)(d \in \mathrm{Acc}_\theta).$$

From which we may use ii) to obtain $a + b \in \mathrm{Acc}_\theta$, completing the proof.

iv) Again a function $\varphi : T(\theta) \times T(\theta) \rightarrow T(\theta)$ may be defined in a θ-primitve recursive fashion. It is our aim to show $(\forall x, y \in \mathrm{Acc}_\theta)(\varphi xy \in \mathrm{Acc}_\theta)$, to this end let

$$F(a) := (\forall b \in \mathrm{Acc}_\theta)(\varphi ab \in \mathrm{Acc}_\theta)$$

and assume

(*) $$(\forall z \prec a)F(z)$$

by 5.2 it suffices to verify $F(a)$. So let us assume

(**) $$a, b \in \mathrm{Acc}_\theta \quad \text{and} \quad (\forall y \prec b)(\varphi ay \in \mathrm{Acc}_\theta)$$

now we must verify $\varphi ab \in \mathrm{Acc}_\theta$. To do this we prove that

$$d \prec \varphi ab \Rightarrow d \in \mathrm{Acc}_\theta$$

by induction on Gd; the term complexity of d.

1) If d is strongly critical then $d \preceq a$ or $d \preceq b$ in which case $d \in \mathrm{Acc}_\theta$ by (*) or (**).

2) If $d \equiv \varphi d_0 d_1$ then we have the following subcases:

2.1) If $d_0 \prec a$ and $d_1 \prec \varphi ab$ then since $Gd_1 < Gd$ we get $d_1 \in \mathrm{Acc}_\theta$ from the induction hypothesis. So by (*) we get $d \equiv \varphi d_0 d_1 \in \mathrm{Acc}_\theta$.

2.2) If $d \equiv \varphi a d_1$ and $d_1 \prec b$ then $d \in \mathrm{Acc}_\theta$ by (**).

2.3 If $a \prec d_0$ and $d \prec b$ then $d \in \mathrm{Acc}_\theta$ since $b \in \mathrm{Acc}_\theta$.

3. If $d \equiv d_1 + \ldots + d_n$ and $n > 1$ we get $d_1, \ldots, d_n \in \mathrm{Acc}_\theta$ from the induction hypothesis and thus $d \in \mathrm{Acc}_\theta$ follows from iii).

Thus we have verified that

$$(\forall b \in \mathrm{Acc}_\theta)[(\forall y \prec b)(\varphi ay \in \mathrm{Acc}_\theta) \rightarrow \varphi ab \in \mathrm{Acc}_\theta].$$

So, from Acc_θ-induction we get $(\forall b \in \mathrm{Acc}_\theta)(\varphi ab \in \mathrm{Acc}_\theta)$, ie. $F(a)$ completing the proof.

v) We aim to show that

$$(\forall \beta \leq \theta)[(\forall \gamma < \beta)(\Gamma_\gamma \in \mathrm{Acc}_\theta) \rightarrow \Gamma_\beta \in \mathrm{Acc}_\theta]$$

from which we may use transfinite induction along θ (since θ is an ordinal) to obtain the desired result.

So suppose $\beta \leq \theta$ and $(\forall \delta < \beta)(\Gamma_\delta \in \mathrm{Acc}_\theta)$. Now suppose $b \prec \Gamma_\beta$, by induction on the term complexity of b we verify that $b \in \mathrm{Acc}_\theta$.

If $b \equiv 0$ we are trivially done by ii) or if $b \equiv \Gamma_\delta$ for some $\delta < \beta$ then we know $b \in \mathrm{Acc}_\theta$ by assumption.

If $b \equiv b_0 + \ldots + b_n$ or $b \equiv \varphi b_0 b_1$ then we may use parts iii) and iv) and the induction hypothesis since the components b_i have smaller term complexity.

It cannot be the case that $b \equiv \psi b_0$ since $\psi a \succ \Gamma_\theta$ for every a.

Thus using ii) we get that $\Gamma_\beta \in \mathrm{Acc}_\theta$ and the proof is complete. □

Definition 5.4. By recursion through the construction of ordinal terms in $T(\theta)$ we define the set $SC_{\prec\Omega}(a)$ which lists the most recent strongly critical ordinal below Ω used in the build up of the ordinal term a:

1) $SC_{\prec\Omega}(0) := SC_{\prec\Omega}(\Omega) := \emptyset$.

2) $SC_{\prec\Omega}(a) := \{a\}$ if $a \equiv \Gamma_\beta$ for some $\beta \leq \theta$ or $a \equiv \psi a_0$.

3) $SC_{\prec\Omega}(a_1 + \ldots + a_n) := \bigcup_{1 \leq i \leq n} SC_{\prec\Omega}(a_i)$.

4) $SC_{\prec\Omega}(\varphi a_0 a_1) := SC_{\prec\Omega}(a_0) \cup SC_{\prec\Omega}(a_1)$.

5) $SC_{\prec\Omega}(\psi a) := \{\psi a\}$.

Now let

$$M_\theta := \{a \in T(\theta) \mid SC_{\prec\Omega}(a) \subseteq \mathrm{Acc}_\theta\}$$

and

$$a \prec_{M_\theta} b := a, b \in M_\theta \wedge a \prec b.$$

Finally for a definable class U we define the following formula

$$\mathrm{Prog}_{M_\theta}(U) := (\forall y \in M_\theta)[(\forall z \prec_{M_\theta} y)(z \in U) \rightarrow (y \in U)].$$

Lemma 5.5.

$$\mathrm{Acc}_\theta = M_\theta \cap \Omega := \{a \in M_\theta \mid a \prec \Omega\}.$$

Proof. Suppose that $a \in \mathrm{Acc}_\theta$ and observe that $(\forall x \in SC_{\prec\Omega}(a))(x \preceq a)$, thus $SC_{\prec\Omega}(a) \subseteq \mathrm{Acc}_\theta$ by 5.3i) thus we have verified that $a \in M_\theta \cap \Omega$.

Now let us suppose that $a \in M_\theta \cap \Omega$, so we know that $SC_{\prec\Omega}(a) \subseteq \mathrm{Acc}_\theta$. By induction on the term complexity Ga we verify that $a \in \mathrm{Acc}_\theta$.

Clearly $0 \in \mathrm{Acc}_\theta$ and if $a \equiv \Gamma_\beta$ for some $\beta \leq \theta$ then $a \in \mathrm{Acc}_\theta$ by Lemma 5.3v).

If $a \equiv a_1 + \ldots + a_n$ then we get $a_1, \ldots, a_n \in M_\theta \cap \Omega$ since $SC_{\prec\Omega}(a_i) \subseteq SC_{\prec\Omega}(a)$ for each i. Now using the induction hypothesis we get $a_1, \ldots, a_n \in \mathrm{Acc}_\theta$ and so by Lemma 5.3ii) we have $a \in \mathrm{Acc}_\theta$.

If $a \equiv \varphi bc$ then we get $b, c \in M_\theta \cap \Omega$, so using the induction hypothesis we get $b, c \in \mathrm{Acc}_\theta$ and so by Lemma 5.3iii) we have $a \in \mathrm{Acc}_\theta$.

If $a \equiv \psi a_0$ then $SC_{\prec\Omega}(a) = \{a\}$ so we have $a \in \mathrm{Acc}_\theta$ by assumption. □

Definition 5.6. For a definable class U let

$$U^\delta := \{b \in M_\theta \mid (\forall a \in M_\theta)[M_\theta \cap a \subseteq U \rightarrow M_\theta \cap a + \omega^b \subseteq U]\}$$

where $M_\theta \cap a := \{b \in M_\theta \mid b \prec a\}$.

Lemma 5.7. **KP** $\vdash \mathrm{Prog}_{M_\theta}(U) \rightarrow \mathrm{Prog}_{M_\theta}(U^\delta)$.

Proof. Assume

(1) $\qquad \mathrm{Prog}_{M_\theta}(U)$

(2) $\qquad b \in M_\theta$

(3) $\qquad (\forall x \prec_{M_\theta} b)(z \in U^\delta)$.

Under these assumptions we need to verify that $b \in U^\delta$. Since we already have that $b \in M_\theta$ by (2), it suffices to verify

$$(\forall a \in M_\theta)[M_\theta \cap a \subseteq U \rightarrow M_\theta \cap a + \omega^b \subseteq U]$$

to this end we assume that

(4) $\qquad a \in M_\theta \quad \text{and} \quad M_\theta \cap a \subseteq U.$

Now choose some $d \in M_\theta \cap a + \omega^b$, we must show that $d \in U$ under the assumptions (1)-(4).

If $d \prec a$ then we have $d \in U$ by (4).

If $d = a$ then using (1) and (4) we have $a \in U$.

If $d \succ a$ then since $d \prec a + \omega^b$, we may find $d_1, \ldots, d_k$ such that

$$d = a + \omega^{d_1} + \ldots + \omega^{d_k} \quad \text{and} \quad d_k \preceq \ldots \preceq d_1 \prec b.$$

Since $M_\theta \cap a \subseteq U$ we get $M_\theta \cap a + \omega^{d_1} \subseteq U$ from (3).

In a similar fashion using (3) a further $k-1$ times we obtain

$$M_\theta \cap a + \omega^{d_1} + \ldots + \omega^{d_k} \subseteq U.$$

Finally using one application of $\mathrm{Prog}_{M_\theta}(U)$ (assumption (1)) we have $d \in U$ and thus the proof is complete. □

Definition 5.8. We define the class X_θ in **KP** as

$$X_\theta := \{a \in M_\theta \mid (\exists x \in Ka)(x \succeq a) \vee \psi a \in \mathrm{Acc}_\theta\}.$$

Recall that the function k was defined in Definition 2.11 and can be computed in a θ-primitive recursion fashion. The class X_θ may be thought of as those $a \in M_\theta$ for which either ψa is undefined or $\psi a \in \mathrm{Acc}_\theta$.

Lemma 5.9. **KP** $\vdash \mathrm{Prog}_{M_\theta}(X_\theta)$.

Proof. Assume

(1) $\quad a \in M_\theta$

(2) $\quad (\forall z \prec_{M_\theta} a)(z \in X_\theta)$.

We need to verify that $a \in X_\theta$. If $(\exists x \in Ka)(x \succeq a)$ then we are done, so assume $(\forall x \in Ka)(x \prec a)$ and thus $\psi a \in T(\theta)$ and we must verify that $\psi a \in \mathrm{Acc}_\theta$. To achieve this we verify that

(*) $\quad b \prec \psi a \Rightarrow b \in \mathrm{Acc}_\theta$

from which we would be done by 5.3ii). To verify (*) we proceed by induction on Gb, the term complexity of b.

If $b \equiv 0$ or $b \equiv \Gamma_\beta$ for some $\beta \leq \theta$ we are done by 5.3v).

If $b \equiv b_0 + \ldots + b_n$ or $b \equiv \varphi b_0 b_1$ then the result follows by the induction hypothesis and 5.3ii) or 5.3iii).

So suppose that $b \equiv \psi b_0$. It must be the case that $(\forall x \in K b_0)(x \prec b_0)$ and $b_0 \prec a$. We must now show that $b_0 \in M_\theta$ in order to use (2) to conclude that $b_0 \in X_\theta$. The claim is that

$$(**) \qquad \mathrm{SC}_{\prec\Omega}(b_0) \subseteq \mathrm{Acc}_\theta \quad \text{and thus} \quad b_0 \in M_\theta.$$

Suppose $d \in \mathrm{SC}_{\prec\Omega}(b_0)$ then either $d \equiv \Gamma_\beta$ for some $\beta \leq \theta$ in which case $d \in \mathrm{Acc}_\theta$ by 5.3v) or $d \equiv \psi d_0 \prec \psi a$ for some d_0. But

$$Gd \leq Gb_0 < Gb$$

and thus $d \in \mathrm{Acc}_\theta$ by induction hypothesis. Thus the claim (**) is verified. Now using (2) we obtain $b_0 \in X_\theta$ which implies $b \equiv \psi b_0 \in \mathrm{Acc}_\theta$. □

Lemma 5.10. For any $n < \omega$ and any definable class U

$$\mathbf{KP} \vdash \mathrm{Prog}_{M_\theta}(U) \;\rightarrow\; M_\theta \cap e_n \subseteq U \wedge e_n \in U.$$

Proof. We proceed by induction on n [outside of **KP**].

If $n = 0$ then $\mathrm{Prog}_{M_\theta}(U)$ says that

$$(\forall a \in \mathrm{Acc}_\theta)[(\forall b \prec a)(b \in U) \rightarrow a \in U].$$

So using Acc_θ-induction (Lemma 5.2) we obtain $\mathrm{Acc}_\theta \subseteq U$. Hence from 5.5 we get $M_\theta \cap \Omega \subseteq U$. Now $\Omega, \Omega + 1 \in M_\theta$ so using $\mathrm{Prog}_{M_\theta}(U)$ a further two times we have $\Omega + 1 := e_0 \in U$ as required.

Now suppose the result holds up to n; since the induction hypothesis holds for *all* definable classes we have that that

$$\mathbf{KP} \vdash \mathrm{Prog}_{M_\theta}(U^\delta) \rightarrow M_\theta \cap e_n \subseteq U^\delta \wedge e_n \in U^\delta$$

and by Lemma 5.7 we have

$$\text{(1)} \qquad \mathbf{KP} \vdash \mathrm{Prog}_{M_\theta}(U) \to M_\theta \cap e_n \subseteq U^\delta \wedge e_n \in U^\delta.$$

Now we argue informally in **KP**. Suppose $\mathrm{Prog}_{M_\theta}(U)$, then from (1) we obtain

$$M_\theta \cap e_n \subseteq U^\delta \quad \wedge \quad e_n \in U^\delta.$$

This says that

$$(\forall b \in M_\theta \cap (e_n + 1))(\forall a \in M_\theta)[M_\theta \cap a \subseteq U \to M_\theta \cap a + \omega^b \subseteq U].$$

Now if we put $a = 0$ and $b = e_n$ (noting that $e_n \in M_\theta$) we obtain

$$M_\theta \cap \omega^{e_n} \subseteq U$$

from which $\mathrm{Prog}_{M_\theta}(U)$ implies $\omega^{e_n} \in U$ as required. □

Theorem 5.11. For every $n < \omega$

$$\mathbf{KP} \vdash \forall\theta\, \psi(e_n) \in \mathrm{Acc}_\theta$$

and hence $\mathbf{KP} \vdash \forall\theta\, A_n(\theta)$.

Proof. By 5.9 we have $\mathrm{Prog}_{M_\theta}(X_\theta)$ recalling that

$$X_\theta := \{a \in M_\theta \mid (\exists x \in Ka)(x \succeq a) \vee \psi a \in \mathrm{Acc}_\theta\}.$$

So from 5.10 we get $e_n \in X_\theta$ for any $n < \omega$ and thus $\psi(e_n) \in \mathrm{Acc}_\theta$. □

6 The provably total set functions of KP

At this point we should perhaps remind ourselves that the ordinal $\psi\alpha$ depends on a parameter θ which is the rank of $TC(\{X\})$ as ψ is defined simultaneously with the sets $B_\theta(\alpha)$. After Definition 2.4 we adopted the convention to drop the subscript θ from ψ_θ. For the next application we have to be aware of this dependence. For each $n < \omega$ we define the following recursive set function

$$G_n(X) := L_{\psi_\theta(e_n)}(X)$$

where θ is the rank of $TC(\{X\})$. For a formula $A(a, b)$ of **KP** let

$$\forall x \exists! y A(x, y) := \forall x \forall y_1 \forall y_2 [A(x, y_1) \wedge A(x, y_2) \to y_1 = y_2] \wedge \forall x \exists y A(x, y).$$

Definition 6.1. If T is a theory formulated in the language of set theory, f a set function and $\mathfrak{X}$ a class of formulae. We say that f is $\mathfrak{X}$ definable in T if there is some $\mathfrak{X}$-formula $A_f(a, b)$ with exactly the free variables a, b such that

i) $V \models A_f(x, y) \leftrightarrow f(x) = y$.

ii) $T \vdash \forall x \exists ! y A_f(x, y)$.

Theorem 6.2. Suppose f is a set function that is Σ definable in **KP**, then there is some n (which we may compute from the finite derivation) such that

$$V \models \forall x(f(x) \in G_n(x)).$$

Moreover G_m is Σ definable in **KP** for each $m < \omega$.

Proof. Let $A_f(a, b)$ be the Σ formula expressing f such that **KP** $\vdash \forall x \exists ! y A_f(x, y)$ and fix an arbitrary set X. Let θ be the rank of X. Applying Theorem 4.10 we can compute some $k < \omega$ such that

$$\mathcal{H}_0 \, \overset{\Omega \cdot \omega^k}{\underset{\Omega + k}{\vdash}} \, \forall x \exists ! y A_f(x, y)\,.$$

Applying Lemma 3.15 iv) twice we get

$$\mathcal{H}_0 \, \overset{\Omega \cdot \omega^k}{\underset{\Omega + k}{\vdash}} \, \exists y A_f(X, y)\,.$$

Applying Theorem 3.17 (predicative cut elimination) we get

$$\mathcal{H}_0 \, \overset{e_{k+1}}{\underset{\Omega + 1}{\vdash}} \, \exists y A_f(X, y)\,.$$

Now by Theorem 3.23 (collapsing) we have

$$\mathcal{H}_{e_{k+2}} \, \overset{\psi_\theta(e_{k+2})}{\underset{\psi_\theta(e_{k+2})}{\vdash}} \, \exists y A_f(X, y)\,.$$

Applying Theorem 3.17 (predicative cut elimination) again yields

$$\mathcal{H}_\gamma \, \overset{\varphi(\psi_\theta \gamma)(\psi_\theta \gamma)}{\underset{0}{\vdash}} \, \exists y A_f(X, y) \quad \text{where } \gamma := e_{k+2}.$$

Now by Lemma 3.18 (boundedness) we obtain

(1) $$\mathcal{H}_\gamma \, \overset{\alpha}{\underset{0}{\vdash}} \, (\exists y \in \mathbb{L}_\alpha(X)) A_f(X, y)^{\mathbb{L}_\alpha(X)} \quad \text{where } \alpha := \varphi(\psi_\theta \gamma)(\psi_\theta \gamma).$$

Since (1) contains no instances of (Cut) or (Σ-$\mathrm{Ref}_\Omega(X)$), it follows by induction on α that

$$L_\alpha(X) \models \exists y A_f(X, y).$$

It remains to note that $L_\alpha(X) \subseteq G_{k+3}(X)$ to complete this direction of the proof.

For the other direction we argue informally in **KP**. Let X be an arbitrary set, we may specify the rank of X in a Δ_0 manner([3] p. 29). By Theorem 5.11 we can find an ordinal of the same order type as $\psi_\theta(e_n)$ with θ being the rank of $TC(\{X\})$. We can now generate $L_{\psi_\theta(e_n)}(X)$ by Σ-recursion ([3] p. 26 Theorem 6.4). □

The comparison of Theorem 1.2 with Theorem 6.2 provides a pleasing relation between the arithmetic and set theoretic worlds.

Remark 6.3. In fact the first part of 6.2 can be carried out *inside* **KP**, i.e. If f is Σ definable in **KP** then we can compute some n such that $\mathbf{KP} \vdash \forall x(\exists! y \in G_n(x))A_f(x, y)$. This is not immediately obvious since it appears we need induction up to $\psi_\theta(\varepsilon_{\Omega+1})$, which we do not have access to in **KP**. The way to get around this is to note that we could, in fact, have managed with an infinitary system based on an ordinal representation built out of $B_\theta(e_m)$, provided m is high enough, and we may compute how high m needs to be from the finite derivation. We do have access to induction up to $\psi(e_m)$ for any ordinal θ in **KP** by Theorem 5.11.

7 Applications to semi-intuitionistic KP

PA is conservative over its intuitionistic cousin (called Heyting arithmetic, **HA**) for Π^0_2-statements. One might wonder whether a corresponding result holds in set theory for Π_2-statements. As it turns out, such a result does not obtain for **KP** and its intuitionistic version **IKP**,[2] however, adding the law of excluded middle for atomic formulas to **IKP** yields conservativity for Π_2 theorems.

A semi-intuitionistic version of **IKP** is obtained by assuming the law of excluded middle for atomic formulas, i.e.,

$$\forall x \forall y\, (x \in y \ \vee \ \neg x \in y). \tag{3}$$

Semi-intuitionistic versions of **KP** have become important in Feferman's work in connection with discussions of definiteness of concepts and the continuum hypothesis (cf. [17, 18, 19, 20, 51]).

[2]See [1, 2] for a definition of **IKP**.

Theorem 7.1. **KP** *is* Π_2 *conservative over the semi-intuitionistic theory* **IKP** *plus (3).*

Proof. Let T be the theory **IKP** augmented by (3). Assume that $\mathbf{KP} \vdash \forall x \exists y A(x, y)$, where $A(a, b)$ is Δ_0. We now argue in T. Let X be an arbitrary set. As in the proof of Theorem 4.10 we can determine an α (uniformly depending on the rank of $TC(\{X\})$ such that

$$\mathcal{H}_\gamma \overset{\alpha}{\underset{0}{\vdash}} (\exists y \in \mathbb{L}_\alpha) A_f(X, y)^{\mathbb{L}_\alpha}. \tag{4}$$

To see that we can do this inside T note that the m in Remark 6.3 does not depend on θ. Since (4) contains no instances of (Cut) or (Σ-Ref$_\Omega(X)$), it follows by induction on α that

$$L_\alpha(X) \models \exists y A(X, y).$$

Excluded middle for atomic formulas is required at several points. For instance it is needed in Lemma 4.3i), Case 1. Also when showing that all sequents Λ occurring in the derivation (4) are true in $L_\alpha(X)$[3] one needs to invoke the law of excluded middle for Δ_0-formulas. The latter follows from (3) with the help of Δ_0-Separation. □

8 A relativised ordinal analysis of KP($\mathcal{P}$)

With the help of [56] and the foregoing machinery one can also characterize the provable power recursive set functions of Power Kripke-Platek set theory, **KP**($\mathcal{P}$). For background on **KP**($\mathcal{P}$) see [56]. To introduce its axioms we need the notion of subset bounded formula.

Definition 8.1. We use subset bounded quantifiers $\exists x \subseteq y \ldots$ and $\forall x \subseteq y \ldots$ as abbreviations for $\exists x(x \subseteq y \wedge \ldots)$ and $\forall x(x \subseteq y \to \ldots)$, respectively.

The $\Delta_0^{\mathcal{P}}$-formulae are the smallest class of formulae containing the atomic formulae closed under $\wedge, \vee, \to, \neg$ and the quantifiers

$$\forall x \in a,\ \exists x \in a,\ \forall x \subseteq a,\ \exists x \subseteq a.$$

A formula is in $\Sigma^{\mathcal{P}}$ if belongs to the smallest collection of formulae which contains the $\Delta_0^{\mathcal{P}}$-formulae and is closed under $\wedge, \vee$ and the quantifiers $\forall x \in a$, $\exists x \in a$, $\forall x \subseteq a$ and $\exists x$. A formula is $\Pi^{\mathcal{P}}$ if belongs to the smallest collection of formulae which contains the $\Delta_0^{\mathcal{P}}$-formulae and is closed under $\wedge, \vee$, the quantifiers $\forall x \in a$, $\exists x \in a$, $\forall x \subseteq a$ and $\forall x$.

[3]This means that the disjunction over all formulas in Λ is true in $L_\alpha(X)$.

Definition 8.2. $\mathbf{KP}(\mathcal{P})$ has the same language as **ZF**. Its axioms are the following: Extensionality, Pairing, Union, Infinity, Powerset, $\Delta_0^{\mathcal{P}}$-Separation, $\Delta_0^{\mathcal{P}}$-Collection and Set Induction (or Class Foundation).

The transitive models of $\mathbf{KP}(\mathcal{P})$ have been termed **power admissible** sets in [22].

Remark 8.3. Alternatively, $\mathbf{KP}(\mathcal{P})$ can be obtained from **KP** by adding a function symbol $\mathcal{P}$ for the powerset function as a primitive symbol to the language and the axiom

$$\forall y\, [y \in \mathcal{P}(x) \leftrightarrow y \subseteq x]$$

and extending the schemes of Δ_0 Separation and Collection to the Δ_0-formulae of this new language.

Lemma 8.4. $\mathbf{KP}(\mathcal{P})$ *is* **not** *the same theory as* $\mathbf{KP} + \mathbf{Pow}$, *where* **Pow** *denotes the Powerset Axiom. Indeed,* $\mathbf{KP} + \mathbf{Pow}$ *is a much weaker theory than* $\mathbf{KP}(\mathcal{P})$ *in which one cannot prove the existence of* $V_{\omega+\omega}$.

Proof. [56, Lemma 2.4]. □

8.1 The infinitary proof system $\mathbf{RS}_{\Omega}^{\mathcal{P}}(X)$

The infinitary proof system $RS_{\Omega}^{\mathcal{P}}$ of [56] is based on a formal analogue of the von Neumann hierarchy along the Bachmann-Howard ordinal. For our purposes both have to be relativised to a given set X.

Definition 8.5. Let X be any set. We may relativise the von Neumann hierarchy to X as follows:

$$\begin{aligned} V_0(X) &:= TC(\{X\}) \quad \text{the } \textit{transitive closure} \text{ of } \{X\} \\ V_{\alpha+1}(X) &:= \{B : B \subseteq V_\alpha(X)\} \\ V_\theta(X) &:= \bigcup_{\xi<\theta} V_\xi(X) \quad \text{when } \theta \text{ is a limit.} \end{aligned}$$

Let X be an arbitrary (well founded) set and let θ be the set-theoretic rank of X (hereby referred to as the $\in$-rank). Henceforth all ordinals are assumed to belong to the ordinal notation system $T(\theta)$ developed in section 3. The system $\mathbf{RS}_{\Omega}^{\mathcal{P}}(X)$ will be the relativised version of the infinitary proof system $RS_{\Omega}^{\mathcal{P}}$ from [56].

Definition 8.6. We give an inductive definition of the set $\mathcal{T}^{\mathcal{P}}$ of $\mathbf{RS}_{\Omega}^{\mathcal{P}}(X)$ terms. To each term $t \in \mathcal{T}^{\mathcal{P}}$ we assign an ordinal level $|\, t\, |$.

(i) For every $u \in TC(\{X\})$, $\bar{u} \in \mathcal{T}^{\mathcal{P}}$ and $|\, \bar{u}\, | := \Gamma_{\mathrm{rank}(u)}$.

(ii) For every $\alpha < \Omega$, $\mathbb{V}_\alpha(X) \in \mathcal{T}^{\mathcal{P}}$ and $|\mathbb{V}_\alpha(X)| := \Gamma_{\theta+1} + \alpha$.

(iii) For each $\alpha < \Omega$, we have infinitely many free variables $a_1^\alpha, a_2^\alpha, a_3^\alpha, \ldots$ which are terms of level $\Gamma_{\theta+1} + \alpha$.

(iv) If $\alpha < \Omega$, $A(a, b_1, \ldots, b_n)$ is a $\Delta_0^{\mathcal{P}}$ formula of **KP**($\mathcal{P}$) with all free variables displayed and $s_1, \ldots, s_n$ are terms in $\mathcal{T}^{\mathcal{P}}$ then

$$[x \in \mathbb{V}_\alpha(X) | A(x, s_1, \ldots, s_n)]$$

is a term of level $\Gamma_{\theta+1} + \alpha$.

The $\mathbf{RS}_\Omega^{\mathcal{P}}(X)$*-formulae* are the expressions of the form $F(s_1, \ldots, s_n)$, where $F(a_1, \ldots, a_n)$ is a formula of **KP**($\mathcal{P}$) with all free variables exhibited and $s_1, \ldots, s_n$ are $\mathbf{RS}_\Omega^{\mathcal{P}}(X)$-terms. We set

$$|F(s_1, \ldots, s_n)| \quad = \quad \{|s_1|, \ldots, |s_n|\}.$$

For a sequent $\Gamma = \{A_1, \ldots, A_n\}$ we define

$$|\Gamma| \quad := \quad |A_1| \cup \ldots \cup |A_n|.$$

A formula is a $\Delta_0^{\mathcal{P}}$**-formula** of $\mathbf{RS}_\Omega^{\mathcal{P}}(X)$ if it is of the form $F(s_1, \ldots, s_n)$ with $F(a_1, \ldots, a_n)$ being a $\Delta_0^{\mathcal{P}}$-formula of **KP**($\mathcal{P}$) and $s_1, \ldots, s_n$ $\mathbf{RS}_\Omega^{\mathcal{P}}(X)$-terms.

As in the case of the Tait-style version of **KP**($\mathcal{P}$) in [56, Sec. 3], we let $\neg A$ be the formula which arises from A by (i) putting $\neg$ in front of each atomic formula, (ii) replacing $\wedge, \vee, (\forall x \in s), (\exists x \in s), (\forall x \subseteq s), (\exists x \subseteq s), \forall x, \exists x$ by $\vee, \wedge, (\exists x \in s), (\forall x \in s), (\exists x \subseteq s), (\forall x \subseteq s), \exists x, \forall x$, respectively, and (iii) dropping double negations. $A \to B$ stands for $\neg A \vee B$.

Remark 8.7. There is a crucial difference between Definition 3.3 and Definition 8.6 when it comes to measuring the level of a comprehension term. The level of $[x \in \mathbb{V}_\alpha(X) | A(x, s_1, \ldots, s_n)]$ does not take the terms $s_1, \ldots, s_n$ into account. They may be of arbitrary (especially higher) level.

Since we also want to keep track of the complexity of cuts appearing in derivations, we endow each formula with an ordinal rank.

Definition 8.8. The *rank* of a term or formula is determined as follows.

1. $rk(\bar{u}) := \Gamma_{rank(u)}$ *for* u *in the transitive closure of* X.

2. $rk(\mathbb{V}_\alpha(X)) := \Gamma_{\theta+1} + \omega \cdot \alpha$.

3. $rk([x \in \mathbb{V}_\alpha(X) \mid F(x)]) := \max\{\Gamma_{\theta+1} + \omega \cdot \alpha + 1, rk(F(\bar{0})) + 2\}$.

4. $rk(s \in t) := rk(s \notin t) := \max\{|\, s\,| + 6, |\, t\,| + 1\}$.

5. $rk((\exists x \in t)F(x)) := rk((\forall x \in t)F(x)) := max\{rk(t) + 3, rk(F(\bar{0})) + 2\}$.

6. $rk((\exists x \subseteq t)F(x)) := rk((\forall x \subseteq t)F(x)) := \max\{rk(t) + 3, rk(F(\bar{0})) + 2\}$.

7. $rk(\exists x\, F(x)) := rk(\forall x\, F(x)) := \max\{\Omega, rk(F(\bar{0})) + 2\}$.

8. $rk(A \wedge B) := rk(A \vee B) := max\{rk(A), rk(B)\} + 1$.

Definition 8.9. The *axioms* of $\mathbf{RS}_\Omega^\mathcal{P}(X)$ are:

(X1) $\Gamma,\ \bar{u} \in \bar{v}$ if $u, v \in TC(X)$ and $u \in v$.

(X2) $\Gamma,\ \bar{u} \notin \bar{v}$ if $u, v \in TC(X)$ and $u \notin v$.

(A1) $\Gamma,\ A,\ \neg A$ for A in $\Delta_0^\mathcal{P}$.

(A2) $\Gamma,\ t = t$.

(A3) $\Gamma,\ s_1 \neq t_1, \ldots, s_n \neq t_n, \neg A(s_1, \ldots, s_n),\ A(t_1, \ldots, t_n)$
for $A(s_1, \ldots, s_n)$ in $\Delta_0^\mathcal{P}$.

(A4) $\Gamma,\ s \in \mathbb{V}_\alpha(X)$ if $|\, s\,| < |\, \mathbb{V}_\alpha(X)\,|$.

(A5) $\Gamma,\ s \subseteq \mathbb{V}_\alpha(X)$ if $|\, s\,| \leq |\, \mathbb{V}_\alpha(X)\,|$.

(A6) $\Gamma, t \notin [x \in \mathbb{V}_\alpha(X) \mid F(x, \vec{s})],\ F(t, \vec{s})$
whenever $F(t, \vec{s})$ is $\Delta_0^\mathcal{P}$ and $|\, t\,| < |\, \mathbb{V}_\alpha(X)\,|$.

(A7) $\Gamma, \neg F(t, \vec{s}),\ t \in [x \in \mathbb{V}_\alpha(X) \mid F(x, \vec{s})]$
whenever $F(t, \vec{s})$ is $\Delta_0^\mathcal{P}$ and $|\, t\,| < |\, \mathbb{V}_\alpha(X)\,|$.

We adopt the notion of operator from Definition 3.6. If s is an $\mathbf{RS}_\Omega^\mathcal{P}(X)$-term, the operator $\mathcal{H}[s]$ is defined by

$$\mathcal{H}[s](X) \quad = \quad \mathcal{H}(X \cup \{|\, s\,|\}).$$

Likewise, if $\mathfrak{X}$ is a formula or a sequent we define

$$\mathcal{H}[\mathfrak{X}](X) \quad = \quad \mathcal{H}(X \cup |\, \mathfrak{X}\,|).$$

Definition 8.10. Let $\mathcal{H}$ be an operator and let Λ be a finite set of $\mathbf{RS}_\Omega^{\mathcal{P}}(X)$–formulae. $\mathcal{H} \overset{\alpha}{\underset{\rho}{\vdash}} \Lambda$ is defined by recursion on α.

If Λ is an **axiom** and $|\Lambda| \cup \{\alpha\} \subseteq \mathcal{H}(\emptyset)$, then $\mathcal{H} \overset{\alpha}{\underset{\rho}{\vdash}} \Lambda$.

Moreover, we have inductive clauses pertaining to the inference rules of $\mathbf{RS}_\Omega^{\mathcal{P}}(X)$, which all come with the additional requirement that

$$|\Lambda| \cup \{\alpha\} \subseteq \mathcal{H}(\emptyset)$$

where Λ is the sequent of the conclusion. We shall not repeat this requirement below.

Below the third column gives the requirements that the ordinals have to satisfy for each of the inferences. For instance in the case of $(\forall)_\infty$, to be able to conclude that $\mathcal{H} \overset{\alpha}{\underset{\rho}{\vdash}} \Gamma, \forall x F(x)$, it is required that for all terms s there exists α_s such that $\mathcal{H}[s] \overset{\alpha_s}{\underset{\rho}{\vdash}} \Gamma, F(s)$ and $|s| < \alpha_s + 1 < \alpha$. The side conditions for the rules $(b\forall)_\infty, (pb\forall)_\infty, (\notin)_\infty, (\not\subseteq)_\infty$ below have to be read in the same vein.

Below we shall write $|s| \dot{<} |t|$ and $|s| \dot{\leq} |t|$ for $|s| < \max(\Gamma_{\theta+1}, |t|)$ and $|s| \leq \max(\Gamma_{\theta+1}, |t|)$, respectively.

The clauses are the following:

$(\wedge)$	$\dfrac{\mathcal{H} \overset{\alpha_0}{\underset{\rho}{\vdash}} \Gamma, A_0 \qquad \mathcal{H} \overset{\alpha_0}{\underset{\rho}{\vdash}} \Gamma, A_1}{\mathcal{H} \overset{\alpha}{\underset{\rho}{\vdash}} \Gamma, A_0 \wedge A_1}$	$\alpha_0 < \alpha$
$(\vee)$	$\dfrac{\mathcal{H} \overset{\alpha_0}{\underset{\rho}{\vdash}} \Lambda, A_i}{\mathcal{H} \overset{\alpha}{\underset{\rho}{\vdash}} \Gamma, A_0 \vee A_1}$	$\alpha_0 < \alpha$ $i \in \{0,1\}$
(Cut)	$\dfrac{\mathcal{H} \overset{\alpha_0}{\underset{\rho}{\vdash}} \Lambda, B \qquad \mathcal{H} \overset{\alpha_0}{\underset{\rho}{\vdash}} \Lambda, \neg B}{\mathcal{H} \overset{\alpha}{\underset{\rho}{\vdash}} \Lambda}$	$\alpha_0 < \alpha$ $rk(B) < \rho$
$(b\forall)_\infty$	$\dfrac{\mathcal{H}[s] \overset{\alpha_s}{\underset{\rho}{\vdash}} \Gamma, s \in t \rightarrow F(s) \text{ for all } |s| < |t|}{\mathcal{H} \overset{\alpha}{\underset{\rho}{\vdash}} \Gamma, (\forall x \in t)F(x)}$	$|s| \leq \alpha_s < \alpha$
$(b\exists)$	$\dfrac{\mathcal{H} \overset{\alpha_0}{\underset{\rho}{\vdash}} \Gamma, s \in t \wedge F(s)}{\mathcal{H} \overset{\alpha}{\underset{\rho}{\vdash}} \Gamma, (\exists x \in t)F(x)}$	$\alpha_0 < \alpha$ $|s| < |t|$ $|s| < \alpha$

$$(pb\forall)_\infty \quad \frac{\mathcal{H}[s] \overset{\alpha_s}{\underset{\rho}{\vdash}} \Gamma, s \subseteq t \to F(s) \text{ for all } |s| \dot{\leq} |t|}{\mathcal{H} \overset{\alpha}{\underset{\rho}{\vdash}} \Gamma, (\forall x \subseteq t)F(x)} \quad |s| \leq \alpha_s < \alpha$$

$$(pb\exists) \quad \frac{\mathcal{H} \overset{\alpha_0}{\underset{\rho}{\vdash}} \Gamma, s \subseteq t \wedge F(s)}{\mathcal{H} \overset{\alpha}{\underset{\rho}{\vdash}} \Gamma, (\exists x \subseteq t)F(x)} \quad \begin{array}{r} \alpha_0 < \alpha \\ |s| \dot{\leq} |t| \\ |s| < \alpha \end{array}$$

$$(\forall)_\infty \quad \frac{\mathcal{H}[s] \overset{\alpha_s}{\underset{\rho}{\vdash}} \Gamma, F(s) \text{ for all } s}{\mathcal{H} \overset{\alpha}{\underset{\rho}{\vdash}} \Gamma, \forall x F(x)} \quad |s| < \alpha_s + 1 < \alpha$$

$$(\exists) \quad \frac{\mathcal{H} \overset{\alpha_0}{\underset{\rho}{\vdash}} \Gamma, F(s)}{\mathcal{H} \overset{\alpha}{\underset{\rho}{\vdash}} \Gamma, \exists x F(x)} \quad \begin{array}{r} \alpha_0 + 1 < \alpha \\ |s| < \alpha \end{array}$$

$$(\not\in)_\infty \quad \frac{\mathcal{H}[r] \overset{\alpha_r}{\underset{\rho}{\vdash}} \Gamma, r \in t \to r \neq s \text{ for all } |r| < |t|}{\mathcal{H} \overset{\alpha}{\underset{\rho}{\vdash}} \Gamma, s \notin t} \quad |r| \leq \alpha_r < \alpha$$

$$(\in) \quad \frac{\mathcal{H} \overset{\alpha_0}{\underset{\rho}{\vdash}} \Gamma, r \in t \wedge r = s}{\mathcal{H} \overset{\alpha}{\underset{\rho}{\vdash}} \Gamma, s \in t} \quad \begin{array}{r} \alpha_0 < \alpha \\ |r| < |t| \\ |r| < \alpha \end{array}$$

$$(\not\subseteq)_\infty \quad \frac{\mathcal{H}[r] \overset{\alpha_r}{\underset{\rho}{\vdash}} \Gamma, r \subseteq t \to r \neq s \text{ for all } |r| \dot{\leq} |t|}{\mathcal{H} \overset{\alpha}{\underset{\rho}{\vdash}} \Gamma, s \not\subseteq t} \quad |r| \leq \alpha_r < \alpha$$

$$(\subseteq) \quad \frac{\mathcal{H} \overset{\alpha_0}{\underset{\rho}{\vdash}} \Gamma, r \subseteq t \wedge r = s}{\mathcal{H} \overset{\alpha}{\underset{\rho}{\vdash}} \Gamma, s \subseteq t} \quad \begin{array}{r} \alpha_0 < \alpha \\ |r| \dot{\leq} |t| \\ |r| < \alpha \end{array}$$

$$(\Sigma^{\mathcal{P}}\text{-}Ref) \quad \frac{\mathcal{H} \overset{\alpha_0}{\underset{\rho}{\vdash}} \Gamma, A}{\mathcal{H} \overset{\alpha}{\underset{\rho}{\vdash}} \Gamma, \exists z\, A^z} \quad \begin{array}{r} \alpha_0 + 1, \Omega < \alpha \\ A \in \Sigma^{\mathcal{P}} \end{array}$$

Remark 8.11. Suppose $\mathcal{H} \overset{\alpha}{\underset{\rho}{\vdash}} \Gamma(s_1, \ldots, s_n)$ where $\Gamma(a_1, \ldots, a_n)$ is a sequent of $\mathbf{KP}(\mathcal{P})$ such that all variables $a_1, \ldots, a_n$ do occur in $\Gamma(a_1, \ldots, a_n)$ and $s_1, \ldots, s_n$ are $\mathbf{RS}_\Omega^{\mathcal{P}}(X)$-terms. Then we have that $|s_1|, \ldots, |s_n| \in \mathcal{H}(\emptyset)$. Standing in sharp contrast to the ordinal analysis of **KP**, however, the terms s_i may and often will contain subterms that the operator $\mathcal{H}$ does **not** control, that is, subterms t with $|t| \notin \mathcal{H}(\emptyset)$.

The embedding of $\mathbf{KP}(\mathcal{P})$ into $\mathbf{RS}_\Omega^\mathcal{P}(X)$ and the ordinal analysis of $\mathbf{RS}_\Omega^\mathcal{P}(X)$ can be carried out in much the same way as for $RS_\Omega^\mathcal{P}$ in [56] with only minor amendments necessary to deal with terms and axioms pertaining to the given set X. Below we list the main steps.

Theorem 8.12. If $\mathbf{KP}(\mathcal{P}) \vdash \Gamma(a_1,\ldots,a_n)$ where $\Gamma(a_1,\ldots,a_n)$ is a finite set of formulae whose free variables are amongst $a_1,\ldots,a_n$, then there is some $m < \omega$ (which we may compute from the derivation) such that

$$\mathcal{H}[s_1,\ldots,s_n] \overset{\Omega\cdot\omega^m}{\underset{\Omega+m}{\vdash}} \Gamma(s_1,\ldots,s_n)$$

for any operator $\mathcal{H}$ and any $\mathbf{RS}_\Omega^\mathcal{P}(X)$ terms $s_1,\ldots,s_n$.

Proof. This can be proved in the same way as [56, Theorem 6.9]. □

Theorem 8.13 (Cut elimination I).

$$\mathcal{H} \overset{\alpha}{\underset{\Omega+n+1}{\vdash}} \Gamma \quad \Rightarrow \quad \mathcal{H} \overset{\omega_n(\alpha)}{\underset{\Omega+1}{\vdash}} \Gamma$$

where $\omega_0(\beta) := \beta$ *and* $\omega_{k+1}(\beta) := \omega^{\omega_k(\beta)}$.

Proof: The proof is the special case of Theorem 3.17 when $\rho = \Omega+n$ and $\alpha = 0$. See also [56, Theorem 7.1]. □

For a formula C of $\mathbf{RS}_\Omega^\mathcal{P}(X)$, $C^{\mathbb{V}_\delta(X)}$ is obtained from C by replacing all unbounded quantifiers Qz in C by $(Qz \in \mathbb{V}_\delta(X))$.

Lemma 8.14 (Boundedness for $\mathbf{RS}_\Omega^\mathcal{P}(X)$)**.** If C is a $\Sigma^\mathcal{P}$ formula, $\alpha \leq \beta < \Omega$, $\beta \in \mathcal{H}$ and $\mathcal{H} \overset{\alpha}{\underset{\rho}{\vdash}} \Gamma, C$ then $\mathcal{H} \overset{\alpha}{\underset{\rho}{\vdash}} \Gamma, C^{\mathbb{V}_\beta(X)}$.

Proof. Similar to Lemma 3.18. □

Theorem 8.15 (Collapsing for $\mathbf{RS}_\Omega^\mathcal{P}(X)$)**.** Suppose Γ is a set of $\Sigma^\mathcal{P}$ formulae such that $|\Gamma| \subseteq B(\eta)$ and $\eta \in B(\eta)$.

$$\text{If} \quad \mathcal{H}_\eta \overset{\alpha}{\underset{\Omega+1}{\vdash}} \Gamma \quad \text{then} \quad \mathcal{H}_{\hat{\alpha}} \overset{\psi\hat{\alpha}}{\underset{\psi\hat{\alpha}}{\vdash}} \Gamma$$

where $\hat{\alpha} = \eta + \omega^{\Omega+\alpha}$.

Proof. The proof is essentially the same as that of [56, Theorem 7.4]. $\square$

For the characterisation theorem for $\mathbf{KP}(\mathcal{P})$, we need to show that derivability in $\mathbf{RS}_{\Omega}^{\mathcal{P}}(X)$ entails truth for $\Sigma^{\mathcal{P}}$-formulae. Since $\mathbf{RS}_{\Omega}^{\mathcal{P}}(X)$-formulae contain variables we need the notion of assignment. Let VAR be the set of free variables of $\mathbf{RS}_{\Omega}^{\mathcal{P}}(X)$. A variable assignment ℓ is a function

$$\ell : VAR \longrightarrow V_{\psi(\varepsilon_{\Omega+1})}$$

satisfying $\ell(a^{\alpha}) \in V_{\alpha+1}(X)$. ℓ can be canonically lifted to all $\mathbf{RS}_{\Omega}^{\mathcal{P}}(X)$-terms as follows:

$$\begin{aligned} \ell(\bar{u}) &= u \text{ for } u \text{ in } TC(\{X\}) \\ \ell(\mathbb{V}_{\alpha}(X)) &= V_{\alpha}(X) \\ \ell([x \in \mathbb{V}_{\alpha}(X) \mid F(x, s_1, \ldots, s_n)]) &= \{x \in V_{\alpha}(X) : F(x, \ell(s_1), \ldots, \ell(s_n))\} . \end{aligned}$$

Note that $\ell(s) \in V_{\psi(\varepsilon_{\Omega+1})}(X)$ holds for all $\mathbf{RS}_{\Omega}^{\mathcal{P}}(X)$-terms s. Moreover, we have $\ell(s) \in V_{|s|+1}(X)$.

Theorem 8.16 (Soundness). *Let $\mathcal{H}$ be an operator with $\mathcal{H}(\emptyset) \subseteq B(\varepsilon_{\Omega+1})$ and $\alpha, \rho < \psi(\varepsilon_{\Omega+1})$. Let $\Gamma(s_1, \ldots, s_n)$ be a sequent consisting only of $\Sigma^{\mathcal{P}}$-formulae with constants from $TC(\{X\})$. Suppose*

$$\mathcal{H} \overset{\alpha}{\underset{\rho}{\vdash}} \Gamma(s_1, \ldots, s_n) .$$

Then, for all variable assignments ℓ,

$$V_{\psi(\varepsilon_{\Omega+1})}(X) \models \Gamma(\ell(s_1), \ldots, \ell(s_n)) ,$$

where the latter, of course, means that $V_{\psi(\varepsilon_{\Omega+1})}$ is a model of the disjunction of the formulae in $\Gamma(\ell(s_1), \ldots, \ell(s_n))$.

Proof: The proof is basically the same as for [56, Theorem 8.1]. It proceeds by induction on α. Note that, owing to $\alpha, \rho < \Omega$, the proof tree pertaining to $\mathcal{H} \overset{\alpha}{\underset{\rho}{\vdash}} \Gamma(s_1, \ldots, s_n)$ neither contains any instances of $(\Sigma^{\mathcal{P}}\text{-}Ref)$ nor of $(\forall)_{\infty}$, and that all cuts are performed with $\Delta_0^{\mathcal{P}}$-formulae. The proof is straightforward as all the axioms of $RS_{\Omega}^{\mathcal{P}}$ are true under the interpretation and all other rules are truth preserving with respect to this interpretation. Observe that we make essential use of the free variables when showing the soundness of $(b\forall)_{\infty}$, $(pb\forall)_{\infty}$, $(\notin)_{\infty}$ and $(\not\subseteq)_{\infty}$. We treat $(pb\forall)_{\infty}$ as an example. So assume $(\forall x \subseteq s_i)F(x, \vec{s}) \in \Gamma(\vec{s})$ and

$$\mathcal{H}[r] \overset{\alpha_r}{\underset{\rho}{\vdash}} \Gamma(s_1, \ldots, s_n), r \subseteq s_i \rightarrow F(r, \vec{s})$$

holds for all terms r with $|r| \leq |s_i|$ for some $\alpha_r < \alpha$. In particular we have

$$\mathcal{H}[a^\beta] \frac{\alpha'}{\rho} \Gamma(s_1, \ldots, s_n), a^\beta \subseteq s_i \to F(a^\beta, \vec{s})$$

where $\beta = |s_i|$ and a^β is a free variable not occurring in $\Gamma(s_1, \ldots, s_n)$ and $\alpha' = \alpha_{a^\beta}$. By the induction hypothesis we have

$$V_{\psi_\Omega(\varepsilon_{\Omega+1})} \models \Gamma(\ell(s_1), \ldots, \ell(s_n)), \ell'(a^\beta) \subseteq \ell(s_i) \to F(\ell'(a^\beta), \ell(s_1), \ldots, \ell(s_n))$$

where ℓ' is an arbitrary variable assignment. This entails that either

$$V_{\psi_\Omega(\varepsilon_{\Omega+1})} \models \Gamma(\ell(s_1), \ldots, \ell(s_n))$$

or

$$V_{\psi_\Omega(\varepsilon_{\Omega+1})} \models \ell'(a^\beta) \subseteq \ell(s_i) \to F(\ell'(a^\beta), \ell(s_1), \ldots, \ell(s_n))$$

for all assignments ℓ'. In the former case we have found what we want and in the latter case we arrive at $V_{\psi_\Omega(\varepsilon_{\Omega+1})} \models (\forall x \subseteq \ell(s_i))F(x, \ell(s_1), \ldots, \ell(s_n))$ and therefore also have $V_{\psi_\Omega(\varepsilon_{\Omega+1})} \models \Gamma(\ell(s_1), \ldots, \ell(s_n))$. □

8.2 The provably total set functions of KP($\mathcal{P}$)

For each $n < \omega$ we define the following recursive set function

$$G_n^{\mathcal{P}}(X) := V_{\psi_\theta(e_n)}(X)$$

where e_n was defined in (2) and θ stands for the rank of the transitive closure of X.

Theorem 8.17. Suppose f is a set function that is $\Sigma^{\mathcal{P}}$ definable in **KP**($\mathcal{P}$), then there is some n (which we may compute from the finite derivation) such that

$$V \models \forall x(f(x) \in G_n^{\mathcal{P}}(x)).$$

Moreover $G_m^{\mathcal{P}}$ is $\Sigma^{\mathcal{P}}$ definable in **KP**($\mathcal{P}$) for each $m < \omega$.

Proof. Let $A_f(a, b)$ be the $\Sigma^{\mathcal{P}}$ formula expressing f such that **KP**($\mathcal{P}$)$\vdash \forall x \exists! y A_f(x, y)$ and fix an arbitrary set X. Let θ be the rank of X. Applying Theorem 8.12 we can compute some $k < \omega$ such that

$$\mathcal{H}_0 \frac{\Omega \cdot \omega^k}{\Omega + k} \forall x \exists! y A_f(x, y) \, .$$

Applying inversion as in Lemma 3.15 iv) twice we get

$$\mathcal{H}_0 \frac{\Omega \cdot \omega^k}{\Omega + k} \exists y A_f(X, y) \, .$$

Applying Theorem 8.13 we get

$$\mathcal{H}_0 \overset{e_{k+1}}{\underset{\Omega+1}{\vdash}} \exists y A_f(X,y)\,.$$

Now by Theorem 8.15 (collapsing) we have

$$\mathcal{H}_{e_{k+2}} \overset{\psi_\theta(e_{k+2})}{\underset{\psi_\theta(e_{k+2})}{\vdash}} \exists y A_f(X,y)\,.$$

Now by Lemma 8.14 (boundedness) we obtain

$$(5) \qquad \mathcal{H}_\gamma \overset{\psi_\theta(\gamma)}{\underset{\psi_\theta(\gamma)}{\vdash}} (\exists y \in \mathbb{V}_{\psi_\theta(\gamma)}(X)) A_f(X,y)^{\mathbb{V}_{\psi_\theta(\gamma)}} \quad \text{where } \gamma := e_{k+2}.$$

The Soundness Theorem 8.16 applied to (5) now yields that

$$\mathbb{V}_{\psi_\theta(\gamma)} \models \exists y\, A_f(X,y).$$

It remains to note that $V_\alpha(X) \subseteq G^{\mathcal{P}}_{k+3}(X)$ to complete this direction of the proof.

For the other direction we argue informally in $\mathbf{KP}(\mathcal{P})$. Let X be an arbitrary set. By Theorem 5.11 we can find an ordinal of the same order type as $\psi_\theta(e_n)$. We can now generate $V_{\psi_\theta(e_n)}(X)$ by $\Sigma^{\mathcal{P}}$-recursion (similar to [3] p. 26 Theorem 6.4). □

Remark 8.18. As was the case for **KP**, the first part of 6.2 can be carried out *inside* $\mathbf{KP}(\mathcal{P})$, i.e. If f is $\Sigma^{\mathcal{P}}$ definable in $\mathbf{KP}(\mathcal{P})$ then we can compute some n such that

$$\mathbf{KP}(\mathcal{P}) \vdash \forall x(\exists! y \in G^{\mathcal{P}}_n(x)) A_f(x,y)\,.$$

This is not immediately obvious since it appears we need induction up to $\psi_\theta(\varepsilon_{\Omega+1})$, which we do not have access to in $\mathbf{KP}(\mathcal{P})$. The way to get around this is to note that we could, in fact, have managed with an infinitary system based on an ordinal representation built out of $B_\theta(e_m)$, provided m is high enough, and we may compute how high m needs to be from the finite derivation. We do have access to induction up to $\psi_\theta(e_m)$ in $\mathbf{KP}(\mathcal{P})$ by Theorem 5.11.

9 Adding global choice: $\mathbf{KP}(\mathcal{P}) + \mathbf{AC}_{global}$

Here we extend the relativised ordinal analysis to $\mathbf{KP}(\mathcal{P})$ with global choice. Since the global axiom of choice, $\mathbf{AC}_{global}$, is less familiar, let us spell out the details. By $\mathbf{KP}(\mathcal{P}) + \mathbf{AC}_{global}$ we mean an extension of $\mathbf{KP}(\mathcal{P})$ where the language contains a new binary relation symbol R and the axiom schemes of **KP**($\mathcal{P}$) are extended to this richer language and the following axioms pertaining to R are added:

(6) $\quad (i) \quad \forall x \forall y \forall z[\mathsf{R}(x,y) \wedge \mathsf{R}(x,z) \rightarrow y = z]$

(7) $\quad (ii) \quad \forall x[x \neq \emptyset \rightarrow \exists y \in x\, \mathsf{R}(x,y)]$.

Section 3 of [58] describes an extension of $RS_\Omega^{\mathcal{P}}$ that incorporates the new symbol R. We can now relativise this system to a given set X as we did with $RS_\Omega^{\mathcal{P}}$ in the previous section. Let us call the relativized version $\mathbf{RS}_\Omega^{\mathcal{P}}(\mathsf{R}, X)$. The ordinal analysis of $\mathbf{RS}_\Omega^{\mathcal{P}}(\mathsf{R}, X)$ can be performed with almost no changes as for $\mathbf{RS}_\Omega^{\mathcal{P}}(X)$ in the foregoing section. On account of the relativization we arrive at stronger versions of [58, Corollary 3.1] and [58, Theorem] which incorporate the parameter X. A $\Pi_2^{\mathcal{P}}$-formula is a formula of the form $\forall y\, A(y)$ with $A(y)$ in $\Sigma^{\mathcal{P}}$.

Theorem 9.1. *Let B be $\Pi_2^{\mathcal{P}}$-sentence of the language without the predicate* R. *If* $\mathbf{KP}(\mathcal{P}) + \mathbf{AC}_{global} \vdash B$, *then* $\mathbf{KP}(\mathcal{P}) + \mathbf{AC} \vdash B$.

Proof. Basically as in [58, Theorem 3.2]. □

The acronym **CZF** stands for Constructive Zermelo-Fraenkel set theory. For details see [1, 2].

Corollary 9.2. *(i)* $\mathbf{KP}(\mathcal{P}) + \mathbf{AC}_{global}$, $\mathbf{KP}(\mathcal{P}) + \mathbf{AC}$, *and* $\mathbf{CZF} + \mathbf{AC}$ *prove the same $\Pi_2^{\mathcal{P}}$-sentences.*

(ii) The three theories are of the same proof-theoretic strength as **KP**($\mathcal{P}$). *More precisely, they prove the same Π_4^1-sentences of the language of second order arithmetic when identified with their canonical translation into the language of set theory.*

Proof. (i) For $\mathbf{KP}(\mathcal{P}) + \mathbf{AC}_{global}$ and $\mathbf{KP}(\mathcal{P}) + \mathbf{AC}$ this follows from the foregoing Theorem. A question left open in [55] was that of the strength of constructive Zermelo-Fraenkel set theory with the axiom of choice. There $\mathbf{CZF} + \mathbf{AC}$ was interpreted in $\mathbf{KP}(\mathcal{P}) + V = L$ ([55, Theorem 3.5]). However, the realizability interpretation works with $\mathbf{AC}_{global}$ as well. Moreover, for this notion of realizability, realizability of a $\Pi_2^{\mathcal{P}}$-sentence B entails its truth. Therefore if $\mathbf{CZF} + \mathbf{AC} \vdash B$, then $\mathbf{KP}(\mathcal{P}) + \mathbf{AC}_{global} \vdash B$.

Conversely note that $\mathbf{CZF} + \mathbf{AC}$ proves the law of excluded middle for $\Delta_0^{\mathcal{P}}$-formulae. This amount of classical logic suffices to prove the power set axiom from the subset collection axiom. The proof-theoretic ordinal of **CZF** is also the Bachmann-Howard ordinal. Moreover, in Theorem 5.11, **KP** can be replaced by

CZF + **AC**. As a result, the ordinal analysis for B utilizing $\mathbf{RS}^{\mathcal{P}}_{\Omega}(\mathsf{R}, X)$, can be carried out in **CZF** + **AC** itself and the proof of the pertaining soundness is also formalisable in **CZF** + **AC**, whence the latter theory proves B.

(ii) follows from (i) viewed in conjunction with [54, Corollary 3.5]. □

Finally, we remark that the three theories of Corollary 9.2 can be added to the list of proof-theoretically equivalent theories presented in [57, Theorem 15.1].

10 The provably total set functions of other theories

Part of the machinery developed here could also be used to give a characterization of the total set functions of extensions of **KP** such as the theories **KPi** and **KPM** that are describing a recursively inaccessible and a recursively Mahlo universe of sets, respectively (see [30, 43, 49]). This however would also require an interpretation of collapsing functions as acting on set-theoretic ordinals along the lines of [46, 44].

References

[1] P. Aczel, M. Rathjen: *Notes on constructive set theory.* Technical Report 40, Institut Mittag-Leffler (The Royal Swedish Academy of Sciences,Stockholm,2001). *http://www.ml.kva.se/preprints/archive2000-2001.php*

[2] P. Aczel, M. Rathjen: *Constructive set theory.* book draft, August 2010.

[3] J Barwise: *Admissible Sets and Structures.* (Springer, Berlin 1975).

[4] M. Beeson: *Foundations of Constructive Mathematics.* Springer, Berlin (1985).

[5] B. Blankertz, W. Weiermann: *A uniform approach for characterizing the provably total number-theoretic functions of KPM and (some of) its subsystems.* Preprint 1994.

[6] B. Blankertz: *Beweistheoretische Techniken zur Bestimmung der* Π^0_2*-Skolem Funktionen.* Dissertation, Westfälische Wilhelms-Universität (Münster, 1997).

[7] B. Blankertz, W. Weiermann: *How to chracterize provably total functions by the Buchholz operator method.* Lecture Notes in Logic 6 (Springer, Heidelberg, 1996).

[8] W. Buchholz, S. Feferman, W. Pohlers, W. Sieg: *Iterated inductive definitions and subsystems of analysis.* (Springer, Berlin, 1981).

[9] W. Buchholz: *A new system of proof-theoretic ordinal functions.* Annals of Pure and Applied Logic 32 (1986) 195-207

[10] W. Buchholz, S. Wainer *Provably computable functions and the fast growing hierarchy.* in: Logic and Combinatorics, Contemporary Mathematics 65 (American Mathematical Society, Providence,1987) 179-198.

[11] W. Buchholz: *A simplified version of local predicativity.* in: P. Aczel, H. Simmons, S. Wainer (eds.), *Leeds Proof Theory 90* (Cambridge University Press, Cambridge, 1993) 115-147.

[12] W. Buchholz: *Notation systems for infinitary derivations.* Arch. Math. Logic 30 (1991) 277–296.

[13] S. Feferman: *Systems of predicative analysis.* Journal of Symbolic Logic 29 (1964) 1-30.

[14] S. Feferman: *predicative provability is set theory.* American Mathematical Society, Volume 72, Number 3 (1966), 486-489.

[15] S. Feferman: *Systems of predicative analysis II. Representations of ordinals.* Journal of Symbolic Logic 33 (1968) 193-220.

[16] S. Feferman: *Proof theory: a personal report.* in: G. Takeuti *Proof Theory*, 2nd edition (North-Holland, Amsterdam, 1987) 445-485.

[17] S. Feferman: *On the strength of some semi-constructive theories.* In: U.Berger, P. Schuster, M. Seisenberger (Eds.): *Logic, Construction, Computation* (Ontos Verlag, Frankfurt, 2012) 201–225.

[18] S. Feferman: *Is the continuum hypothesis a definite mathematical problem?* Draft of paper for the lecture to the Philosophy Dept., Harvard University, Oct. 5, 2011 in the *Exploring the Frontiers of Incompleteness* project series, Havard 2011–2012.

[19] S. Feferman: *Three Problems for Mathematics: Lecture 2: Is the Continuum Hypothesis a definite mathematical problem?* Slides for inaugural Paul Bernays Lectures, ETH, Zürich, Sept. 12, 2012.

[20] S. Feferman: *Why isn't the Continuum Problem on the Millennium ($1,000,000) Prize list?* Slides for CSLI Workshop on Logic, Rationality and Intelligent Interaction, Stanford, June 1, 2013.

[21] G. Frege: *Die Grundlagen der Arithmetik.* (Verlag Wilhelm Koebner, Breslau, 1884).

[22] H. Friedman: *Countable models of set theories.* In: A. Mathias and H. Rogers (eds.): *Cambridge Summer School in Mathematical Logic*, volume 337, Lectures Notes in Mathematics (Springer, Berlin, 1973) 539-573.

[23] H. Friedman and S. Sheard: *Elementary descent recursion and proof theory*, Annals of Pure and Applied Logic 71 (1995) 1–45.

[24] G. Gentzen: *Die Widerspruchsfreiheit der reinen Zahlentheorie.* Mathematische Annalen 112 (1936) 493-565

[25] G. Gentzen: *Neue Fassung des Widerspruchsfreiheitsbeweises für die reine Zahlentheorie*, Forschungen zur Logik und zur Grundlegung der exacten Wissenschaften, Neue Folge 4 (Hirzel, Leipzig, 1938) 19–44.

[26] G. Gentzen: *Beweisbarkeit und Unbeweisbarkeit von Anfangsfällen der transfiniten Induktion in der reinen Zahlentheorie.* Mathematische Annalen 119 (1943) 140–161.

[27] G. Jäger: *Beweistheorie von KPN.* Archiv f. Math. Logik 2 (1980) 53-64.

[28] G. Jäger: *Zur Beweistheorie der Kripke–Platek Mengenlehre über den natürlichen Zahlen.* Archiv f. Math. Logik 22 (1982) 121-139.

[29] G. Jäger: *Theories for admissible sets: a unifying approach to proof theory.* Bibliopolis,

Naples, 1986

[30] G. Jäger and W. Pohlers: *Eine beweistheoretische Untersuchung von* $\mathbf{\Delta}^1_2$-$\mathbf{CA} + \mathbf{BI}$ *und verwandter Systeme*, Sitzungsberichte der Bayerischen Akademie der Wissenschaften, Mathematisch–Naturwissenschaftliche Klasse (1982).

[31] G. Kreisel: *On the interpretation of non-finitist proofs I.* Journal of Symbolic Logic 16 (1951) 241–267.

[32] G. Kreisel: *On the interpretation of non-finitist proofs II.* Journal of Symbolic Logic 17 (1952) 43–58.

[33] G. Kreisel: *A survey of proof theory.* Journal of Symbolic Logic 33 (1968) 321-388.

[34] G. Kreisel, G. Mints, S. Simpson: *The use of abstract language in elementary metamathematics: Some pedagogic examples.* in: Lecture Notes in Mathematics, vol. 453 (Springer, Berlin, 1975) 38-131.

[35] S. Kripke: *Transfinite recursion on admissible ordinals.* Journal of Symbolic logic 29 (1964) 161-162

[36] E. Lopez-Escobar: *An extremely restricted* ω*-rule*, Fundamenta Mathematicae 90 (1976) 159-172.

[37] M. Michelbrink: *A Buchholz derivation system for the ordinal analysis of* $KP + \Pi_3$*-reflection.* Journal of Symbolic Logic 71 (2006) 1237–1283.

[38] Y.N. Moschovakis: *Recursion in the universe of sets*, mimeographed note, 1976.

[39] D. Normann: *Set recursion*, in: Fenstad et al. (eds.): *Generalized recursion theory II* (North-Holland, Amsterdam, 1978) 303-320.

[40] R. A. Platek: *Foundations of recursion theory* PhD Thesis, Stanford University, 1966 (219pp).

[41] W. Pohlers: *Proof Theory*, Unversitext (Springer 2009).

[42] W. Pohlers, J.-C. Stegert: *Provably recursive functions of reflection.* In: U.Berger, P. Schuster, M. Seisenberger (Eds.): *Logic, Construction, Computation* (Ontos Verlag, Frankfurt, 2012) 381–474.

[43] M. Rathjen: *Proof-Theoretic Analysis of KPM*, Arch. Math. Logic 30 (1991) 377–403.

[44] M. Rathjen: *Collapsing functions based on recursively large ordinals: A well–ordering proof for KPM.* Archive for Mathematical Logic 33 (1994) 35–55.

[45] Michael Rathjen: *A Proof-Theoretic Characterization of the Primitive Recursive Set Functions.* The Journal of Symbolic Logic Vol. 57, No. 3 (1992), pp. 954-969

[46] M. Rathjen: *How to develop proof–theoretic ordinal functions on the basis of admissible ordinals.* Mathematical Logic Quarterly 39 (1993) 47-54.

[47] M. Rathjen: *Proof theory of reflection.* Annals of Pure and Applied Logic 68 (1994) 181-224

[48] M. Rathjen *Recent advances in ordinal analysis:* $\Pi^1_2 - CA$ *and related systems.* Bulletin of Symbolic Logic 1, (1995) 468-485

[49] M. Rathjen: *The realm of ordinal analysis.* In: S.B. Cooper and J.K. Truss (eds.): *Sets and Proofs.* (Cambridge University Press, 1999) 219-279.

[50] M. Rathjen: *Unpublished Lecture Notes on Proof Theory.* (1999)

[51] M. Rathjen: *Indefiniteness in semi-intuitionistic set theories: On a conjecture of Feferman.* Journal of Symbolic Logic 81 (2016) 742–754.

[52] M. Rathjen: *An ordinal analysis of stability.* Archive for Mathematical Logic 44 (2005) 1-62

[53] M. Rathjen: *An ordinal analysis of parameter-free* Π^1_2 *comprehension.* Archive for Mathematical Logic 44 (2005) 263-362.

[54] M. Rathjen: *Replacement versus collection in constructive Zermelo-Fraenkel set theory.* Annals of Pure and Applied Logic 136 (2005), 156–174.

[55] M. Rathjen: *Choice principles in constructive and classical set theories.* In: Z. Chatzidakis, P. Koepke, W. Pohlers (eds.): *Logic Colloquium 2002*, Lecture Notes in Logic 27 (A.K. Peters, 2006) 299–326.

[56] M. Rathjen: *Relativized ordinal analysis: The case of Power Kripke-Platek set theory.* Annals of Pure and Applied Logic 165 (2014) 316-393.

[57] M. Rathjen: *Constructive Zermelo-Fraenkel Set Theory, Power Set, and the Calculus of Constructions.* In: P. Dybjer, S. Lindström, E. Palmgren and G. Sundholm: *Epistemology versus Ontology: Essays on the Philosophy and Foundations of Mathematics in Honour of Per Martin-Löf*, (Springer, Dordrecht, Heidelberg, 2012) 313–349.

[58] M. Rathjen: *Power Kripke-Platek set theory and the axiom of choice.* Submitted. Also published in the Isaac Newton Institute for Mathematical Sciences preprint series for the programme *'Mathematical, Foundational and Computational Aspects of the Higher Infinite'*.

[59] G.E Sacks: *Higher recursion theory.* (Springer, Berlin, 1990)

[60] K. Schütte: *Eine Grenze für die Beweisbarkeit der transfiniten Induktion in der verzweigten Typenlogik.* Archiv für Mathematische Logik und Grundlagenforschung 67 (1964) 45-60.

[61] K. Schütte: *Predicative well-orderings*, in: Crossley, Dummett (eds.), *Formal systems and recursive functions* (North Holland, 1965) 176–184.

[62] K. Schütte: *Proof Theory.* (Springer, Berlin 1977).

[63] H. Schwichtenberg: *Proof theory: Some applications of cut-elimination.* In: J. Barwise (ed.): *Handbook of Mathematical Logic.* (North Holland, Amsterdam, 1977) 867-895.

[64] H. Schwichtenberg, S.S. Wainer: *Proofs and Computations.* (Cambridge University Press, Cambridge, 2012).

[65] G. Takeuti: *Proof theory, second edition.* (North Holland, Amsterdam, 1987).

[66] A. Weiermann: *How to characterize provably total functions by local predicativity.* Journal of Symbolic Logic 61 (1996) 52-69.

Received 15 September 2015

www.ingramcontent.com/pod-product-compliance
Lightning Source LLC
Chambersburg PA
CBHW080659110426
42739CB00034B/3337
* 9 7 8 1 8 4 8 9 0 2 1 9 0 *